# International Conference: Water Resources Assessment and Seasonal Prediction



Proceedings of the International Association of Hydrological Sciences (PIAHS) Volume 374, 2016

Koblenz, Germany
13-16 October 2016

**Editors:**

**W. Grabs** **S. Demuth**

ISBN: 978-1-5108-3910-6

**Printed from e-media with permission by:**

Curran Associates, Inc.
57 Morehouse Lane
Red Hook, NY 12571

**Some format issues inherent in the e-media version may also appear in this print version.**



Printed by Curran Associates, Inc. (2017)

International Association of Hydrological Sciences
Centre for Ecology and Hydrology
Wallingford, Oxfordshire OX10 8BB
United Kingdom

Phone: 441 491 692 442
Fax: 441 491 692 448

http://iahs.info/

**Additional copies of this publication are available from:**

Curran Associates, Inc.
57 Morehouse Lane
Red Hook, NY 12571 USA
Phone: 845-758-0400
Fax: 845-758-2633
Email: curran@proceedings.com
Web: www.proceedings.com

# TABLE OF CONTENTS

Proc. IAHS, 374, 1–1, 2016
proc-iahs.net/374/1/2016/
doi:10.5194/piahs-374-1-2016


# Preface: International conference: Water Resources Assessment and Seasonal Prediction

**Wolfgang Grabs[1] and Siegfried Demuth[2]**

[1]Chief, International Water Affairs, German Federal Institute of Hydrology, Koblenz, Germany
[2]Director, UNESCO International Centre for Water Resources and Global Change at German Federal Institute of Hydrology, Koblenz, Germany

*Correspondence to:* Wolfgang Grabs (grabs@bafg.de)

Published: 17 October 2016

Changes in water quantity, and quality are occurring mainly as consequences of global and local changes including environmental factors, climate change and human-induced changes. These changes impact the water availability which is putting stress on different sectors. Science-based knowledge to assess water resources over a range of scales both in space and time is essential to develop methodologies for water resources assessments and to improve sustainable water management practices. Linked to the dynamic assessment of water resources is the ability to provide reliable seasonal predictions of water availability and water quality. All of these need to be embedded into a framework of scientific and institutional cooperation within a larger scope of policy setting and water diplomacy.

The conference had been attended by seventy experts from twenty-eight countries who exchanged their research results and findings in the different thematic areas of the conference. A number of the presentations have been expanded into full papers and selected for this publication.

The selected papers provide an overview of research efforts and good practices focusing on changes of the hydrological cycle, predictions of freshwater availability under climate change, prediction of variations of global and regional hydrological processes and assessment of water resources linked to seasonal and inter-annual variations of hydrological components of the water cycle.

Main results of the conference have been the recognition to improve earth observing systems, improvement of methods and models to assess and predict water resources availability including improvements in seasonal predictions; recognizing that water quality is a limiting factor in water use and the necessity to enhance transboundary management of water resources in the context of water diplomacy.

The conference is a milestone in documenting current efforts to assess and predict water resources availability in the context of sustainable development.

The editors also wish to gratefully acknowledge the significant contributions made by the co-sponsoring organisations to enable the planning and conduct of the conference. Thanks go in particular to the German Federal Institute of Hydrology (BfG) who hosted the conference, World Meteorological Organization (WMO), UNESCO, the Group on Earth Observations (GEO) and IAHS.

Proc. IAHS, 374, 3–7, 2016
proc-iahs.net/374/3/2016/
doi:10.5194/piahs-374-3-2016


Open Access
PIAHS

# A regional approach to climate adaptation in the Nile Basin

**Michael B. Butts[1], Carlo Buontempo[2], Jens K. Lørup[1], Karina Williams[2], Camilla Mathison[2], Oluf Z. Jessen[1], Niels D. Riegels[1], Paul Glennie[1], Carol McSweeney[2], Mark Wilson[2], Richard Jones[2], and Abdulkarim H. Seid[3]**

[1]DHI, Agern Alle 5, DK 2970, Hoersholm, Denmark
[2]UK Met Office, FitzRoy Road, Exeter, Devon, EX1 3PB, UK
[3]Nile Basin Initiative Secretariat (Nile-Sec), Plot 12 Mpigi Road, Entebbe, Uganda

*Correspondence to:* Michael B. Butts (mib@dhigroup.com)

Published: 17 October 2016

**Abstract.** The Nile Basin is one of the most important shared basins in Africa. Managing and developing the water resources within the basin must not only address different water uses but also the trade-off between developments upstream and water use downstream, often between different countries. Furthermore, decision-makers in the region need to evaluate and implement climate adaptation measures.

Previous work has shown that the Nile flows can be highly sensitive to climate change and that there is considerable uncertainty in climate projections in the region with no clear consensus as to the direction of change. Modelling current and future changes in river runoff must address a number of challenges; including the large size of the basin, the relative scarcity of data, and the corresponding dramatic variety of climatic conditions and diversity in hydrological characteristics. In this paper, we present a methodology, to support climate adaptation on a regional scale, for assessing climate change impacts and adaptation potential for floods, droughts and water scarcity within the basin.

## 1 Introduction

The Nile River is the longest river in the world (UNEP, 2010) and the Nile Basin (Fig. 1) is one of the most critical and most important shared basins in Africa, hosting 25 % of Africa's population (SEDAC, 2010) while accounting for only 10 % of its landmass. Within the basin, agriculture, energy production and livelihoods all depend strongly on the flows. The region is facing rising levels of water scarcity, high population growth, watershed degradation and loss of environmental services (UNEP, 2010). In addition, the water resources are highly sensitive to climate change (Conway et al., 2007).

The Nile basin's climate varies significantly from extreme aridity in the north, including Egypt and Sudan, to the tropical rainforests in Central and East Africa and parts of Ethiopia. The distribution of precipitation can be categorised into two distinct regions; the Equatorial (or East African) lakes and the Ethiopian highlands (Fig. 1). There are significant differences in the wet and dry period distribution with some areas in the tropical region of the Nile basin having two rainfall seasons. In addition, there is significant inter-annual and inter-decadal variability, which also has important implications for the management of water resources in the Nile (Conway, 2005). In addition, many past studies have demonstrated that modelling current and future changes in river runoff presents a number of challenges: the large size of the basin, the relative scarcity of data, its geographical location and the corresponding dramatic variety of climatic conditions and diversity in hydrological characteristics. On top of these hydrological challenges, the climate projections are inherently uncertain because of the natural variability of the climate system, the uncertainty in future anthropogenic emissions and the uncertainty resulting from our incomplete understanding of how to represent important climate system processes in climate models (Buontempo et al., 2011). Nevertheless, decision-makers in the region need to evaluate and implement climate adaptation measures. Such climate adap-

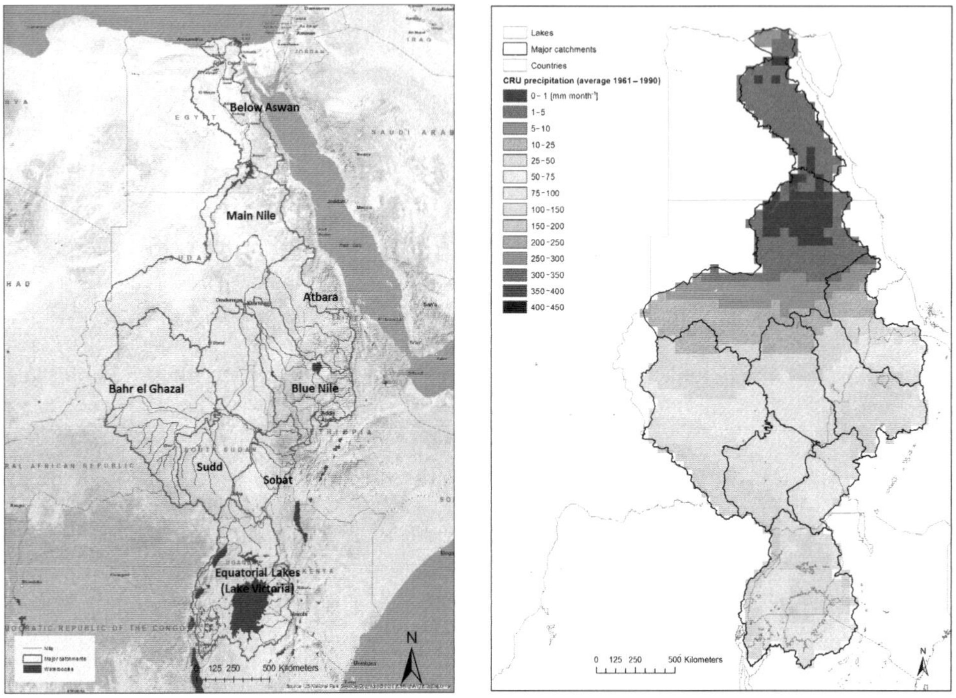


**Figure 1.** An overview of the Nile Basin (left) and distribution of annual average rainfall in mm/month for the period 1961–1990 (right), from CRU (Harris et al., 2014).

tation measures can be implemented at the local, sub-basin or national scale. However, within trans-boundary basins like the Nile the implementation of such measures locally may have important impacts regionally particularly downstream of those measures. Therefore, tools and information are required to understand climate change and adaptation impacts at the regional scale.

## 2 Regional climate modelling

In this study, the Perturbed Physics Ensemble (PPE) approach is used. This PPE approach (Collins et al., 2006) represents the uncertainties or variability in climate projections by perturbing uncertain parameters to create new versions of the climate model. Each of these versions is characterised by different values for a set of parameters that describe the basic unresolved physical processes (Collins and Knight, 2007; Palmer and Williams, 2008). This approach was used to generate a large Global Climate Model (GCM) ensemble and here we apply a recently published systematic methodology (McSweeney et al., 2012) to select a subset of 5 GCM ensemble members. The selection is based on their ability to reproduce important features of the present-day climate over Africa while capturing the range of outcomes from the GCM ensemble (Buontempo et al., 2015). The ability of the selected ensemble to reproduce the rainfall patterns over Africa is illustrated in Fig. 2. To obtain high-resolution information suitable for assessing climate change impacts on flows in the Nile, these GCMs were dynamically downscaled, i.e. lateral boundary conditions from the GCMs are used to drive the regional climate model, HadRM3P, over the 50 km resolution Africa CORDEX (Giorgi et al., 2009) domain. Based on these new RCM simulations, change factors were derived for all 5 ensembles, for temperature, potential evapotranspiration and precipitation and for two periods 2020–2049 and 2070–2099.

## 3 Regional hydrological modelling

To represent the flows and water resources on the regional scale, both for climate change impact assessment and for climate adaptation scenarios, distributed hydrological modelling is required. In this study, a sub-basin based modelling approach using MIKE HYDRO (DHI, 2013) was adopted. The rainfall-runoff processes are represented using the NAM model (Butts et al., 2004, 2007). The runoff from the sub-basins then becomes the inflow to the river or channel network representation used in MIKE HYDRO (Fig. 3). The MIKE HYDRO BASIN model represents rivers and their main tributaries as a network consisting of branches and calculation nodes. Branches represent the individual river sections, while the nodes represent either a confluence or a location where certain water activities occur. These include, for instance, multipurpose reservoirs, withdrawals for wa-

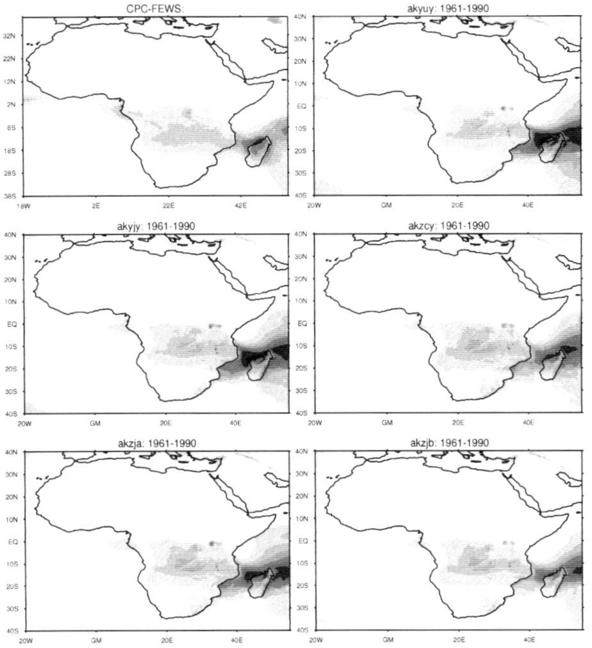


**Figure 2.** Comparison of the observed and simulated precipitation for Africa for the period June–September (JJAS). The observations are from CPC-FEWS (Love et al., 2004) cover the period 1983–2012 while the simulations cover the 1961–1990.

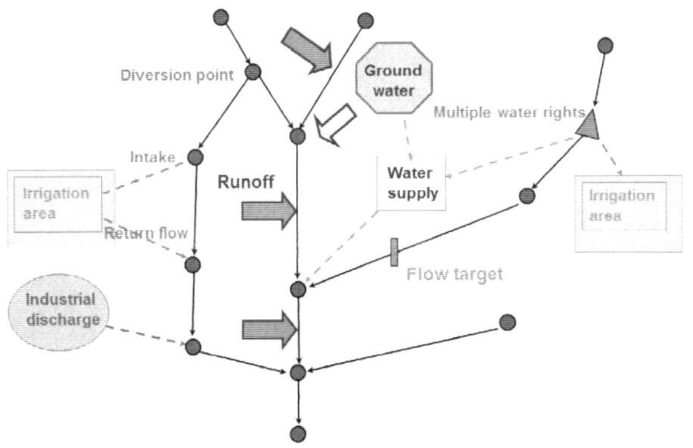


**Figure 3.** Schematic of the MIKE HYDRO network model including different water activities and water users.

ter supply or irrigation, effluent discharges, diversion canals and systems, gauging stations, low flow control points and priority-based allocations. Simple routing along the branches is used between these nodes. Withdrawals from the river or reservoir are specified as irrigation water users and/or regular water users (e.g. domestic or industrial). These capabilities make MIKE HYDRO well-suited not only for regional scale assessments of the impacts of climate change on regional water resources but also for subsequent assessment of regional scale climate adaptation options.

## 4 Results

The regional hydrological model has been developed and calibrated against available discharge data within the period 1960 to 1980. This period of 21 years is assumed in this analysis to be representative of the variability for the reference period (1961–1990). The MIKE HYDRO model was then run using bias-corrected values for rainfall and PET over the same period. All other factors such as the operation and operation strategies of the reservoirs, the extractions for irrigation, etc. were kept fixed. The resulting changes are therefore solely because of projected changes in the climate.

Figure 4 presents the projected flows at the Jinja station for the two periods; near future (2020–2049) and far future (2070–2099). The Jinja station is located at the outlet of Lake Victoria. It is an important control point with the Owen Falls dam located just downstream. The releases follow the so-called "Agreed Curve" to provide releases that resemble the natural outflows from the lake prior to its construction. The outflow control is shown clearly in the flow duration curve for the low flows. One of the key features that determine the hydrology of the lake is the high contribution (85 %) of rainfall directly over the lake to total lake inflow. This direct precipitation suggests that lake levels and the long-term outflow will be highly sensitive to climatic change, as the water does not flow via the terrestrial water cycle (Williams et al., 2014). The flow projections indicate reductions in the flows for the near future (2020–2049) but with different magnitudes among the 5 ensemble members. For the far future (2070–2099), both increasing and decreasing trends are evident.

## 5 Discussion

A regional-scale modelling framework has been developed to address the need for regional scale tools and information that can support and inform decision-making for large transboundary river basins like the Nile. The framework incorporates the state of the art Perturbed Physics Ensemble (PPE) approach for climate change modelling and appropriate hydrological modelling using MIKE HYDRO. As a result, the framework can provide an assessment of climate change impacts, an indication of the uncertainties and can be used for regional scale adaptation. This framework can be readily applied to similar large or transboundary river basins in Africa and in other regions (Asia, South America) or used to at smaller scales to evaluate national or local climate impact assessments for water resources, water resource management and the evaluation of climate adaptation measures.

**Acknowledgements.** This study has been carried out by DHI and the UK Met Office in collaboration with NBI within the project "Adaptation to climate change induced water stress in the Nile River Basin". This project was launched in March 2010 as a

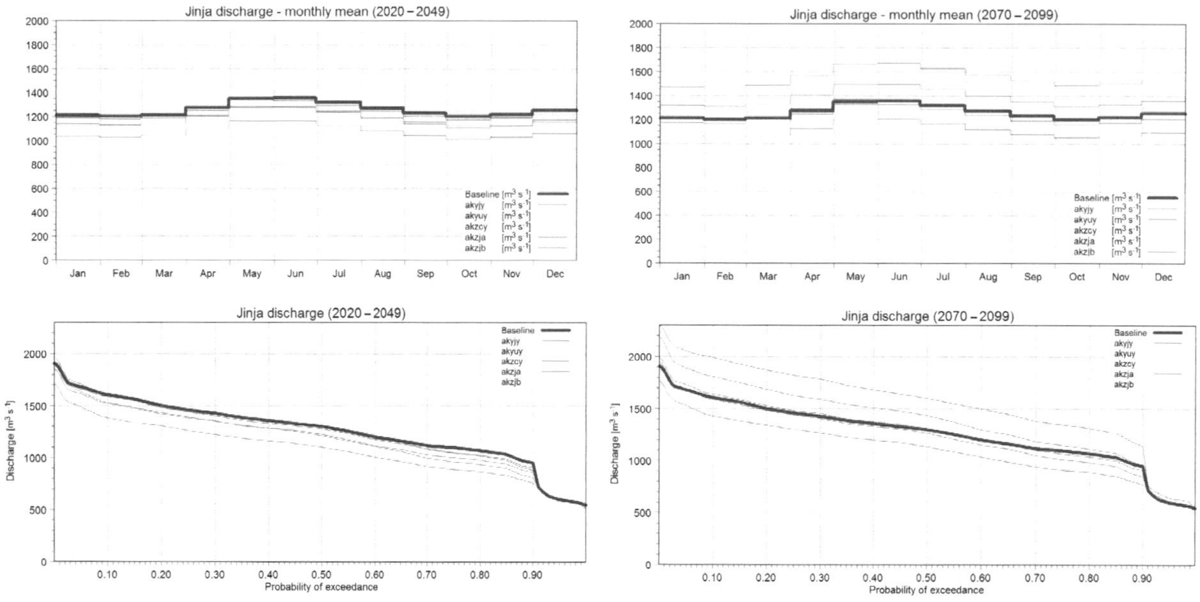


**Figure 4.** Projected changes in the average monthly flows (upper) and flow duration curve (lower) for 2020–2049 (left) and 2070–2099 (right) for the Jinja station. The solid (red) line shows the baseline levels (1961–1990) and the thin green lines show the variability among the 5 RCM ensemble members.

partnership between the United Nations Environment Programme UNEP and the Nile Basin Initiative (NBI), sponsored by the Swedish International Development Co-operation Agency (SIDA). This study was carried out under Project Co-operation Agreement (PCA/2010/DEPI/CCAU/005) between UNEP and DHI. The authors would also like to acknowledge the support of UNEP-DHI and their respective organisations.

## References

Buontempo, C., Lørup, J. K., Sanderson, M., Butts, M., Palin, E., McCarthy, R., Jones, R., Betts, R., and Antar, M.: The impact of uncertainties in climate impacts assessments: the case of the Nile basin, in: Coping with Global Environmental Change: Climate Change, Soil and Desertification, Water Management, Food and Health, Hexagon Series on Human, Environmental Security and Peace (HESP), edited by: Brauch, H.-G., Spring, Ú. O., Mesjasz, C., Grin, J., Kameri-Mbote, P., Chourou, B., Dunay, P., and Birkmann, J., Berlin – Heidelberg – New York, Springer-Verlag, 5, 765–772, ISBN: 978-3-642-17775-0, doi:10.1007/s00382-006-0121-0, 2011.

Buontempo, C., Mathison, C., Jones, R., Williams, K., Wang, C., and McSweeney, C.: An ensemble climate projection for Africa, Clim. Dynam., 44, 2097–2118, doi:10.1007/s00382-014-2286-2, 2015.

Butts, M. B., Payne, J. T., Kristensen, M., and Madsen, H.: An evaluation of the impact of model structure on hydrological modelling uncertainty for streamflow prediction, J. Hydrol., 298, 242–266, doi:10.1016/j.jhydrol.2004.03.042, 2004.

Butts, M. B., Dubicki, A., Stronska, K., Jørgensen, G., Nalberczynski, A., Lewandowski, A., and van Kalken, T.: Flood forecasting for the Upper and Middle Odra River Basin. Flood Risk Management in Europe: Innovation in Policy and Practice Series: Advances in Natural and Technological Hazards Research, edited by: Begum, S., Stive, M. J. F., and Hall, J. W., Vol. 25, Springer Science & Business Media, ISBN: 1-4020-4199-3, doi:10.1007/978-1-4020-4200-3_19, 2007.

Collins, M. and Knight, S. K. (Eds.): Ensembles and probabilities: A new era in the prediction of climate change, Philos. T. R. Soc. A, 365, 1955–2191, ISSN: 1364-503X, 2007.

Collins, M., Booth, B. B. B., Harris, G. R., Murphy, J. M., Sexton, D. M. H., and Webb, M. J.: Towards quantifying uncertainty in transient climate change, Clim. Dynam., 27, 127–147, 2006.

Conway, D.: From headwater tributaries to international river: Observing and adapting to climate variability and climate change in the Nile Basin, Global Environ. Chan., 15, 99–114, doi:10.1016/j.gloenvcha.2005.01.003, 2005.

Conway, D., Hanson, C., Doherty, R., and Persechino, R.: GCM simulations of the Indian Ocean dipole influence on East African rainfall: present and future, Geophys. Res. Lett., 34, L03705, doi:10.1029/2006GL027597, 2007.

DHI: MIKE HYDRO User Manual, Hørsholm, Denmark, 2013.

Giorgi, F., Jones, C., and Asrar, G. R.: Addressing climate change needs at the regional level: the CORDEX framework, WMO Bulletin, 58, 175–183, 2009.

Harris, I., Jones, P. D., Osborn, T. J., and Lister, D. H.: Updated high-resolution grids of monthly climatic observations – the CRU TS3.10 Dataset, Int. J. Climatol., 34, 623–642, doi:10.1002/joc.3711, 2014.

Love, T. B., Kumar, V., Xie, P., and Thiaw, W.: A 20-year daily Africa precipitation climatology using satellite and gauge data, in: Proceedings of the 84th AMS Annual Meeting, P5.4. Conference on Applied Climatology, Seattle, WA, 2004.

McSweeney, C. F., Jones, R. G., and Booth, B. B. B.: Selecting ensemble members to provide regional climate change information, J. Climate, 25, 7100–7121, doi:10.1175/JCLI-D-11-00526.1, 2012.

Palmer, T. N. and Williams, P. D.: Introduction. Stochastic physics and climate modelling, Philos. T. R. Soc. A, 366, 2421–2427, doi:10.1098/rsta.2008.0059, 2008.

SEDAC: Gridded Population of the World: Future Estimates. Socioeconomic Data and Applications Center (SEDAC); collaboration with CIESIN, UN-FAO, CIAT, Retrieved: 10 August 2010, available at: http://sedac.ciesin.columbia.edu/data/collection/gpw-v4, 2010.

UNEP: Africa Water Atlas, Division of Early Warning and Assessment (DEWA), United Nations Environment Programme (UNEP), Nairobi, Kenya, 2010.

Williams, K., Chamberlain, J., Buontempo, C., and Bain, C.: Regional climate model performance in the Lake Victoria basin, Clim. Dynam., 44, 1699–1713, doi:10.1007/s00382-014-2201-x, 2014.

Proc. IAHS, 374, 9–15, 2016
proc-iahs.net/374/9/2016/
doi:10.5194/piahs-374-9-2016


Open Access
PIAHS

# Developing new scenarios for water allocation negotiations: a case study of the Euphrates River Basin

**Mohammad Reza Jarkeh[1], Ameneh Mianabadi[2], and Hojjat Mianabadi[3]**

[1]Department of Water engineering, University of Zabol, Zabol, 9861335586, Iran
[2]Department of Water engineering, Ferdowsi University of Mashhad, Mashhad, 9177948974, Iran
[3]Department of Water Resources, Faculty of Civil Engineering and Geosciences, Delft University of Technology, Stevinweg 1, 2628 CN, Delft, the Netherlands

*Correspondence to:* M. R. Jarkeh (mohammad.reza.jargeh@uoz.ac.ir, m.jargeh@gmail.com)

Published: 17 October 2016

**Abstract.** Mismanagement and uneven distribution of water may lead to or increase conflict among countries. Allocation of water among trans-boundary river neighbours is a key issue in utilization of shared water resources. The bankruptcy theory is a cooperative Game Theory method which is used when the amount of demand of riparian states is larger than total available water. In this study, we survey the application of seven methods of Classical Bankruptcy Rules (CBRs) including Proportional (CBR-PRO), Adjusted Proportional (CBR-AP), Constrained Equal Awards (CBR-CEA), Constrained Equal Losses (CBR-CEL), Piniles (CBR-Piniles), Minimal Overlap (CBR-MO), Talmud (CBR-Talmud) and four Sequential Sharing Rules (SSRs) including Proportional (SSR-PRO), Constrained Equal Awards (SSR-CEA), Constrained Equal Losses (SSR-CEL) and Talmud (SSR-Talmud) methods in allocation of the Euphrates River among three riparian countries: Turkey, Syria and Iraq. However, there is not a certain documented method to find more equitable allocation rule. Therefore, in this paper, a new method is established for choosing the most appropriate allocating rule which seems to be more equitable than other allocation rules to satisfy the stakeholders. The results reveal that, based on the new propose model, the CBR-AP seems to be more equitable to allocate the Euphrates River water among Turkey, Syria and Iraq.

## 1 Introduction

About 40 % of world population lives in transboundary river basins which are shared among two or more countries (Prins, 2013). Based on the Transboundary Freshwater Dispute Database (TFDD), 276 transboundary freshwater river basins are shared among almost 148 countries (De Stefano et al., 2012). These basins cover nearly half of the Earth's land surface (United Nations, 2014). In the Middle East, water – as a scarce resource – has become increasingly vital for economic and agricultural development (Gleick, 1993). Mismanagemen uneven distribution of water and the increase in demands as well as distrust among riparian countries, may lead to conflict in shared river basins (Homer-Dixon, 1994). The important role of water in supporting the demands of growing population, crop production, soil degradation and other environmental effects, can cause the freshwater to be a source of conflict in the twentieth century (Westing, 1986). However, it is argued that the outbreak of war is unexpected by two reasons. First, there are almost 450 agreements on international waters signed from 1820 to 2007 (United Nations, 2014) which confirm the tendency of nations to dispute resolution instead of war. Second, starting a war needs many factors such as political, economic, cultural, religious and historical factors (Gleick, 1993). Thus, shared water resource is a source of both conflict and cooperation. Although an international agreement among riparian countries can easily overcome the territorial water resource problem, the basic challenge happens when water demands of riparian countries are more than available water.

Equitable allocation of shared water among riparian states is a complex process and one of the most important is-

Published by Copernicus Publications on behalf of the International Association of Hydrological Sciences.

sues in quantitative conflict resolution, specifically at international scale. Therefore, water scarcity should be analysed with respect to five prospective in the literature of hydropolitics: economic, legal, technical, environmental and security issues (Dolatyar and Gray, 2000). Negotiation is necessary for riparian countries to reach an agreement peacefully over shared water resources. However, the complex problem in consensus building to reach an equitable and reasonable agreement (Rieu-Clarke et al., 2012) is how to allocate water among riparian countries.

Bankruptcy theory is a branch of cooperative Game Theory which can be used in dispute resolution and resources allocation when demand or claim of countries is more than the total available resources (Ansink and Weikard, 2012). The bankruptcy theory was introduced in the seminal papers by O'Neill (1982) and Aumann and Maschler (1985) and some aspects of this theory have been studied by several researchers (Alcalde et al., 2014; Aumann and Maschler, 1985; Hendrickx et al., 2005; Lorenzo-Freire et al., 2010; O'Neill, 1982; Pérez et al., 2010; Thomson, 2003, 2012). Few researchers tried to apply the bankruptcy rules to allocate the shared available water among riparian countries (Mianabadi et al., 2014, 2015; Madani and Zarezadeh, 2012). However, due to different definition of fairness, there is no certain documented method to choose the most appropriate allocation rule. Therefore, in this paper we established a new method to choose the most appropriate allocating rule which seems to be more equitable and reasonable than other allocation rules to satisfy the riparian countries. To evaluate this new proposed method, we applied seven Classical Bankruptcy Rules (CBRs) including Proportional (CBR-Pro), Constrained Equal Awards (CBR-CEA), Constrained Equal Losses (CBR-CEL), CBR-Talmud, Adjusted Proportional (CBR-AP), CBR-Piniles and Minimal Overlap (CBR-MO) and four Sequential Sharing Rules (SSRs) including SSR-Pro, SSR-CEA, SSR-CEL and SSR-Talmud in allocating the Euphrates River water among three riparian countries.

## 2 Euphrates River basin

The Euphrates River is the longest river in the west of Asia (Meyers, 1997) with three riparian countries: Iraq, Syria and Turkey (ESCWA, 2013). It originates from the mountains in the east of Turkey and after following into Syria and Iraq, it connects to the Arvand Rud which goes into the Persian Gulf. The length of river, from the source of Murat Su to the confluence with the Tigris River is 3000 km; 1230 km in Turkey, 710 km in Syria and 1060 km in Iraq (Frenken, 2009) (Fig. 1). After constructing the Keban dam in 1974 and Karakaya dam in 1987, related to the Southeastern Anatolia Project, or Guneydogu Anadolu Projesi, GAP, project, water conflicts raised between Turkey and Syria (Kaya, 1998). Although the GAP project had been scheduled to be completed in 2010, it has been delayed because the World Bank has withheld funding the project due to the lack of an official agreement on water sharing between Turkey and the downstream states on the Euphrates and Tigris Rivers (Jongerden, 2010). The contribution of the riparian countries for draining to the Euphrates River and their claim on this river are illustrated in Table 1 (Ibrahim et al., 2004; Beaumont, 1998).

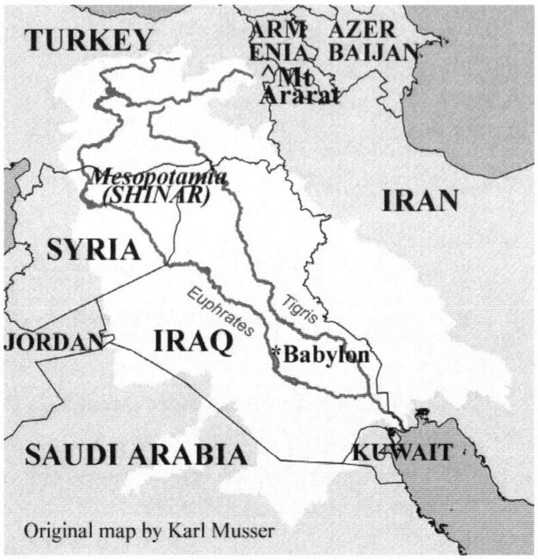


**Figure 1.** The Euphrates-Tigris Rivers basin (created by Karl Musser).

## 3 Bankruptcy allocation rules

### 3.1 Classical Bankruptcy problem

Bankruptcy theory aims to allocate available resource (asset) among stakeholders (creditors) to reach an equitable share of resources, when the asset is not sufficient to satisfy all stakeholders. To determine the meaning of "*equitable*", a large variety of factors, including population, geography, availability of alternative resources, social and economic needs, etc., should be considered during negotiations over the allocation of resources (Gleick, 1993).

#### 3.1.1 Definition

A bankruptcy problem with set of $n$ claimants, $\boldsymbol{N} = \{1, \ldots, i, \ldots, n\}$ is an ordered pair of $(E, \underline{c}) \in R \times R^n$, where $0 \leq c_1 \leq c_2 \leq \ldots \leq c_n$ and $0 \leq E \leq D$ such that $D$ is the total demand:

$$D = \sum\nolimits_{i=1}^{n} c_i .$$

$E$ is the total assets which have to be divided among $n$ claimants with claims of $c_i$ on $E$. An allocation rule is a function that assigns a solution of $f(E, \underline{c}) = (f_1(E, \underline{c}), f_2(E, \underline{c}), \ldots, f_n(E, \underline{c}))$ to a bankruptcy problem

**Table 1.** The rate of contribution and claim of riparian countries on the Euphrates River (Beaumont, 1998; Ibrahim et al., 2004).

| Riparian Countries | Claim (MCM/year) | Claim (%) | Contribution (MCM/year) | Contribution (%) |
|---|---|---|---|---|
| Turkey | 14 000 | 25.6 | 31 580 | 88.8 |
| Syria | 12 600 | 23 | 4000 | 11.2 |
| Iraq | 28 100 | 51.4 | 0 | 0 |
| Total | 54 700 | 100 | 35 580 | 100 |

$(E, \underline{c})$ such that (Curiel et al., 1987):

$f_1(E, \underline{c}) \geq 0$ for every $i \in N$ (Individual rationality)
$\sum_{i=1}^{n} f_1(E, \underline{c}) = E$ (Efficiency).

In this paper allocation vector of $f(E, \underline{c})$ will be shown by $\underline{x} = \langle x_1, x_2, \ldots, x_n \rangle$.

#### 3.1.2 Proportional rule (PRO)

Proportional method can be described as follow:

$$x_i^{\text{pro}} = \mu c_i \text{ where } \mu = E/D. \tag{1}$$

The $\mu$ in this rule is the proportional coefficient.

#### 3.1.3 Constrain Equal Award (CEA)

Using this method, for all $c_i$ there exists $\mu > 0$, such that:

$$x_i^{\text{CEA}} = \min\{\mu, c_i\}. \tag{2}$$

The $\mu$ is calculated by $\sum_{i \in N} \min\{\mu, c_i\} = E$.

#### 3.1.4 Constrained Equal Losses (CEL)

Using this method, for all $c_i$ there exists $\mu > 0$, such that

$$x_i^{\text{CEL}} = \max\{0, c_i - \mu\} \tag{3}$$

where $\sum_{i \in N} \max\{0, c_i - \mu\} = E$.

#### 3.1.5 Talmud

This rule is the combination of CEA and CEL rules and is described as follow:

$$x_i^{\text{TAL}} = \begin{cases} \text{CEA}\left\{\frac{1}{2}c_i, E\right\} & \text{if } E \leq \frac{1}{2}C \\ \frac{1}{2}c_i + \text{CEL}\left\{\frac{1}{2}c_i, E - \frac{1}{2}C\right\} & \text{otherwise.} \end{cases} \tag{4}$$

#### 3.1.6 Piniles

For each $c_i$, $x_i^{\text{Pin}}$ is calculated as follow: (Bosmans and Lauwers, 2007)

$$x_i^{\text{Pin}} = \begin{cases} x_i^{\text{CEA}}\left(\frac{1}{2}\underline{c}, E\right) & \text{if } E \leq \frac{D}{2} \\ \frac{1}{2}\underline{c} + x_i^{\text{CEA}}\left(\frac{1}{2}\underline{c}, E - \frac{D}{2}\right) & \text{if } E \geq \frac{D}{2}. \end{cases} \tag{5}$$

#### 3.1.7 Minimal Overlap (MO) rules

Ibn Ezra's rule

Consider the class of $\psi = \{(E, \underline{c}) : c_n \geq E\}$. The Ibn Ezra's rule is a function of $f^{\text{IE}}(E, \underline{c}) : \psi \to R_+^n$ which is calculated by

$$f_i^{\text{IE}}(E, \underline{c}) = x_i^{\text{IE}} = \sum_{j=1}^{i} \frac{\min\{E, c_j\} - \min\{E, c_{j-1}\}}{n - j + 1} \tag{6}$$

where $c_0 = 0$.

However, this rule is applicable only for a restricted class of bankruptcy problem.

A formal description for Minimal Overlap rule which is a modification of Ibn Ezra's rule for all classes of bankruptcy problem, was proposed by Chun and Thomson (2005). Awards vectors are chosen by minimizing the extent of conflict over each available unit (Chun and Thomson, 2005). For each $c_i$, claims on specific parts of $E$ are arranged in such a way that starting from the highest claim, and in decreasing order. So there is a minimal overlap between them; then, for each unit, equal division among all agents will be applied (Alcalde et al., 2008). Two conditions should be considered (Alcalde et al., 2014).

If $E \geq c_n$

$$x_i^{\text{MO}} = \sum_{j=1}^{i} \frac{\min\{c_j, t\} - \min\{c_{j-1}, t\}}{n - j + 1} + \max\{c_i - t, 0\} \tag{7}$$

where $c_0 = 0$ and $t$ is the unique solution for

$$\sum_{k=1}^{n} \max\{c_k - t, 0\} = E - t. \tag{8}$$

If $E < c_n$

$$x_i^{\text{MO}} = x_i^{\text{IE}} = \sum_{j=1}^{i} \frac{\min\{E, c_j\} - \min\{E, c_{j-1}\}}{n - j + 1}. \tag{9}$$

Adjusted Proportional (AP)

The adjusted proportional (AP) rule, which is a generalization of 2-player contested garment principle (appears in Babylonian Talmud, Baba Metzia, 2a) (Dagan and Volij, 1993), recommends to allocate the asset among claimants as follow (Curiel et al., 1987):

$$x_i^{\mathrm{AP}} = \begin{cases} v_i + \left[\frac{c_i - v_i}{\sum_{j\in N}(c_j - v_j)} \times \left(E - \sum_{j\in N} v_j\right)\right] & \text{if } 0 < E < D \\ c_i & \text{if } E = D \\ 0 & \text{if } E = 0 \end{cases} \tag{10}$$

where

$$v_i = \max\left\{0, E - \sum_{j\neq i} c_j\right\}. \tag{11}$$

This method assigns more allocation to stakeholders with more claims.

### 3.2 Sequential Sharing Rules (SSRs)

In SSR, Ansik and Weikard (2012) considered an ordered set $N$ of $n \geq 2$ stakeholders in which country 1 is the most upstream and country $n$ is the most downstream. Stakeholder $i$ is upstream of $j$ if $i < j$. The set of stakeholders downstream of $i$ is denoted by $D_i = \{j \in N; j > i\}$ and the set of stakeholders upstream of $i$ is denoted by $U_i = \{j \in N; j < i\}$. Contribution of country i to increase flow in the river is shown by $e_i \geq 0$; $\underline{\boldsymbol{e}} = \langle e_1, \ldots, e_n \rangle$. The claimant vector is described by $\underline{\boldsymbol{c}} = \langle c_1, \ldots, c_n \rangle$.

A SSR problem is a triple vector of $\boldsymbol{\omega} = \langle \boldsymbol{N}, \underline{\boldsymbol{e}}, \underline{\boldsymbol{c}} \rangle$. The total available water on the territory of agent $i$ is

$$E_i = e_i + \sum_{j\in U_i} \left(e_j - x_j\right) \tag{12}$$

while $\sum_{j\in U_i}\left(e_j - x_j\right)$ refers to the unallocated water to all upstream countries of country $i$.

Downstream excess claims are defined as follow:

$$c_{D_i} = \sum_{j\in D_i} \left(c_j - e_j\right). \tag{13}$$

In order to solve SSR, a bankruptcy rule is applied to the sequence of $(\omega_1, \ldots, \omega_i, \ldots, \omega_n)$, where $\omega_i = \langle \{i, D_i\}, (E_i, 0), (c_i, c_{D_i}) \rangle$.

There are two major differences between CBRs and SSRs. First, in CBRs all agents have equal position, but in SSRs, the agents are ordered linearly, in the direction of river flow. Second, in CBRs resource is completely separated from the agents while in SSRs the resource is initially endowed to the agents (Ansink and Weikard, 2012).

#### 3.2.1 SSR based on PRO

For all $\omega_i = \langle \{i, D_i\}, (E_i, 0), (c_i, c_{D_i}) \rangle$ there exists $\lambda > 0$, so that $x_i^{\mathrm{Pro}} = \lambda c_i$ and $x_{D_i}^{\mathrm{Pro}} = \lambda c_{D_i}$. Then $\lambda$ is

$$\lambda = \frac{E_i}{c_i + c_{D_i}}. \tag{14}$$

#### 3.2.2 SSR based on CEA

For all $\omega_i = \langle \{i, D_i\}, (E_i, 0), (c_i, c_{D_i}) \rangle$ there exists $\lambda > 0$, so that

$$\begin{cases} x_i^{\mathrm{CEA}} = \min\{c_i, \lambda\} \\ x_{D_i}^{\mathrm{CEA}} = \min\left\{c_{D_i}, \lambda\right\}. \end{cases} \tag{15}$$

#### 3.2.3 SSR based on CEL

For all $\omega_i = \langle \{i, D_i\}, (E_i, 0), (c_i, c_{D_i}) \rangle$ there exists $\lambda > 0$, where

$$\begin{cases} x_i^{\mathrm{CEL}} = \min\{0, c_i - \lambda\} \\ x_{D_i}^{\mathrm{CEL}} = \min\left\{0, c_{D_i} - \lambda\right\}. \end{cases} \tag{16}$$

#### 3.2.4 SSR based on Talmud

For all $\omega_i = \langle \{i, D_i\}, (E_i, 0), (c_i, c_{D_i}) \rangle$ there exists $\lambda > 0$, so that

$$\begin{cases} x_i^{\mathrm{TAL}} = \begin{cases} \min\left\{\frac{1}{2}c_i, \lambda\right\} & \text{if } E_i \leq \frac{1}{2}\left(c_i + c_{D_i}\right) \\ c_i - \min\left\{\frac{1}{2}c_i, \lambda\right\} & \text{otherwise} \end{cases} \\ x_{D_i}^{\mathrm{CEL}} = \begin{cases} \min\left\{\frac{1}{2}c_{D_i}, \lambda\right\} & \text{if } E_i \leq \frac{1}{2}\left(c_i + c_{D_i}\right) \\ c_i - \min\left\{\frac{1}{2}c_{D_i}, \lambda\right\} & \text{otherwise.} \end{cases} \end{cases} \tag{17}$$

## 4 Choosing the best rule

Due to different definition of "equity", there is no certain documented method to choose the most appropriate allocation rule. While the upstream countries prefer the rules closer to the Absolute Territorial Sovereignty (ATS) doctrine (Rieu-Clarke et al., 2012), the downstream riparian countries prefer the rules with more water allocation to them based on the Absolute Territorial Integrity (ATI) doctrine (Rieu-Clarke et al., 2012). In this paper, a new method is introduced to select the most appropriate allocation rule. Three assumptions should be considered for establishing this method. The first is that all stakeholders have the same power (including military power, economic, social welfares and so on) in negotiation and none of the countries considers its power in consensus building process. The second assumption tells that each country prefers to choose the highest rate of allocation. In other word, each country insists on its claim as other countries. The third assumption is that there is no other solution on the table to allocate the shared water resource.

This method chooses a rule which all stakeholders have the lowest dispersion about their total preferences on that rule. So firstly, the allocations are ranked ascending for each stakeholder, separately. Hence, the priority vectors set, $\boldsymbol{\Omega}$, with elements of $\boldsymbol{\omega}_i$ is formed, in which $\boldsymbol{\omega}_i$ is a vector with elements of $\vartheta_j^i$. The vectors $\boldsymbol{\omega}_i$ are the preferences vectors.

**Table 2.** The allocated water of The Euphrates River among the riparian countries as a percentage of their water demand (%).

| Riparian | CBR | | | | | | | SSR | | | |
|---|---|---|---|---|---|---|---|---|---|---|---|
| | Pro | CEA | CEL | Talmud | Piniles | MO | AP | Pro | CEA | CEL | Talmud |
| Turkey | 65 | 85 | 54 | 54 | 70 | 54 | 58 | 62 | 100 | 32 | 50 |
| Syria | 65 | 94 | 49 | 50 | 72 | 49 | 58 | 66 | 86 | 62 | 50 |
| Iraq | 65 | 42 | 77 | 77 | 60 | 77 | 72 | 66 | 38 | 83 | 79 |

**Table 3.** Priority vectors, Priority Index and ranking of eleven CBR and SSR rules.

| Country | CBR_Pro | CBR_CEA | CBR_CEL | CBR_Talmud | CBR_Piniles | CBR_MO | CBR_AP | SSR_Pro | SSR_CEA | SSR_CEL | SSR_Talmud |
|---|---|---|---|---|---|---|---|---|---|---|---|
| Turkey | 2.78 | 5.44 | 0.11 | 2.78 | 4.00 | 0.11 | 0.11 | 0.11 | 13.44 | 25.00 | 9.00 |
| Syria | 0.44 | 11.11 | 11.11 | 0.44 | 4.00 | 11.11 | 0.44 | 1.78 | 7.11 | 0.00 | 4.00 |
| Iraq | 5.44 | 32.11 | 13.44 | 5.44 | 16.00 | 13.44 | 0.11 | 2.78 | 40.11 | 25.00 | 25.00 |
| $\delta_i$ | 8.67 | 48.67 | 24.67 | 8.67 | 24.00 | 24.67 | 0.67 | 4.67 | 60.67 | 50.00 | 38.00 |
| Rank | 3 | 9 | 6 | 3 | 5 | 6 | 1 | 2 | 11 | 10 | 8 |

$i$ and $j$ refers to number of rules and stakeholders, respectively. ($1 \leq i \leq$ Number of rules, $1 \leq j \leq n$ where $n$ is the number of stakeholders). In the current study, $1 \leq i \leq 11$ and $1 \leq j \leq 3$.

In our study, the priority vectors set, $\boldsymbol{\Omega}$ is $\boldsymbol{\Omega} = \{\boldsymbol{\omega}_1, \boldsymbol{\omega}_2, \ldots, \boldsymbol{\omega}_{11}\}$, in which $\boldsymbol{\omega}_1 = \langle 4,5,8 \rangle$, $\boldsymbol{\omega}_2 = \langle 2,1,10 \rangle$ and $\boldsymbol{\omega}_{11} = \langle 10,9,2 \rangle$. By this task each bankruptcy rule corresponds to a priority vector $\boldsymbol{\omega}_i$. The best priority vector is the vector with the lowest distance around an intermediate value e.g. average value in this paper, $\overline{\boldsymbol{\omega}}$. The dispersion around the mean of vector $\boldsymbol{i}$, $\delta_i$, is calculated by

$$\delta_i = \frac{\sum_{j=1}^{n} \left(\vartheta_j^i - \overline{\boldsymbol{\omega}_i}\right)^2}{n} = \frac{\sum_{j=1}^{n} \left(\vartheta_j^i - \frac{\sum_{j=1}^{n} \vartheta_j^i}{n}\right)^2}{n}.$$

For example for CBR-MO rule we have:

$$\overline{\omega_6} = \frac{8+11+4}{3} \approx 7.7$$

$$\delta_6 = \delta_{\mathrm{CBP\_MO}} = \frac{(8-7.7)^2 + (11-7.7)^2 + (4-7.7)^2}{3} \approx 24.67.$$

Table 3 presents $\delta_i$ for all rules. Each rule which has the lowest $\delta_i$, will be chosen as the most appropriate allocation rule.

## 5 Results

Figure 2 illustrates the allocation of the Euphrates River among three riparian countries by seven CBRs and four SSRs. The allocated water of the Euphrates River among the countries as a percentage of their water demands are presented in Table 2.

The results shown in Fig. 2 indicate that the method of the CBR-CEA allocates the lowest rate of water to Iraq, the most downstream agent, while the SSR-CEL and SSR-Talmud allocate the highest rate of water to Iraq. For the upstream country, Turkey, the most and the least rate of water were allocated by the SSR-CEA and the SSR-CEL, respectively. Based on ATS principle, the SSR-CEA is preferable for upstream countries, while the SSR-CEL is preferable for downstream countries based on ATI principle. Morever the highest and lowest allocations to Syria are assigned by CBR-CEA, CBR-CEL and CBR-MO rules, respectively.

The preferences orders for the three riparian states rules are as follow:

- *Turkey*: SSR-CEA > CBR-CEA > CBR-Piniles > CBR-Pro > SSR-Pro > CBR-AP > CBR-CEL = CBR-MO > CBR-Talmud > SSR-Talmud > SSR-CEL
- *Syria*: CBR-CEA > SSR-CEA > CBR-Piniles > SSR-Pro > CBR-Pro > SSR-CEL > CBR-AP > SSR-Talmud = CBR-Talmud > CBR-CEL = CBR-MO
- *Iraq*: SSR-CEL > SSR-Talmud > CBR-CEL = CBR-MO > CBR-Talmud > CBR-AP > SSR-Pro > CBR-Pro > CPR-Piniles > CBR-CEA > SSR-CEA

According to Table 3 and based on new proposed method CBR_AP and SSR_CEA are proposed as best and worse allocation rules, respectively. Although, allocation to Turkey is the highest rate by SSR-CEA rule and this rule has the highest rank for this country (and second place for Syria) but, as it is illustrated in Table 3, it has the lowest rank for Iraq and only 38 % of Iraq claims is allocated by this method (Table 2). So the more equitable rule is the rule which satisfies the stakeholders in the more appropriate way.

## 6 Discussion

Eleven bankruptcy rules in 2 different views, classical bankruptcy problems and sharing river problem, have been presented in this paper. A new choosing method base on minimal priority index is proposed for selecting the most appropriate rule among these eleven bankruptcy rules. According

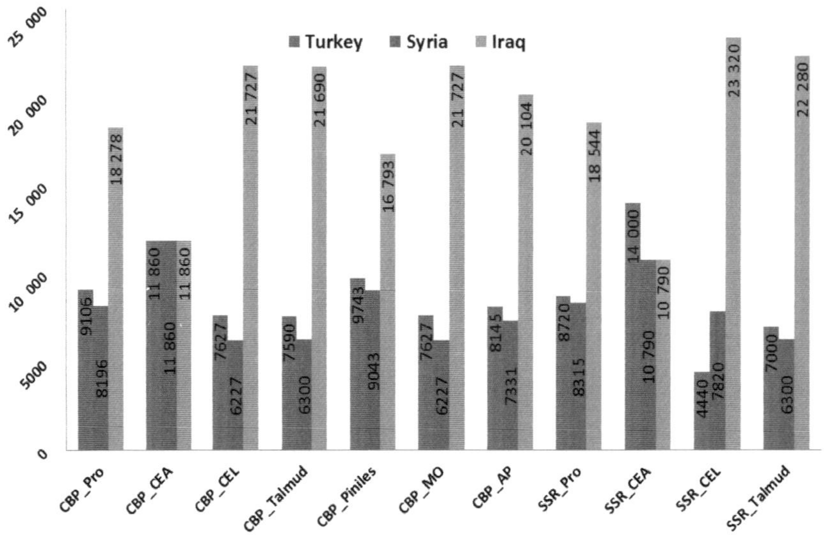


**Figure 2.** Reallocation of the Euphrates River among three riparian countries by Classic Bankruptcy Rules (CBR) and Sequential Sharing Rules (SSR) methods.

to these indices CBR_AP rule select as the best allocation rule for sharing Euphrates River among the three countries.

No doubt, consensus building depends on several factors including allocation percentage, rate of contribution, the damage to the downstream states due to the development in the upstream countries, compensation to the downstream countries, socio-political conditions, and the benefits of the development for the upstream countries, etc. Some factors that should be considered for equitable and reasonable utilisation of an international watercourse are established at UN Watercourses Convention, Part II (General Principle), article 6 (factor relevant to equitable and reasonable utilization) (Rieu-Clarke et al., 2012). Almost all the factors emphasis that any kind of development along the basin must consider existing use of the watercourse which is referred to "historical uses" (Ansink and Weikard, 2009). However, it should be consider that "Historical Uses" is differs from "Historical rights" (Grover, 2007).

In this research, we established a new method to find the most appropriate allocation rule to satisfy the three countries of Turkey, Syria and Iraq more equitably for water allocation of the Euphrates River. Although the allocation rules can give an appropriate vision to the conflict management of the transboundary water resources, but it should also be noted that reallocation of shared water resources is a complicated problem that it cannot be solved only by some mathematical methods, especially when the power distribution is asymmetric (Susskind, 2006; Madani et al., 2014). Therefore, water diplomacy and triple negotiating among Turkey, Syria and Iraq are suggested to build a consensus and to reach an agreement on this river.

## 7 Data availability

Data are collected from two different articles. (1) doi:10.1007/978-3-662-10866-6_28 by Ibrahim et al. (2004) (2) Restructuring of Water Usage in the Tigris-Euphrates Basin: The Impact of Modern Water Management Policies by Peter Beaumont website (Beaumont et al., 1998): http://environment.yale.edu/publication-series/796.html.

## References

Alcalde, J., Marco, M., and Silva, J.: The minimal overlap rule revisited, Social Choice and Welfare, 31, 109–128, doi:10.1007/s00355-007-0269-5, 2008.

Alcalde, J., del Carmen Marco-Gil, M., and Silva-Reus, J. A.: The minimal overlap rule: restrictions on mergers for creditors' consensus, Top, 22, 363–383, 2014.

Ansink, E. and Weikard, H.-P.: Contested water rights, European Journal of Political Economy, 25, 247–260, doi:10.1016/j.ejpoleco.2008.09.007, 2009.

Ansink, E. and Weikard, H.-P.: Sequential sharing rules for river sharing problems, Social Choice and Welfare, 38, 187–210, 2012.

Aumann, R. J. and Maschler, M.: Game theoretic analysis of a bankruptcy problem from the Talmud, J. Econ. Theor., 36, 195–213, 1985.

Beaumont, P.: Restructuring of water usage in the Tigris-Euphrates basin: the impact of modern water management policies, in: Transformation of Middle Eastern natural environments: Legacies and lessons, edited by: Coppock, J. and Miller, J. A., Publisher: Yale School of forestry & environmental studies, available at: http://environment.yale.edu/publication-series/796.html, 168–186, 1998.

Bosmans, K. and Lauwers, L.: Lorenz comparisons of nine rules for the adjudication of conflicting claims, SSRN 958691, 2007.

Chun, Y. and Thomson, W.: Convergence under replication of rules to adjudicate conflicting claims, Games and Economic Behavior, 50, 129–142, 2005.

Curiel, I. J., Maschler, M., and Tijs, S. H.: Bankruptcy games, Zeitschrift für Operations Research, 31, A143–A159, doi:10.1007/BF02109593, 1987.

Dagan, N. and Volij, O.: The bankruptcy problem: a cooperative bargaining approach, Mathematical Social Sciences, 26, 287–297, 1993.

De Stefano, L., Duncan, J., Dinar, S., Stahl, K., Strzepek, K. M., and Wolf, A. T.: Climate change and the institutional resilience of international river basins, J. Peace Res., 49, 193–209, 2012.

Dolatyar, M. and Gray, T.: Water Politics in the Middle East: A Context for Conflict Or Co-operation?, Macmillan, 2000.

ESCWA: Inventory of Shared Water Resources in Western Asia ESCWA, 2013.

Frenken, K.: Irrigation in the Middle East region in figures AQUASTAT Survey-2008, Water Reports, 2009.

Gleick, P. H.: Water and Conflict: Fresh Water Resources and International Security, International Security, 18, 79–112, doi:10.2307/2539033, 1993.

Grover, V. I.: Water: a source of conflict or cooperation?, Science Publishers, Inc., 2007.

Hendrickx, R., Borm, P., Van Elk, R., and Quant, M.: Minimal overlap rules for bankruptcy, International Mathematical Forum, 61, 3001–3012, 2005.

Homer-Dixon, T. F.: Environmental scarcities and violent conflict: evidence from cases, International Security, 19, 5–40, 1994.

Ibrahim, G., Sonmez, B., Bogardi, J., and Castelein, S.: Water issues among the riparian states of Euphrates and Tigris transboundary rivers, From Conflict to Co-operation in International Water Resources Management: Challenges and Opportunities, International Conference, Delft, the Netherlands, 20–22 November 2002, 344–354, 2004.

Jongerden, J.: Dams and Politics in Turkey: Utilizing Water, Developing Conflict, Middle East Policy, 17, 137–143, doi:10.1111/j.1475-4967.2010.00432.x, 2010.

Kaya, I.: The Euphrates-Tigris basin: An overview and opportunities for cooperation under international law, Arid Lands Newsletter, 44, available at: http://ag.arizona.edu/OALS/ALN/aln44/kaya.html#international (last access: August 2015), 1998.

Lorenzo-Freire, S., Casas-Méndez, B., and Hendrickx, R.: The two-stage constrained equal awards and losses rules for multi-issue allocation situations, Top, 18, 465–480, 2010.

Madani, K. and Zarezadeh, M.: Bankruptcy methods for resolving water resources conflicts, 2012 World Environmental and Water Resources Congress, 2247–2252, 2012.

Madani, K., Zarezadeh, M., and Morid, S.: A new framework for resolving conflicts over transboundary rivers using bankruptcy methods, Hydrol. Earth Syst. Sci., 18, 3055–3068, doi:10.5194/hess-18-3055-2014, 2014.

Meyers, E. M.: The Oxford encyclopedia of archaeology in the Near East, Oxford University Press, 1997.

Mianabadi, H., Mostert, E., Zarghami, M., and van de Giesen, N.: A new bankruptcy method for conflict resolution in water resources allocation, J. Environ. Manage., 144, 152–159, doi:10.1016/j.jenvman.2014.05.018, 2014.

Mianabadi, H., Mostert, E., Pande, S., and van de Giesen, N.: Weighted Bankruptcy Rules and Transboundary Water Resources Allocation, Water Resour. Manage., 29, 2303–2321, 2015.

O'Neill, B.: A problem of rights arbitration from the Talmud, Mathematical Social Sciences, 2, 345–371, 1982.

Pérez, J. A., Reus, J. A. S., and Gil, M. d. C. M.: A new prospect of additivity in bankruptcy problems, Instituto Valenciano de Investigaciones Económicas, SA (Ivie), 2010.

Prins, G.: Threats without enemies: facing environmental insecurity, Routledge, 2013.

Rieu-Clarke, A., Moynihan, R., and Magsig, B.-O.: UN Watercourses Convention, IHP-HELP Centre for Water Law, Policy and Science (under the auspices of UNESCO), Dundee, 2012.

Susskind, L.: Breaking Robert's rules, Negotiation Journal, 22, 351–355, 2006.

Thomson, W.: Axiomatic and game-theoretic analysis of bankruptcy and taxation problems: a survey, Mathematical Social Sciences, 45, 249–297, 2003.

Thomson, W.: Lorenz rankings of rules for the adjudication of conflicting claims, Economic Theory, 50, 547–569, 2012.

UN Water: UN-Water factsheet on transboundary waters, available at: http://www.un.org/waterforlifedecade/transboundary_waters.shtml, last access: 6 August 2014.

Westing, A. H.: Global resources and international conflict: environmental factors in strategic policy and action, Oxford University Press, 1986.

Proc. IAHS, 374, 17–21, 2016
proc-iahs.net/374/17/2016/
doi:10.5194/piahs-374-17-2016


Open Access PIAHS

# Post conflict water management: learning from the past for recovery planning in the Orontes River basin

**Myriam Saadé-Sbeih**[1], **François Zwahlen**[2], **Ahmed Haj Asaad**[1], **Raoul Gonzalez**[1], **and Ronald Jaubert**[1]

[1]Graduate Institute of International and Development Studies, Chemin Eugène-Rigot 2, 1202 Geneva, Switzerland
[2]Centre for Hydrogeology and Geothermics (CHYN), University of Neuchâtel, Rue Emile-Argand 11, 2000 Neuchâtel, Switzerland

*Correspondence to:* Myriam Saadé-Sbeih (myriam.saade@gmail.com)

Published: 17 October 2016

**Abstract.** Water management is a fundamental issue in post-conflict planning in Syria. Based on historical water balance assessment, this study identifies the drivers of the profound changes that took place in the Lebanese and Syrian parts of the Orontes River basin since the 1930s. Both drastic effects of the conflict on the hydro-system and the strong uncontrolled anthropization of the river basin prior to the crisis have to be considered in the design of recovery interventions.

## 1 Introduction

Current human–hydrosystem interactions are shaped by past water uses, hydraulic development and hydro-social configurations (Swyngedouw, 2009). Historical reconstruction of water use is considered a powerful approach to identify drivers of changes in human-water relationships in order to design realistic future water management scenarios and relevant policies (Grouillet et al., 2015; Lu et al., 2015). While general trends such as population growth or increase of irrigated areas have been considered as major drivers of changes, little attention has been paid to radical changes, for example in the case of armed conflicts. In those cases, evidence-based decision making is requested for sound emergency interventions and post-conflict planning. However, decision makers are faced with poor and asymmetric information challenging their ability to define appropriate responses (Gibson et al., 2005).

## 2 Study area

The transboundary Orontes (Al Asi) River basin, shared between Lebanon, Syria and Turkey, is representative of the global changes in water use that impacted hydrosystems during the last century in the southern and eastern parts of the Mediterranean basin. Since the 1950s and especially in the 1990s, the area witnessed an intensive surface and ground water development, mostly for agricultural purposes, leading to a sharp decrease in the river discharge in the middle course of the basin; the drying up of numerous springs and localized groundwater overexploitation in several areas (Zwahlen et al., 2014). After the outbreak of the Syrian crisis in 2011, the study area is a key region in the conflict (Haj Asaad and Jaubert, 2014) and witnessed a drastic spatial redistribution of the water consumption and a radical reconfiguration of surface and ground water flows. As a consequence of the decrease in water consumption in the middle reach of the basin, the Orontes River discharge increased by two-fold at the Syrian-Turkish border in 2014; whereas water consumption increased in "shelter" areas, where displaced populations have taken refuge.

## 3 Method

This study aims to provide guidance for recovery planning in the post conflict transition period in the Lebanese and Syrian parts of the Orontes River basin. It identifies through a historical water balance assessment, the socioeconomic drivers (Reynard et al., 2014) of the profound changes that took place at the regional catchment level. The multidisciplinary

Published by Copernicus Publications on behalf of the International Association of Hydrological Sciences.

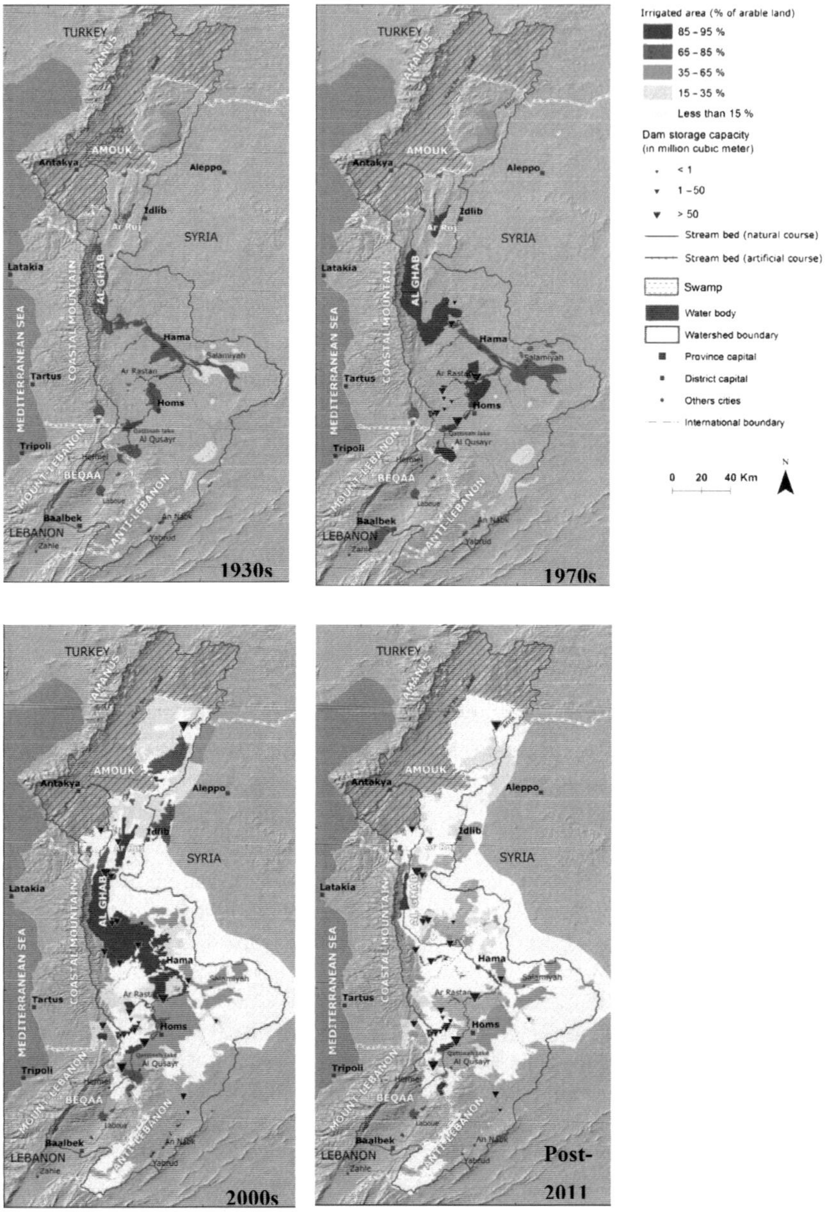


**Figure 1.** Evolution of irrigated areas for the 1930s; 1970s; 2010s and after 2011.

approach consists of assessing the human-hydrosystem co-evolution through the calculation of water balances at two scales for four representative periods in terms of water exploitation: the 1930s; 1970s; 2000s and post-2011. The lack of accurate and comprehensive long term datasets is tackled by confronting different estimation methods of historical surface and ground water flows: literature reviews, historical records, statistical analysis, expert interviews, maps and satellite images analysis to assess land use changes. Water balance, as a synthesis tool, is used to check the plausibility of the estimated parameters.

## 4 Transformation of the hydro-system between the 1930s and the 2000s

In terms of magnitude, irrigation is the main driver of changes in the Orontes River basin water balance: in the last century, water consumption has been significantly modified by an intensive anthropization, especially the development of

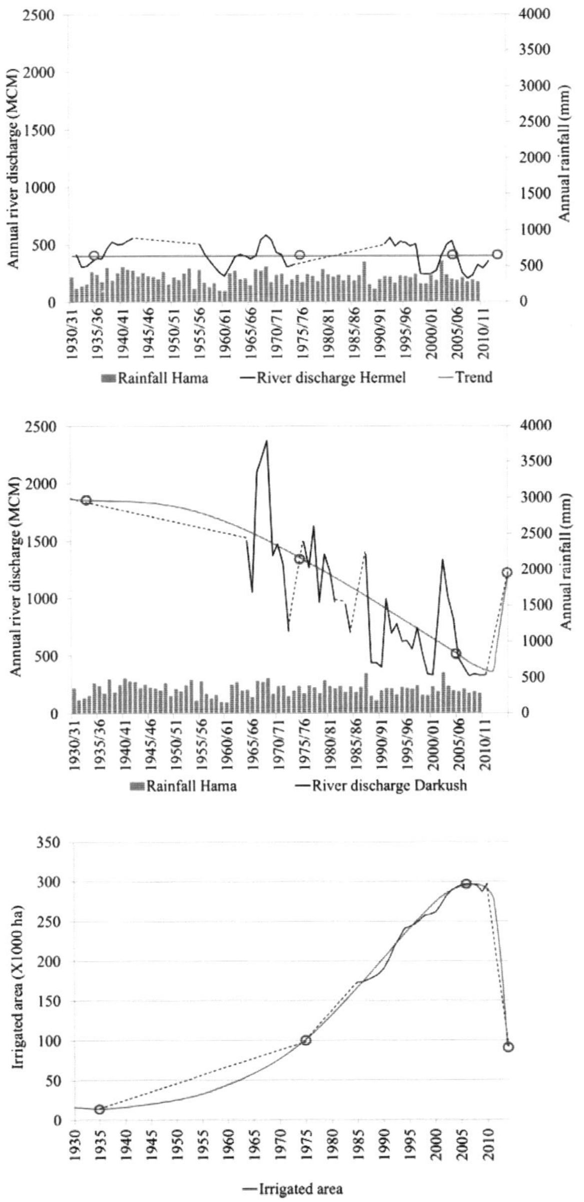


**Figure 2.** Changes in river discharge and irrigated areas 1930/1931–2013/2014. River discharge series from UN-ESCWA and BGR (2013).

irrigation (Fig. 1). The caused by human activities increased by a factor 9 between the 1930s and the 2000s (from 235 to 2210 MCM $yr^{-1}$), with an increase in consumption by a factor 5 between the 1930s and the 1970s; and about a factor 2 (1.8) between the 1970s and the 2000s (Fig. 2).

As a consequence, the Orontes River discharge decreased sharply in the downstream part of the basin: at the Darkush gauging station, the annual discharge decreases from 1850 MCM $yr^{-1}$ in the 1930s to 1250 in the 1970s and 600 in the 2000s. According to our last and more fitting water balance reconstitutions, a first negative variation of the groundwater storage is estimated for the 1970s, with localized groundwater extraction in the Salamyah and Karnaz areas. Then an important variation of the storage is estimated in the 2000s ($-155$ MCM $yr^{-1}$ for the global basin) due to generalized intensive groundwater extraction in different parts of the study area: mainly Al Qaa – Qusayr area in the Muhardah sub-catchment ($-50$ MCM $yr^{-1}$), North Muhardah area in the Jisr ash Shughur sub-catchment ($-85$ MCM $yr^{-1}$) and ar Ruj area in the Darkush sub-catchment ($-20$ MCM $yr^{-1}$).

The assessment at different temporal and spatial scales shows that each sub-catchment has different dynamics and evolution trends. The upstream sub-catchment remained relatively stable during the last century. The downstream sub-catchments faced major changes in terms of surface and ground water flows and storage: a sharp decrease in river and spring discharges and localized negative variation of groundwater storage (Fig. 3). For example, in the Jisr ash Shughur sub-catchment, 54 % of the total average inflow in the 1930s was provided by surface water (direct runoff or river discharge), it decreases to 43 % in the 1970s and to 36 % in the 2000s. Future water management policies have to consider irrigation as a main target and to be adapted to the specific conditions of each sub-catchment.

## 5 Drastic changes after 2011

The Syrian crisis led to a drastic redistribution of water use patterns and a sharp decrease in water consumption for irrigation, because of combats, population displacement, destruction of water infrastructures and the lack of energy and agricultural inputs. In the first years of the crisis water consumption fell by a factor 3.3 (from 2240 to 670 MCM $yr^{-1}$), the river discharge increased in the downstream part, by 2-fold at the Syrian-Turkish border. A positive variation in groundwater storage is estimated for downstream sub-catchments. There is a disintensification of water consumption at large scale, but it should be noticed that, in some places, there is a local intensification of water consumption (Al Qaa area for example).

## 6 Conclusions

Water management is a fundamental element in post-conflict planning in Syria and the Orontes River basin. To design relevant water recovery program and policy and prevent water mismanagement, decision makers will have to consider the drastic effects of the conflict on the hydro-system since 2011, as well as the strong uncontrolled anthropization of the river basin prior to the crisis. They should take into consideration expected drivers, such as irrigation and energy policies, as well as unexpected drivers, such as the political crisis. As-

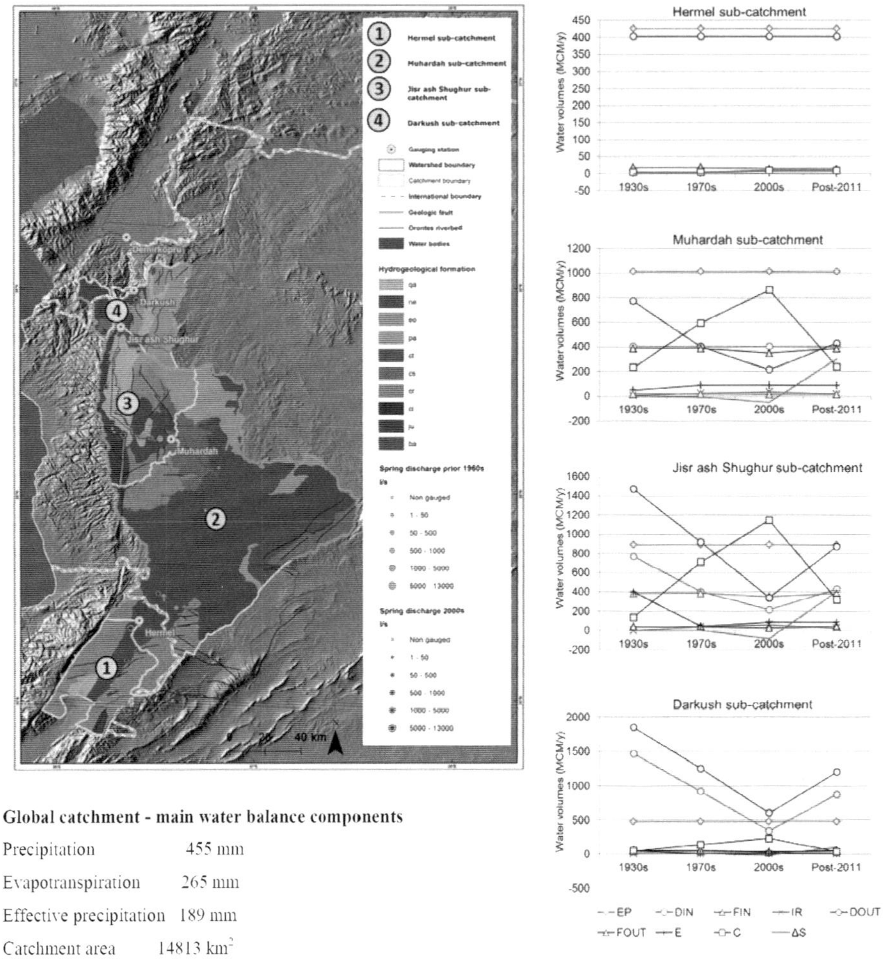


**Figure 3.** Historical water balance in the Lebanese and Syrian parts of the Orontes River basin. EP: Effective precipitation; DIN: River discharge (inflow); FIN: Ground water or surface water inflow; DOUT: River discharge (outflow); FOUT: Ground water or surface water outflow; $E$: Water surface evaporation; $C$: Anthropic evapotranspiration; $\Delta S$ change in groundwater storage.

sessing present and past situations is a precondition to design sound policies.

## 7 Data availability

For Fig. 2, data are coming from the following sources:

River discharge series from UN-ESCWA and GBR (2013) http://waterinventory.org/sites/waterinventory.org/files/chapters/Chapter-07-Orontes-River-Basin-web_1.pdf.

Irrigated areas based on NAPC Syrian Agriculture Database http://www.agriportal.gov.sy/napcsyr/sadb.htm.

For Figs. 1 and 3, the methodology and data sets will be detailed in further publications and made available on the Orontes Project website https://www.water-security.org/ (planned to be available beginning of next year).

**Acknowledgements.** This study is part of the Orontes River Basin Research Program led by the Graduate Institute of International and Development Studies with the support of the Swiss Development and Cooperation Agency. It aims to provide baseline elements to support emergency interventions and post-conflict planning in Syria and to contribute to coordinated management of the Orontes River Basin's shared water resources in the long term (Jaubert et al., 2014). The study is funded by the Swiss Federal Department of Foreign Affairs.

## References

Gibson, C. C., Andersson, K., Ostrom, E., and Shivakumar, S.: The Samaritan's Dilemma: The Political Economy of Development Aid, Oxford University Press, 288 p., 2005.

Grouillet, B., Fabre, J., Ruelland, D., and Dezetter, A.: Historical reconstruction and 2050 projections of water demand under anthropogenic and climate changes in two contrasted Mediterranean catchments, J. Hydrol., 522, 684–696, doi:10.1016/j.jhydrol.2015.01.029, 2015.

Haj Asaad, A. and Jaubert, R.: Geostrategic stakes and the impact of the conflict in the Orontes River basin, Confluences méditerranée, 89, 173–184, doi:10.3917/come.089.0173, 2014.

Jaubert, R., Saadé-Sbeih, M., Al Dbiyat, M., and Haj Asaad, A.: E-Atlas of the Orontes River basin, https://www.water-security.org/, 2014.

Lu, Z., Wei, Y., Xiao, H., Zou, S., Xie, J., Ren, J., and Western, A.: Evolution of the human-water relationships in the Heihe River basin in the past 2000 years, Hydrol. Earth Syst. Sci., 19, 2261–2273, doi:10.5194/hess-19-2261-2015, 2015.

Reynard, E., Bonriposo, M., Graefe, O., Homewood, C., Huss, M., Kauzlaric, M., Liniger, H., Rey, E., Rist, S., Schädler, B., Schneider, F., and Weingartner, R.: Interdisciplinary assessment of complex regional water systems and their future evolution: how socioeconomic drivers can matter more than climate, Wiley Interdiscip. Rev. Water, 1, 413–426, doi:10.1002/wat2.1032, 2014.

SADB (Syrian Agriculture Database): http://www.agriportal.gov.sy/napcsyr/sadb.htm, last access: 9 September 2014.

Swyngedouw, E.: The Political Economy and Political Ecology of the Hydro-Social Cycle, J. Contemp. Water Res. Educ., 142, 56–60, doi:10.1111/j.1936-704X.2009.00054.x, 2009.

UN-ESCWA and BGR (United Nations Economic and Social Commission for Western Asia; Bundesanstalt für Geowissenschaften und Rohstoffe), Inventory of Shared Water Resources in Western Asia, Beirut, 2013.

Zwahlen, F., Bakalowicz, M., Gonzalez, R., Haj Asaad, A., Saadé-Sbeih, M., and Jaubert, R.: Groundwater flows in the Orontes River basin and groundwater in the Syria-Lebanon water sharing agreement, International Association of Hydrogeologists, Selected Papers on Hydrogeology, 53–62, 2016.

Proc. IAHS, 374, 23–28, 2016
proc-iahs.net/374/23/2016/
doi:10.5194/piahs-374-23-2016


Open Access
PIAHS

# To what extent do they sway Australian water management decision making?

**Maureen Papas**

Honorary Fellow, Faculty of Law, University of Western Australia, 35 Stirling Highway, Crawley, Perth, Western Australia, Australia

*Correspondence to:* Maureen Papas (maureen.papas@uwa.edu.au)

Published: 17 October 2016

**Abstract.** At a time when the reliability of freshwater resources has become highly unpredictable, as a result of climate change and increased droughts frequency, the role of scientific evidence in forecasting the availability of seasonal water has become more critical. Australia is one of the driest inhabited continents. Its freshwater availability is highly variable, which poses unique problems for the management of the nation's water resources. Under Australia's federal system, water management challenges have been progressively dealt with through political institutions that rely on best available science to inform policy development. However, it could be argued that evidenced-based policy making is an impossible aim in a highly complex and uncertain political environment: that such a rational approach would be defeated by competing values and vested interests across stakeholders. This article demonstrates that, while science has a fundamental role to play in effective water resource management, the reality *on the ground* often diverges from the intended aim and does not always reflect efforts at reform. This article briefly reviews the *Water Act 2007* (Cth) and comments on why policy makers need to manage rather than try to eliminate uncertainty to promote change.

## 1 Introduction

Decision making in water resource management in Australia has long been driven by the need to adapt to changes in water availability and to respond to increasing water scarcity. Australia is regarded as the world's driest continent, much of it flat, and with a legacy of high levels of salinity buried deep within it (Cullen et al., 2012). Salinity is an inherent feature of the Australian landscape, which if left unmanaged, has serious implications for water quality, biodiversity, land productivity and the supply of water generally (Thompson, 2014). Similarly, a flat continent suggests very low or no mountain ranges, with no permanent snow or glaciers, and therefore, no freshwater source other than rivers and aquifers (i.e. receptacles for groundwater). To compound the problem, Australia's weather is influenced by a highly variable climate with an annual cycle of wet and dry periods (Bureau of Meteorology, 2013). It could be argued that given these factors, science – based decision-making would improve sustainable water management practices. When the Federal *Water Act 2007* (Cth)[1] was introduced in response to Australia's millennium drought (2001–2009) (van Dijk et al., 2013), and in an attempt to prevent the further decline of water resources in the Murray Darling Basin (MDB) (also referred to as the Basin), one of the key reform measures was to prepare a Basin Plan to cap water extractions to levels that scientific evidence indicated would be sustainable. In short, the Act was in response to a severe drought period, unprecedented in the instrumental record, and called on water management policy to recognise the impact of climate change. However, the Basin-wide hydrological modelling required determining the level of water extraction from the environment, which was vigorously contested (explained further in Sect. 3) and proved challenging to guiding management decisions. This suggests that science-based knowledge had a limited role in stakeholder engagement outcomes. To demonstrate this point, the article proceeds in three sections. The first provides a very brief overview of the geography and manage-

[1]Cth refers to Commonwealth indicating that the Act is valid Australia wide.

Published by Copernicus Publications on behalf of the International Association of Hydrological Sciences.

ment framework in the MDB, the second section critically evaluates a key objective of the *Water Act 2007*. The final section explores the role of science in policy formulation and its effect on decision making relating to water resources more generally. This article concludes that although science is key to policy formulation, it is not in itself sufficient to reconcile competing stakeholder interests or to decide which objectives should be achieved and at what cost.

## 2 Background overview: brief geography and governance of the MDB

Located in south eastern Australia, the MDB is considered one of the flattest catchments in the world largely comprised of vast low-lying dry plains (Murray-Darling Basin Authority, 2012a). Although the MDB average annual rainfall supplies is in excess of 530 000 GL of freshwater, 90 % evaporates or transpires back into the atmosphere, leaving less than 10 % to drain as run off into rivers, lakes, streams and into groundwater aquifers (MDBA, 2012b). Despite these challenges, the MDB is Australia's agricultural heartland, producing one-third of the nation's food supply. It also plays an important role in ecological diversity and supports a wide range of animals, plants and ecosystems. The MDB covers an area of more that one million square kilometres (about 14 % of Australia's land mass) (MDBA, 2012c) and spans four states (Queensland, New South Wales, Victoria and South Australia) and the Australian Capital Territory. Good governance and reliable seasonal predictions of water availability for effective use of water is paramount for the MDB.

Historically, a co-operative framework governed water sharing in the MDB (whereby states and Commonwealth governments had to reach consensus, to manage the Basin resources), rather than a constitutional framework to ensure a right of supply (Connell and Grafton, 2011). This co-operative framework was fundamental and crucial to the MDB because it required federal and state governments to agree on an overall plan, with each jurisdiction responsible for its implementation (Connell, 2007). Yet, in the last twenty years, concerns about over-allocated water and the continuing decline of water resources in the river system remains a serious issue, severely affecting not just environment quality but also the security of water entitlements for consumptive users (Scanlon, 2006; Papas, 2007; Connell, 2011). These problems have arisen despite a number of management programmes and joint initiatives between states and the Commonwealth – namely, the National Water Initiative and the Living Murray – that had been implemented expressly to address the highly variable water conditions that characterise the MDB (Papas, 2007, p. 90). These reforms were implemented to ensure the ecological system is restored to good health and retains an optimum level of productivity. However, for these measures to work effectively, inter-jurisdictional arrangements require that each government maintain a shared responsibility as to how the water is managed and allocated (Papas, 2007). This is especially true of Australia's federal system of government, because primary responsibility for water and environmental management rests with state governments (Kildea and Williams, 2010).

Australia's most recent water reform, the Water Act 2007, was introduced in the context of the most devastating drought in recorded history (Skinner and Langford, 2013, p. 871). The reform occurred within an existing mix of institutional arrangements, which inherently suggests that reforms have to adapt to a set of constraints (Skinner and Langford, 2013, p. 872). Yet, the principles of governance that underpin the *Water Act 2007*, marked a renewed approach to environmental protection and the way scientific knowledge was used to arbitrate the final outcome. Some of the issues are explored below.

## 3 A new governance framework: the Water Act 2007

The *Water Act 2007* established a new governance structure, whereby the MDB was brought under Commonwealth management for the first time since Federation (Kildea and Williams, 2011). The passing of the Act represented a widespread consensus that ongoing unsustainable levels of water extractions in the Basin needed to be seriously tackled through a central statutory planning process that could provide clear guidance in the matter (Fisher et al., 2010). In effect, the Act imposes a national interest in the management of the MDB system, rather than leaving its management at the state level.

The Act also established an independent Murray–Darling Basin Authority (MDBA), an agency of the federal government, to prepare a Basin Plan (MDBA, 2007). The aim of this Basin Plan was to apply best available science to define an environmentally sustainable level of take and to reduce over-allocated water entitlements that threatened water security (Skinner and Langford, 2013). That is, the Basin Plan was to set enforceable limits on the quantity of water (both surface and groundwater) – known as sustainable diversion limits (SDLs) – that could be extracted from the Basin as a whole, without compromising key environmental assets and ecosystem functions. This is the amount of water, expressed as the long-term average annual volume, that may be taken from a given river or aquifer (Hamstead and O'Keefe, 2008). These limits are central to the Basin Plan in securing the long-term health of the MDB, and come into effect in 2019 (MDBA, 2014).

Hydrologic modelling was used to determine SDL across the Basin (MDBA, 2012a). However, the diversion limits that were initially proposed were systematically contested and provoked considerable community anger. Some academics argue that one of the main issues to provoke strong opposition from water users was whether in developing the Basin

Plan, the MDBA was prepared to give future environmental considerations precedence over current social and economic factors (Kildea and Williams, 2011). Their concerns were valid given that in previous decades, water needs for the environment and for irrigation were often treated as mutually exclusive, and served the political purposes of particular interest groups (e.g. Musgrave, 2008; Skinner and Langford, 2013, p. 890). Similarly, the distinctive "battle line" between irrigators and regulators to engage farmers with water regulation was fundamentally weak; the decision–making process was fraught and highly politicised, and this resulted in little progress (Holley and Sinclair, 2012; Doremus and Tarlock, 2008). Consequently, some academics have questioned the reliance of scientific evidence to best inform water resources regulation and facilitate management decision making about socio-economic effects (Liu et al., 2008). Others maintain that given the prevalence of the role of modelling in natural resources and environmental regulation in other countries, policy makers should engage critically with these scientific tools and not neglect their effect as the realm of scientific expertise (Fisher et al., 2010).

### 3.1 The role of science

Science is considered an essential discipline used to promote policy formulation and political decision making (Davis et al., 2015, p. 1). As Frewer and Salter (2002, p. 138) explain, the role of science as an aid to policy development and to guide complex decision making is based on two assumptions:

> ... first, that the advice and, in particular, its predictive content is accurate and, second, that the public sees the advice as authoritative and the decisions and policy flowing from that advise as legitimate.

Despite this, it could be argued that the context in which specialist advice is sought, plays a role in the interpretation of scientific knowledge. As we saw above, science was used to frame the objectives of the SDLs, and thus to formulate policy. However, there was considerable reluctance from particular interest groups to comply with the proposed limits. Their reluctance, besides their concerns about the primacy the Water Act gives environmental needs in decision making (e.g. Godden, 2011), also stemmed from their expressed concern about the MDBA's failure to engage local experts in the hydrological modelling content of the Basin Plan (Parliament of the Commonwealth of Australia, 2011). Pileke (2007) argues that this undermined both the reputation of scientific knowledge to guide policy development and its effectiveness as a regulatory tool to inform decision making. In contrast, other commentators suggest that complex public policy problems, including those faced by catchment water managers, must be made in the best interest of the public, which represents a broad cross-section of views and opinions (Heazle, 2010, p. 6; Skinner and Langford, 2013, p. 873). In this context, while scientific advice was desirable, it remained subsidiary to community participation in decision making.

Another issue worth considering when seeking scientific advice is the notion of uncertainty. Some observers assert that scientists use statements of uncertainty to characterise information that is by nature never "black and white", or to indicate what they do not yet know (Gibbs et al., 2013). In short, uncertainty is normal in scientific research (Gibbs et al., 2013). However, it could be argued that uncertainty has an adverse impact on complex decision making. For example, uncertainty can play a key role when policy makers attempt to justify one policy over another, particularly when both the level of uncertainty and the political stakes are high (Heazle, 2010, p. 135). Heazle (2010, p. 135) notes that uncertainty strips our ability to proclaim the best course of action on the basis of evidenced-based knowledge alone, while simultaneously granting license to speculate on various policy alternatives for those seeking to frame policy issues. Scientific findings are not exact. Therefore, policy makers like to draw their decisions based on various scenarios (or findings) to guide their policy options. Still, this suggests that the scope for scientific uncertainty is a valid factor that influences decision making. Similarly, Skinner and Langford (2013, p. 888) believes that the effect on constituents, ideological perspective and socio-economic implications also actively contribute to what happens on the ground.

More recently, predictions by Australia's Bureau of Meteorology (BoM) of a "substantial" El Niño event in 2015 for the first time in five years (BoM, 2015) points to the need to adapt to worsening conditions. El Niño periods in Australia are generally associated with warmer temperatures across much of the country, below average rainfall and higher risk of bush fires (The Guardian, 2015). Under such extreme forecasts, the reliability of seasonal water is likely to be even less predictable, further undermining efforts to manage water resources sustainably.

### 3.2 The way forward

The Basin Plan, which received final approval from Australia's Federal Minister in 2012, allows the MDBA to amend SDLs up or down within defined parameters (see *Water Act 2007* (s23a) through the SDL mechanisms by 2015 (Murray-Darling Basin Authority, 2015). These mechanisms were implemented to ensure sufficient flexibility in adapting to Australia's highly variable weather conditions and were based on best available scientific knowledge in the context of climate change (Murray-Darling Basin Authority, 2015). This suggests that the potential effect of climate change in the reduction of precipitations and run offs (overland flow) are considered in government policy. However, Pittock (2013) argues that the 2012 Basin Plan makes no specific provision for managing the risks posed by climate change, which raises important questions in terms of compliance and regulatory decision making, by both water users and their community.

The *Water Act 2007* (s22, item 8) states that the Basin Plan must include a mechanism for determining whether there has been compliance with SDLs but provides no clear requirements about the methods the MDBA must adopt to include this mechanism (Environment Defenders Office, 2012). Basin state governments are required to implement the SDLs through their own water resources management plans, which typically operate at the scale of river catchment (Foerster, 2014, p. 264). Therefore, it is reasonable to assume, that a lack of clear guidelines for compliance means that water users may choose to exceed their limits, which would compromise the sustainability of water extractions across the Basin and its states (particularly downstream users) (Gardner et al., 2014, p. 4; see also Holley and Sinclair, 2012, p. 179). Further, the MDBA expounds on its adaptive management approach to addressing the effects of climate change on water resources, but does not suggest that adaptive responses may require a reduction of water entitlements (Gardner et al., 2014, p. 4).

In dry periods, it is conceivable that the attitudes of water users and the community to further reducing their entitlements may well result in wilful non-compliance with the SDLs, and a lack of cooperation. Indeed, governments have been found to struggle to provide an effective and comprehensive compliance and enforcement regime for water (Holley and Sinclair, 2012, p. 149). In addition, the national policy in relation to the risks and costs of future reductions in water entitlements is currently assigned to water-entitlement holders, which means that a reductions of water availability as a result of climate change will not be compensated by the Australian Government (Gardner et al., 2014, p. 5). Gardner argues that this policy is significantly unjust, and suggests that economic and environmental losses from reduced water entitlements could be calculated in pecuniary terms, for the purpose of determining compensation to water holders who suffer such losses (Gardner et al., 2014, p. 5). Nonetheless, the Australian Government has demonstrated some level of engagement with and commitment to enforcing standards, with the introduction of the *National Framework for Compliance and Enforcement Systems for Water Resource Management* in 2012 (Department of the Environment, 2012). This initiative creates a prospect for a great deal of improvement in water management in Australia, and it might be necessary to enforce compliance through financial concessions or more stringent enforcement mechanisms (or both) to promote change.

This discussion leads to a question about the long-term implications for scientific research generally and best science advice more specifically. It points to two major issues: first, while science-based knowledge can provide best available hydrological modelling, and the most accurate data, there are clearly inconsistencies between efforts at reform and on-the-ground outcomes when mechanisms to monitor compliance are poorly enforced. These inconsistencies are a major concern given that the Basin Plan outlines legally binding extraction limits on all Basin states, and that ultimately, the effectiveness of these overarching SDLs is vital to the management regime of the MDB under the Federal Government. Secondly, the former Prime Minister Tony Abbott's[2] decision not to have a dedicated science minister in his government, until a recent cabinet reshuffle, raises an important question about the Coalition Government's attitude to the discipline of science (The Guardian, 2013). Similarly, the 2014-15 cuts in the Federal Budget to science programs and research grants has left many observers deeply concerned about Australia's future prosperity in the sciences (Australian Academic of Science, 2014)[3].

At a time when changing weather patterns due to climate change are predicted to affect the hydrological cycle across various regions more than ever, it is important to recognise the value of research efforts that provide the scientific knowledge necessary to drive future preparedness. Australia is about to face the adverse impacts of another El Niño event, which will be compounded by the effects of climate change. In this context, the federal government of Australia should take the long-term view that policy informed by science-based knowledge must transcend both political lines and stakeholders' interests. A good first step in this direction would be to bring science under a single responsible minister and the policy impetus to encourage science to do what it does best – that is, develop sound methodologies for water resource assessments to inform sustainable water management practices.

## 4 Conclusions

The role of science in policy making demonstrates that specialist knowledge may or may not influence complex decision making. Despite the idea that governments and organisations across the world consider evidence-based policy a key factor to making sound policy, the reality tends to fall short of this idea. This article has demonstrated that the treatment of uncertainty in specialist advice largely determines how well the advice fits with, or contradicts the policy goals dominated by the political and stakeholder interests. In Australia, as elsewhere, the role of science remains vital in formulating the evidence and forecasting necessary to prepare against unprecedented changes to our climate and our water ways. More than ever, water managers and decision makers need to be well informed about the consequences of poor policy implementation and the dangers of climate change on

[2]The Coalition Government led by the Prime Minister Tony Abbott was ousted in a dramatic party coup on 14 September 2015. Malcolm Turnbull won the prime ministership from Tony Abbott and has been sworn in as Australia's new Prime Minister.

[3]Prime Minister Malcolm Turnbull recently announced that science will have a more predominant role in his government (see Department of Industry, Innovation and Science http://www.industry.gov.au/science/Pages/default.aspx).

water resources. The prospect of an El Niño event in Australia suggests that the future of water management will need to adapt to highly variable seasonal precipitations. Science can help prepare for such anticipated changes but decision makers have to be motivated to act accordingly.

**Acknowledgements.** Some elements of this article where first developed in one of the papers' prepared by the author, as part of a PhD by publication, Macquarie University, Sydney, Australia. The author would like to thank the German Federal Institute of Hydrology (BfG) and the International Centre for Water Resources and Global Change for the opportunity to present. The author would like to also acknowledge UNESCO's financial support towards her attendance to the conference in Koblenz, Germany (no. 4500282573).

## References

Australian Academy of Science: 2014-15 Federal Budget Analysis of Measures Relating to Science, Research and Innovation, From the Australian Academy of Science, https://www.science.org.au/files/userfiles/support/submissions/2014/2014-15-federal-budget-initial-analysis.pdf, 2014.

BoM: Mature El Nino in the Tropical Pacific Region, http://www.bom.gov.au/climate/enso/#tabs=Overview, 2015.

Bureau of Meteorology (BoM): Australian Water Resources: Assessment 2012, http://www.bom.gov.au/water/awra/2012/documents/summary-hr.pdf (last access: 1 August 2016), 2013.

Connell, D.: Water Politics in the Murray-Darling Basin, The Federation Press, Sydney, 2007.

Connell, D. and Grafton, Q. R.: Water Reform in the Murray-Darling Basin, Water Resour. Res., 47, W00G03, doi:10.1029/2010WR009820, 2011.

Cullen, P., Flannery, T., Harding, R., Morton, S., Possingham, H., Saunders, D., Thom, B., Williams, J., Young, M., Cosier, P., and Boully, L: Blueprint for a Living Continent: A Way Forward from the Wentworth Group of Concerned Scientists (1 November 2002), http://wentworthgroup.org/publications/, 2012.

Davis, K. E., Kingsbury, B., and Merry, S. E.: The Local-Global Life of Indicators: Law, Power, and Resistance, in: The Quiet Power of Indicators: Measuring Governance, Corruption and the Rule of Law, Cambridge Studies in Law and Society, edited by: Merry, S. E., Davies, K. E., and Kingsbury, B., CUP, 2015.

Doremus, H. and Tarlock, A. D.: Water War in the Klamath Basin: Macho Law Combat Biology and Dirty Politics, Island Press, Washington, DC, 2008.

Environment Defenders Office: Legal analysis of the Proposed Murray-Darling Basin Plan, Environment Defenders office (Victoria), https://envirojustice.org.au/downloads/files/law_reform/EDO_legal_analysis_of_draft_MD_Basin_Plan.pdf (last access: 1 August 2016), 2012.

Fisher, E., Pascual, P., and Wagner, W: Understanding environmental models in their legal and regulatory context, J. Environ. Law, 22, p. 251, 2010.

Foerster, A: Water law: adapting to climate change in South-eastern Australia, in: Water and the Law: Towards Sustainability, edited by: Kidd, M., Feris, L., Forombo, T., and Iza, A., The IUCN Academy of Environmental Law, Edward Elgar Publishing Limited, 2014.

Frewer, L. and Salter, B.: Public attitudes, scientific advice and the politics of regulatory policy: The case of BSE, Science and Public Policy, 29, p. 138, 2002.

Gardner, A., Blaskers, R., and Hartley, M.: Legal Scenarios for Integrated Modelling, Australasian Journal of Natural Resources Policy and Law, 17, p. 4, 2014.

Gibbs, P., Hanlon, M., Hardaker, P., Hawkins, E., McDonald, A., Maskell, K., Mayfield, H., Mclean, A., Morris, E., Mylne, K., Naylor, M., Palmer, T., Rawlins, M., Royse, K., Smith, L., So, E., Spielgelhalter, D., Stainforth, D., Stewart, I., Tyler, C., De Fotch, F., Brown T., and Innocent, T.: Making Sense of Uncertainty: Why uncertainty is part of science, Sense about Science, p. 3, London, UK, available at: http://www.senseaboutscience.org/data/files/resources/127/SAS012_MSU_reprint_compressed.pdf, 2013.

Godden, L.: Submission to the Inquiry into the Provisions of the *Water Act 2007*. The Senate Standing Committee on Legal and Constitutional Affairs, Canberra, Australia, 2011.

Hamstead, M., Baldwin, C., and O'Keefe, V.: Water Allocation Planning in Australia – current practices and lessons learned Waterlines No 6, National Water Commission, Canberra, http://archive.nwc.gov.au/__data/assets/pdf_file/0011/11018/Waterlines_no_6.pdf, 2008.

Heazle, M.: Uncertainty in Policy Making: Values and Evidence in Complex Decisions, Earthscan, 2010.

Holley, C. and Sinclair, D.: Compliance and enforcement of water licenses in NSW: Limitations in law, policy and Institutions, Australasian Journal of Natural Resources Policy and Law, 15, 149–189, 2012.

Kildea, P. and Williams, G.: The Constitution and the management of water in Australia's rivers, Sydney Law Review, 32, 595–616, 2010.

Kildea, P. and Williams, G.: Journal Excerpt: The Water Act and the Murray-Darling Basin Plan, Public Law Review, 9, 2011.

Liu, Y., Gupta, H., Springer, E., and Wagener, T.: Linking science with environmental decision making: Experiences from an integrated modelling approach to supporting sustainable water resources management, Environ. Model. Softw., 23, 846–858, 2008.

Murray-Darling Basin Authority (MDBA): The Water Act, http://www.mdba.gov.au/about-us/governance/water-act (last access: 1 August 2016), 2007.

Murray-Darling Basin Authority (MDBA): Hydrological modelling, http://www.mdba.gov.au/managing-water/hydrological-modelling (last access: 1 August 2016), 2012a.

Murray-Darling Basin Authority (MDBA): About the Basin, http://www.mdba.gov.au/sites/default/files/archived/proposed/Hydro_Modelling_Report.pdf (last access: 1 August 2016), 2012b.

Murray-Darling Basin Authority (MDBA): SDL Adjustment Ecological Elements Methods Development Report Review of Final Project Report by the Independent Review Panel, http://www.mdba.gov.au/publications/independent-reports/sdl-adjustment-ecological-elements-method-development-report-review (last accessed: 1 August 2016), 2014.

Murray-Darling Basin Authority (MDBA): Geography, http://www.mdba.gov.au/discover-basin/landscape/geography (last accessed 1 August 2016), 2012c.

Murray-Darling Basin Authority (MDBA): The SDL adjustment assessment framework for supply measures, http://www.mdba.gov.au/publications/mdba-reports/sdl-adjustment-assessment-framework-supply-measures (last accessed: 1 August 2016), 2015.
Musgrave, W.: Historical development of water resources in Australia, in: Water Policy in Australia: The Impact of Change and Uncertainty, edited by: Crase, L., Resource for the Future, 2008.
Papas, M.: The Proposed Governance Framework for the Murray-Darling Basin, Macquarie Journal of International and Comparative Environmental Law, 4, 77–90, 2007.
Parliament of the Commonwealth of Australia: House of Representatives Standing Committee on Regional Australia, Of Drought and Flooding Rains: Inquiry into the Impact of the Guide to the Murray-Darling Basin Plan, 2011.
Pileke, R.: The Honest Broker: Making Sense of Science in Policy and Politics, Cambridge University Press, Cambridge, 2007.
Pittock, J.: Lessons from adaptation to sustain freshwater environments in the Murray-Darling Basin, Australia, Wiley Interdisciplinary Reviews: Climate Change, 4, 429–438, 2013.
Scanlon, J.: A hundred years of negotiations with no end in sight: Where is the Murray-Darling Basin Initiative leading us?, Environmental & Planning Law Journal, p. 386, 2006.
Skinner, D. and Langford, J.: Legislating for sustainable basin management: the story of Australia's Water Act (2007), Water Policy, 15, 871–894, 2013.
The Guardian: Australia, where is your science Minister?, 17 September 2013.
The Guardian: El Nino on its way to Australia, says Bureau of Meteorology, 12 May 2015.
Thompson, A: Soil – salinity, Soil salinity, dryland and salinity, http://www.natsoc.org.au/our-projects/biosensitivefutures/part-4-facts-and-principles/ecological-issues/soils-salinity (last access: 1 August 2016), 2014.
van Dijk, A. I. J. M., Beck, H. E., Crosbie, R. S., Jeu, R. A. M., Liu, Y. Y., Podger, G. M., Timbal, B., and Viney, N. R.: The millennium drought in Southeast Australia (2001–2009): natural and human causes and implications for water resources, ecosystems, economy and society, Water Resour. Res., 49, 1040–1057, 2013.

Proc. IAHS, 374, 29–34, 2016
proc-iahs.net/374/29/2016/
doi:10.5194/piahs-374-29-2016


Open Access
PIAHS

# The new portfolio of global precipitation data products of the Global Precipitation Climatology Centre suitable to assess and quantify the global water cycle and resources

**Udo Schneider, Markus Ziese, Anja Meyer-Christoffer, Peter Finger, Elke Rustemeier, and Andreas Becker**

Global Precipitation Climatology Centre, Deutscher Wetterdienst, 63067 Offenbach am Main, Germany

*Correspondence to:* Andreas Becker (andreas.becker@dwd.de)

Published: 17 October 2016

**Abstract.** Precipitation plays an important role in the global energy and water cycle. Accurate knowledge of precipitation amounts reaching the land surface is of special importance for fresh water assessment and management related to land use, agriculture and hydrology, incl. risk reduction of flood and drought. High interest in long-term precipitation analyses arises from the needs to assess climate change and its impacts on all spatial scales. In this framework, the Global Precipitation Climatology Centre (GPCC) has been established in 1989 on request of the World Meteorological Organization (WMO). It is operated by Deutscher Wetterdienst (DWD, National Meteorological Service of Germany) as a German contribution to the World Climate Research Programme (WCRP). This paper provides information on the most recent update of GPCC's gridded data product portfolio including example use cases.

## 1 GPCC background and agenda

The GPCC collects and assures quality of world-wide observational in-situ data from rain gauges in order to provide gridded high-quality and high-resolution land surface precipitation analyses as mandated by WMO's World Climate Research Program and the Global Climate Observing System (GCOS). Based on over 25 years of operation the GPCC gridded data products are built on base of the world-wide largest archive of quality controlled in-situ precipitation data (Fig. 1). All gridded GPCC data products are public available in OGC compliant netCDF format to the community (ftp://ftp-anon.dwd.de/pub/data/gpcc/html/download_gate.html).

### 1.1 A product portfolio for archetype requirements

Depending on the climate related application and service aspired, there is a large variety of user needs in terms of timeliness, homogeneity, resolution and accuracy which cannot be addressed by one data product. As a consequence GPCC has issued a suite of products that contains near-real-time as well as non-real-time products in monthly and daily resolution and in spatial resolutions of 1.0° × 1.0° and 2.5° × 2.5° (only monthly) latitude by longitude. The non-real-time products based on the complete GPCC monthly rainfall station database (the largest monthly precipitation station database of the world with data from more than 100 000 different stations) are also available in 0.5° × 0.5° resolution. GPCC's new global precipitation climatology V.2015 (available in 2.5° × 2.5°, 1.0° × 1.0°, 0.5° × 0.5°, and 0.25° × 0.25° resolution, Fig. 2) based on data from the more than 75 000 stations that feature records longer than 10 years of length is used as background climatology for the other GPCC analyses using the anomaly interpolation method also known as climate aided interpolation (CAI).

### 1.2 Data sources of GPCC

Data from national meteorological and hydrological services, regional and global data collections are mainly used to calculate these products, as well as WMO-GTS data. In order to provide the user with a sufficient level of documentation and long term accessibility, GPCC products issued in year 2011 or later are all referenced by digital object identi-

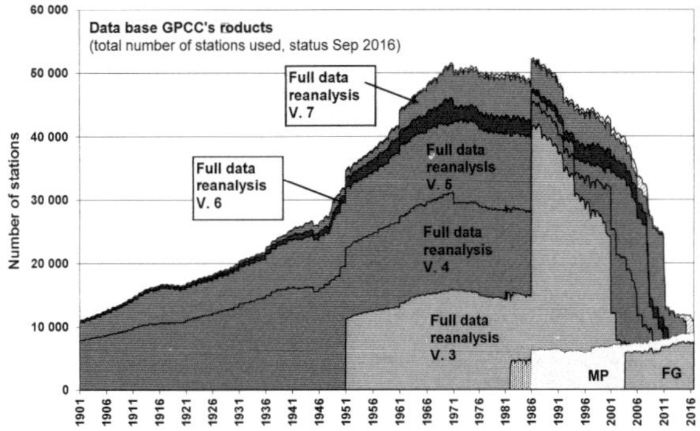


Figure 1. Station data coverage of GPCC's near real-time "Monitoring Product" and "First Guess Product" (indicated by "FG"), as well as for the five most recent Versions of the centennial "Full Data Reanalysis with the most recent Version 7" from June 2015.

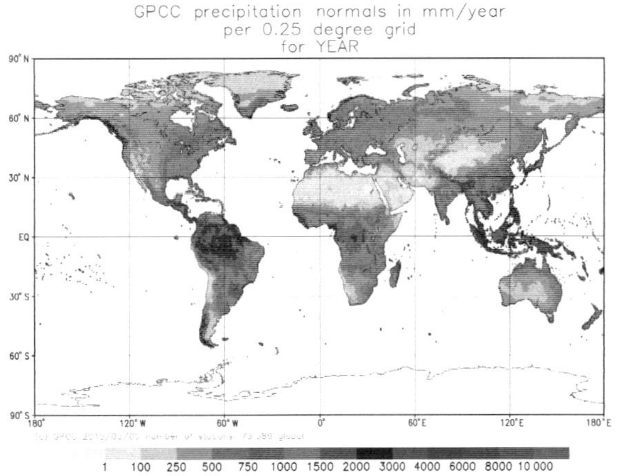


Figure 2. GPCC Climatology V2015, annual totals in mm are shown at 0.25° resolution.

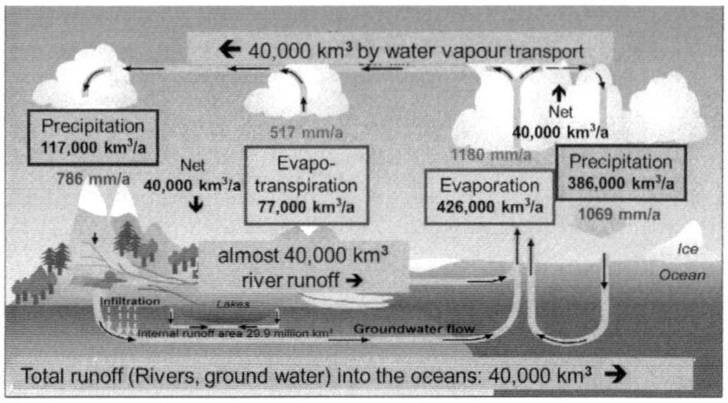


Figure 3. Update of the quantitative assessment of the global water cycle utilizing GPCCs land-surface precipitation analysis (Schneider et al., 2014).

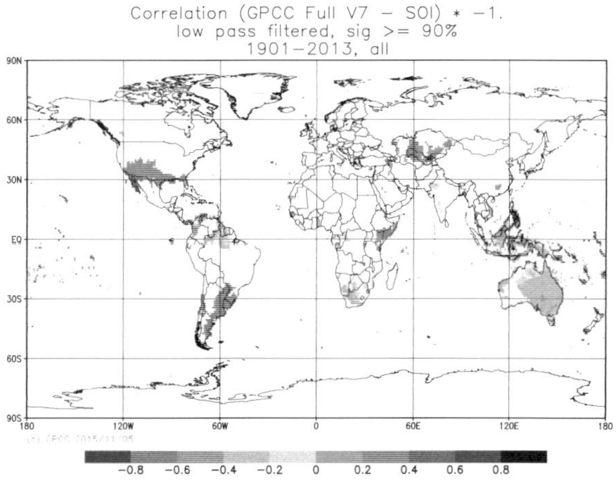


Figure 4. Centennial correlation between precipitation (GPCC Full Data Reanalysis V7) for the period 1901–2013 and the negative ENSO Southern Oscillation index ($-1 \times$ SOI) similar to Becker et al. (2013).

fiers (DOIs, Tables 1 and 2), allowing also for reproducibility and repetition of data utilizations even decades after primary data accesses. Corresponding to international agreements, station data provided by Third Parties are protected. However the gridded GPCC analysis products are freely available via Internet (http://gpcc.dwd.de) and GPCC has also issued a test interpolation data set solely based on the public available station data of NOAAs GHCN.

## 2 Most recent versions of monthly data products issued in June 2015

Most recently updated versions of the Full Data Reanalysis (V7) and Climatology (V2015) were released, replacing their predecessors issued in 2011. The updates are enhanced by almost 8000 additional stations that were added to the quality assured data base of the GPCC. Furthermore the data records of existing stations were extended by the most recent years. All in all, the new versions are based on about 75 100 stations with records exceeding 10 years instead of 67 200 stations reprocessed in year 2011. Due to the additional stations and precipitation data, it was possible to detect errors invisible so far and to improve the analysis in particular across Indonesia, Mexico, Brazil and some other regions. Moreover a new land-sea mask has been introduced, adding many grid cells across islands/atolls that had been missing (sea cells) before. The GPCC Full Data Reanalysis Product is the most recommended for quantitative assessments of the global water cycle (Fig. 3) and resources with regard to the precipitation over land (Schneider et al., 2014). It also contains information to understand natural variabilities (Fig. 4) as well as the hydro-climatological background of water related conflicts (Fig. 5). Since July 2015 a first full data daily product (Fig. 6) has been published together with three other new ones (Table 2).

**Table 1.** Overview on monthly data products of the GPCC and their use cases.

| Name of GPCC data product | | | DOI reference for download | | |
|---|---|---|---|---|---|
| Use cases | Period covered | Time of availability | Regular updates | – Quality<br>– (# of stations)<br>– QC-level | Grid resolution |
| First guess monthly | | | doi:10.5676/DWD_GPCC/FG_M_100 | | |
| – Near real-time drought monitoring<br>– Early monitoring | Oct. 2003 until present | 3–5 days after expiry of month regarded | Yes, monthly | – Limited<br>– (6000 to 7600)<br>– Basic (Auto) | Lat–Lon 1.0° |
| Monthly product V5 (June 2015) | | | doi:10.5676/DWD_GPCC/MP_M_V5_100<br>doi:10.5676/DWD_GPCC/MP_M_V5_100 | | |
| – Regular HQ monitoring<br>– Calibration of satellite data<br>– Sys. error (under catch) correction<br>– Solid precipitation<br>– Annual and monthly anomalies | Jan. 1982 until present | Within two months after end of month regarded | Yes, monthly | – High<br>– (7000 to 8900)<br>– Normal (auto and cross-checks) | Lat–Lon 1.0 and 2.5° |
| Full data reanalysis V7 (June 2015) | | | doi:10.5676/DWD_GPCC/FD_M_V7_050<br>doi:10.5676/DWD_GPCC/FD_M_V7_100<br>doi:10.5676/DWD_GPCC/FD_M_V7_250 | | |
| – Global water cycle and resources<br>– Hydrological studies<br>– Best quality (HR) analysis<br>– Centennial analysis<br>– Climatological analysis<br>– Index (ENSO) sensitive regions<br>– Verification of models from NWP to seasonal<br>– Re-analysis verification | Jan 1901 to Dec 2013 | Upon availability of substantially increased data base of quality controlled precipitation data | No, every three to four years | – Best<br>– (11 000 to 51 000 depending on month; 75 100 in total)<br>– High (in-depth screening and correction) | Lat–Lon 0.5, 1.0, and 2.5° |
| GPCC climatology V2015 (June 2015) | | | doi:10.5676/DWD_GPCC/CLIM_M_V2015_025<br>doi:10.5676/DWD_GPCC/CLIM_M_V2015_050<br>doi:10.5676/DWD_GPCC/CLIM_M_V2015_100<br>doi:10.5676/DWD_GPCC/CLIM_M_V2015_250 | | |
| – Climate aided interpolation (CAI)<br>– School material<br>– Reference material | Reference period 1951–2000 | Similar to full data reanalysis | No, similar to full data reanalysis | – Best<br>– 751 000<br>– High | Lat–Lon 0.25, 0.5, 1.0, and 2.5° |

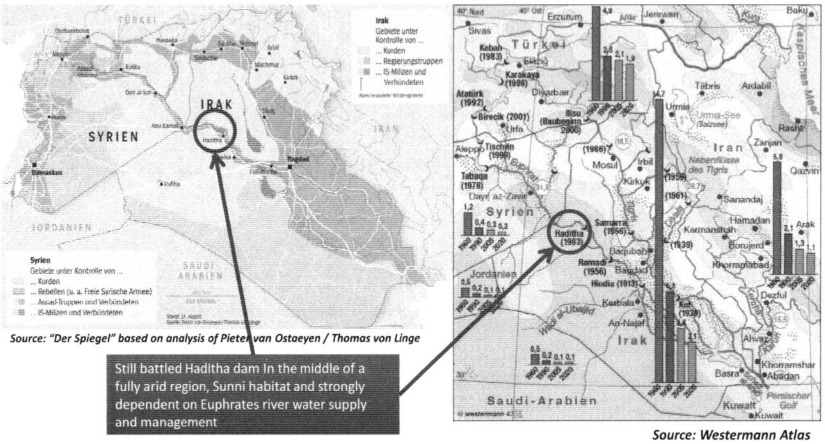


**Figure 5.** GPCC's climatological data sets are utilized as reference material (right panel, here Westermann Atlas) and potentially help to comprehensively understand water modulated conflicts (left panel) and to serve independent reference information for water diplomacy.

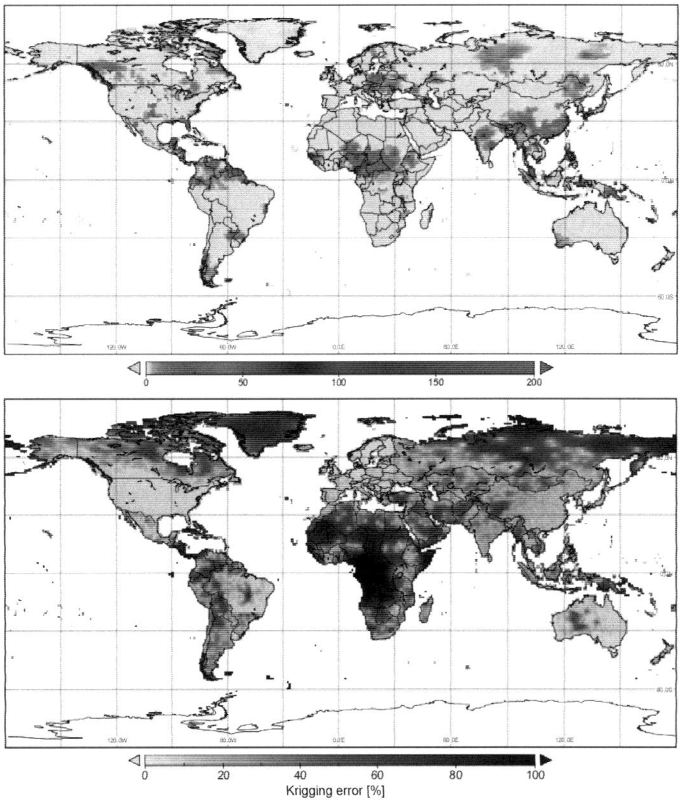


**Figure 6.** Precipitation analysis of 7 July 1997 (top panel) and corresponding interpolation (kriging) error field (bottom panel).

**Table 2.** New GPCC data products issued in July 2015.

| Name of GPCC data product | | | DOI reference for download | | |
|---|---|---|---|---|---|
| Use cases | Period covered | Time of availability | Regular updates | – Quality<br>– (# of stations)<br>– QC-level | Grid resolution |
| GPCC interpolation test dataset | | | doi:10.5676/DWD_GPCC/ITD_M_V1_100 (including monthly station totals of GHCN) | | |
| – Verify GPCC's interpolation method through product based on subset of ∼ 23 000 public available NOAA-GHCN stations | Jan–Dec 1988 | Similar to full data reanalysis | None | – Test<br>– Up to 24 000 stations<br>– GHCN QC | Lat–Lon 1.0° |
| GPCC drought index (monthly; Ziese et al., 2014) | | | doi:10.5676/DWD_GPCC/DI_M_100 | | |
| – Drought monitoring | Jan 2013–present expired month | 10th of following month | Yes, monthly | – Limited<br>– 6000–7600<br>– Basic (Auto) | Lat–Lon 1.0° |
| GPCC first guess daily prod. (Schamm et al., 2014) | | | doi:10.5676/DWD_GPCC/FG_D_100 | | |
| – Early monitoring<br>– Extreme precipitation | Jan 2009–present expired month | 4th of following month | | – Limited<br>– 6000–7600<br>– Basic (Auto) | Lat–Lon 1.0° |
| GPCC full data daily product version 1 (July 2015) | | | doi:10.5676/DWD_GPCC/FD_D_V1_100 | | |
| – Climatological assessment<br>– Climate and weather extremes in precip (WCRP GC and ETCCDI)<br>– Error assessment | Jan 1988–Dec 2013 | Upon availability of full data monthly product | No, 1 month after issuance of full data monthly product | – High<br>– (23 000–30 000)<br>– Auto and vis. check | Lat–Lon 1.0° |

## 3 Data availability

All data sets are provided with free and unlimited access through the DOI references provided. Data will be maintained version dependent on the underlying ftp server operated by Deutscher Wetterdienst.

**Acknowledgements.** We are most appreciative to the data suppliers who are to the largest extent the worldwide spread National Meteorological and/or Hydrological Services, but also some other institutes. These data contributions have put Global Precipitation Climatology Centre (GPCC) into the position to provide the global precipitation analyses described in this document, and we are looking forward to their further contributions, which are crucial in order to maintain and enhance GPCC's level of products in terms of scope and quality.

## References

Becker, A., Finger, P., Meyer-Christoffer, A., Rudolf, B., Schamm, K., Schneider, U., and Ziese, M.: A description of the global land–surface precipitation data products of the Global Precipitation Climatology Centre with sample applications including centennial (trend) analysis from 1901–present, Earth Syst. Sci. Data, 5, 71–99, doi:10.5194/essd-5-71-2013, 2013.

Diercke Weltatlas: Euphrat and Tigris – Water Usage, p. 162, Fig. 2, ISBN 978-3-14-100800-5, 2015.

Schamm, K., Ziese, M., Becker, A., Finger, P., Meyer-Christoffer, A., Schneider, U., Schröder, M., and Stender, P.: Global gridded precipitation over land: A description of the new GPCC First Guess Daily product, Earth Syst. Sci. Data, 6, 49–60, doi:10.5194/essd-6-49-2014, 2014.

Schneider, U., Becker, A., Finger, P., Meyer-Christoffer, A., Ziese, M., and Rudolf, B.: GPCC's new land surface precipitation climatology based on quality-controlled in situ data and its role in quantifying the global water cycle, Theor. Appl. Climatol., 115, 15–40, 2014.

Ziese, M., Schneider, U., Meyer-Christoffer, A., Schamm, K., Vido, J., Finger, P., Bissolli, P., Pietzsch, S., and Becker, A.: The GPCC Drought Index – a new, combined, and gridded global drought index, Earth Syst. Sci. Data, 6, 285–295, doi:10.5194/essd-6-285-2014, 2014.

Proc. IAHS, 374, 35–40, 2016
proc-iahs.net/374/35/2016/
doi:10.5194/piahs-374-35-2016


Open Access
PIAHS

# Rainfall as proxy for evapotranspiration predictions

**Bruno Collischonn**[1] **and Walter Collischonn**[2]
[1]Agencia Nacional de Aguas, Brasilia, 70610-200, Brazil
[2]Instituto de Pesquisas Hidraulicas, Porto Alegre, 91501-970, Brazil
*Correspondence to:* Bruno Collischonn (bruno.collischonn@ana.gov.br)

Published: 17 October 2016

**Abstract.** In this work, we evaluated the relationship between evapotranspiration and precipitation, based on the data recently made available by the Brazilian Meteorological Institute. ETP tend to be lower in rainy periods and vice-versa. This relationship was assessed both in physical and statistical ways, identifying the contribution of each explaining variable of ETP. We derived regression equations between monthly rainfall and ETP, which can be useful in studies where ETP time series are not available, such as reservoir design, irrigation management and flow forecast.

## 1 Introduction

Evaporation and evapotranspiration ($E$/ETP) are components of the water cycle, corresponding to the transfer of water from the terrestrial surface to the atmosphere. Quantitative information about these components is used in several problems involving water resources, such as crop management (both irrigated and non-irrigated), reservoir design and operation, flow forecast and so on. However, $E$/ETP data availability is sparse both in time and space, as well as the data needed to estimate it (solar radiation, wind speed, relative humidity, air temperature and atmospheric pressure) using consecrated methods like Penman-Monteith (Shuttleworth, 2012).

Moreover, $E$/ETP shows a lower variability, compared with precipitation. These factors (data scarcity and low variability) lead to the fact that many studies in water resources in Brazil use simple evaporation vectors, i.e. long-term monthly averages of $E$/ETP, instead of time series. When evaluating reservoir yields, for example, most studies in Brazil use $E$/ETP averages. It is straightforward to conclude, however, that this approach leads to an underestimation of reservoir yields, since during a sequence of dry years, which normally defines the magnitude of reservoir yields, real $E$/ETP is greater than the average. On the other hand, in wet years, when $E$/ETP is lower than average, one can expect frequent reservoir spilling; thus, excess volumes do not compensate interanually.

The Brazilian hydropower network, which accounts for more than 70 % of power supply in the country, is designed to minimize the risk of power shortages in the case of repetition of a so-called critical hydrological period, which was a very dry sequence of years in the country as a whole, from 1949 to 1956. But hydrological data during that period was very scarce, leading to the need of hydrological modelling, mostly using $E$/ETP averages. Better estimates of $E$/ETP would allow a more real assessment of the supply capacity, as well as a better estimation of historical inflows.

Again, given the low variability of $E$/ETP, it is often assumed that the errors incurred by using averages are lower than the estimation of $E$/ETP itself (as well as the estimation of other hydrological variables), making this approach widely accepted.

However small, interannual variability in $E$/ETP does exist, and it is in some ways linked to the variability of precipitation. In general, there is an inverse relationship of rainfall and $E$/ETP in a given time interval. As a consequence, the rainfall-runoff transformation process becomes nonlinear, as stated by Tucci (1998), since an increase of, say, 100 mm in monthly precipitation leads to an increase in runoff higher than 100 mm, because of the decrease of $E$/ETP (besides other non-linearities). The inverse relationship between these two variables has not been very much investigated in the literature, one exception being the work by Mudiare (1985).

Therefore, a way of overcoming $E$/ETP data scarcity, particularly in developing regions, is to establish statistical relations between $E$/ETP and rainfall. Once this relationship exists, $E$/ETP time series are obtained by applying the regression equation with the precipitation time series as input.

Recently, the Brazilian National Institute of Meteorology (INMET) has put its daily hydrometeorological time series available for research purposes. Data since 1961 over 290 stations around the country can be obtained through the internet after registering, attending a long-time demand of water managers and research community in Brazil.

In the present work, some statistics were calculated using this database, in order to identify a more physical explanation for the inverse relationship between rainfall and $E$/ETP, by quantifying the magnitude of the underlying meteorological variables both in dry and wet days.

Moreover, regression equations between monthly $E$/ETP and rainfall at each station were derived. The goal is to obtain a better comprehension of the relationship, as well as generating an $E$/ETP database for the refinement of water resources studies, as an alternative of the use of averaged data.

## 2 Rainfall and $E$/ETP

The Penman-Monteith equation, which is considered to be the most suited method for quantifying $E$/ETP (Smith, 1991), can be rewritten as a function of five meteorological variables:

$$E/\mathrm{ETP} = f(T_{\mathrm{med}}, \mathrm{UR}, \mathrm{PA}, \mathrm{VV}, \mathrm{RAD}) \quad (1)$$

where $T_{\mathrm{med}}$ is the mean air temperature, UR is the relative humidity, PA is the atmospheric pressure, VV is the mean wind speed, and RAD is the incident solar radiation. In our work, incident radiation is calculated from sunshine duration, in hours.

The inverse relationship between rainfall and $E$/ETP can be explained physically, through the variation of its explaining variables. First, rainy days are necessarily cloudy, reducing the incoming radiation. This is the main factor explaining the inverse relationship, since solar radiation (or sunshine duration) is the main driver of the process of evaporation and evapotranspiration (Shuttleworth, 2012).

Rainy days also generally show higher relative humidity. This leads to a decrease of $E$/ETP, since higher humidity causes an increase in vapour pressure (neighbouring air moisture is closer to saturation).

The effects of the remaining variables are not so evident. Mean temperature, for example, is positively correlated with $E$/ETP, since increases vapour saturation pressure (the amount of moisture that can be stored in the surrounding air). However, FAO's recommendation is to use the average between the minimum and maximum daily temperature as an estimate of mean temperature. In rainy days, lower radiation tend to decrease maximum temperatures, but during the night, outgoing longwave radiation is retained in the atmosphere by the clouds, actually increasing minimum temperatures (Allen et al., 1998), at least in temperate climates. Thus, the relevance of mean air temperature in the rainfall-$E$/ETP relationship is relative.

In the case of the other variables (atmospheric pressure and wind velocity), it is harder to establish a correlation intuitively, as made above. The magnitude of each meteorological variable in dry and wet years were statistically tested as follows.

## 3 Statistical analysis of ETP's explaining variables

For each station, daily time series were separated between dry and rainy days, the latter defined as days with precipitation lower or equal than 5 mm. For each of the five variables, we tested the hypothesis that the mean of the variable in dry days is equal to the mean in rainy days (null hypothesis), using Student's $t$ test of the means implemented in Matlab® for a 5 % significance level. If the null hypothesis was rejected, it was verified that the mean in rainy days was higher or lower than in dry days.

Seasonality was also accounted for, by performing the tests for each month of the year. Months where less than 5 % of the days were rainy were discarded (typically the period between July and September in the Southern Hemisphere), in order to avoid non-rejection of the null hypothesis only because of the small size of samples.

INMET stations were grouped according to the five Brazilian climate zones (IBGE, 2013). For each zone, we calculated the frequency of months/stations where magnitude was higher in dry and rainy days, as well as the percentage of months were the null hypothesis could not be rejected (in this case, we considered that no significant difference exists).

### Results – statistical correspondence

#### Wind speed (or: "do rainy days have stronger wind?")

Table 1 shows the condensed results of the hypothesis test for wind velocity. For each climate zone, we calculated the frequency of months/stations where wind was stronger during rainy and dry days, as well as no significant difference.

In most regions, there is no difference between wind velocity during rainy and dry days. In the semi-arid and coastal northeast, stronger wind occurs slightly more frequently during dry days. In the more temperate south, it is the other way round. In general, it was observed that coastal stations showed stronger wind during dry weather, while stations in hillier regions have stronger winds more frequently during rainy days.

Wind velocity influences $E$/ETP by removing moisture from the surface, bringing less saturated air. Thus, higher velocities result in higher $E$/ETP. As seen, dry days, more often than not, show stronger wind velocity, which also helps to explain the negative relationship between rainfall and $E$/ETP.

**Table 1.** Frequency of stronger wind during dry and rainy days.

| Zone | Dry | No diff. | rainy |
|---|---|---|---|
| Temperate | 15 % | 40 % | 46 % |
| Tropical | 27 % | 59 % | 14 % |
| Coastal NE | 41 % | 49 % | 10 % |
| Semiarid | 55 % | 38 % | 7 % |
| Equatorial | 32 % | 56 % | 12 % |
| Brazil | 34 % | 50 % | 15 % |

**Table 2.** Frequency of higher maximum temperature during dry and rainy days.

| Zone | Dry | No diff. | rainy |
|---|---|---|---|
| Temperate | 98 % | 2 % | 0 % |
| Tropical | 95 % | 5 % | 0 % |
| Coastal NE | 80 % | 20 % | 0 % |
| Semiarid | 79 % | 20 % | 1 % |
| Equatorial | 60 % | 38 % | 2 % |
| Brazil | 85 % | 15 % | 1 % |

## Temperature (or: "are dry days warmer?")

Statistical analyses were performed considering minimum, maximum and mean daily temperature. Table 2 shows the analysis for the maximum temperature. In almost 90 % of the cases, maximum temperature was significantly higher during dry days than during rainy days, more so in the temperate south and less so in the equatorial amazon. It confirms the intuitive notion, also stated by Allen (1998), that dry days have maximum temperatures higher than rainy days.

For minimum temperatures, aggregate results are shown in Table 3. Its results show that the assumption by Allen (1998), that rainy/cloudy days have higher minimum temperatures, is valid for temperate climates, but not necessarily for tropical/equatorial regions. Here, very often minimum temperatures were higher in dry days.

In the average, results shown in Table 4 allow the conclusion that dry days tend to be warmer in all regions of the country. As seen, this is in part due to higher maximum temperatures, as expected, but in some parts also due to higher minimum temperatures even in dry days, contrary to some literature, notably Allen et al. (1998).

This also helps to explain the negative relationship between rainfall and $E$/ETP, since, as seen, temperature is positively correlated with $E$/ETP.

## Radiation/Sunshine duration and relative humidity

In the case of sunshine duration (used here to estimate incoming radiation), the correlation is quite intuitive, and is confirmed by the statistical analysis: in almost 100 % of the months/stations, mean sunshine duration was higher in dry days, as seen in Table 5. This is the main factor explaining the inverse relationship between rainfall and $E$/ETP.

**Table 3.** Frequency of higher minimum temperature during dry and rainy days.

| Zone | Dry | No diff. | rainy |
|---|---|---|---|
| Temperate | 3 % | 18 % | 79 % |
| Tropical | 27 % | 38 % | 36 % |
| Coastal NE | 24 % | 60 % | 16 % |
| Semiarid | 26 % | 60 % | 14 % |
| Equatorial | 43 % | 51 % | 6 % |
| Brazil | 26 % | 44 % | 30 % |

**Table 4.** Frequency of higher mean temperature during dry and rainy days.

| Zone | Dry | No diff. | rainy |
|---|---|---|---|
| Temperate | 61 % | 34 % | 5 % |
| Tropical | 79 % | 18 % | 2 % |
| Coastal NE | 46 % | 53 % | 1 % |
| Semiarid | 46 % | 51 % | 2 % |
| Equatorial | 42 % | 56 % | 2 % |
| Brazil | 61 % | 37 % | 3 % |

**Table 5.** Frequency of higher mean sunshine duration during dry and rainy days.

| Zone | Dry | No diff. | rainy |
|---|---|---|---|
| Temperate | 100 % | 0 % | 0 % |
| Tropical | 100 % | 0 % | 0 % |
| Coastal NE | 100 % | 0 % | 0 % |
| Semiarid | 100 % | 0 % | 0 % |
| Equatorial | 98 % | 2 % | 0 % |
| Brazil | 99 % | 1 % | 0 % |

The same happens for relative humidity. Table 6 shows that in 100 % of the cases, relative humidity is higher during rainy days. Since relative humidity is negatively correlated with $E$/ETP, this also explains the inverse relationship investigated in this paper.

Finally, results for atmospheric pressure are shown in Table 7. For higher latitudes, there is a positive correlation between dry weather and atmospheric pressure, as expected. In regions closer do the Equator; however, this correlation is no longer valid, since there is no statistical difference between pressure during dry and rainy days. The variability of atmospheric pressure also plays a minor role in the variability of ETP; for one side, higher pressure increases the specific weight of air, allowing more moisture to be retained in the surrounding air; on the other hand, it increases the psychrometric constant, which comes into the denominator of the Penman-Monteith equation, thus decreasing ETP.

**Table 6.** Frequency of higher relative humidity during dry and rainy days.

| Zone | Dry | No diff. | rainy |
|---|---|---|---|
| Temperate | 0 % | 0 % | 100 % |
| Tropical | 0 % | 0 % | 100 % |
| Coastal NE | 0 % | 0 % | 100 % |
| Semiarid | 0 % | 0 % | 100 % |
| Equatorial | 0 % | 0 % | 100 % |
| Brazil | 0 % | 0 % | 100 % |

**Table 7.** Frequency of higher atmospheric pressure during dry and rainy days.

| Zone | Dry | No diff. | rainy |
|---|---|---|---|
| Temperate | 88 % | 7 % | 5 % |
| Tropical | 65 % | 27 % | 8 % |
| Coastal NE | 9 % | 60 % | 31 % |
| Semiarid | 18 % | 57 % | 24 % |
| Equatorial | 8 % | 71 % | 21 % |
| Brazil | 44 % | 41 % | 15 % |

## 4 Relations between rainfall and ETP

Daily time series for ETP were calculated through the Penman-Monteith, based on Shuttleworth (2012) and synthetized by Collischonn et al. (2007), using the meteorological data recently made available by INMET in 290 meteorological stations in Brazil. Incident radiation data was estimated based on sunshine duration. Reference evapotranspiration was calculated adopting an albedo of 0.23 m, surface resistance of 70 s m$^{-1}$ and crop height of 0.12 m (Allen et al., 1998).

During time intervals where one or more of the variables were missing, ETP was not calculated, the only exception being missing atmospheric pressure. In this case, pressure was estimated based on a seasonal correlation between monthly mean pressure and station elevation. This is considered to be acceptable, given the low influence of this variable over the variability of ETP and the large number of missing data for atmospheric pressure.

ETP data where then accumulated into monthly totals. If missing data accounted for less than 5 days, missing data were filled using the mean ETP of the remaining days. Otherwise, the month was not included in the regression. The same was done for monthly rainfall.

Regression equations were then derived from the paired ETP/rainfall monthly data. Regressions were done for each month of the year, in order to account for seasonality. Given the expected inverse relationship between these variables, the slope of the regression should be negative.

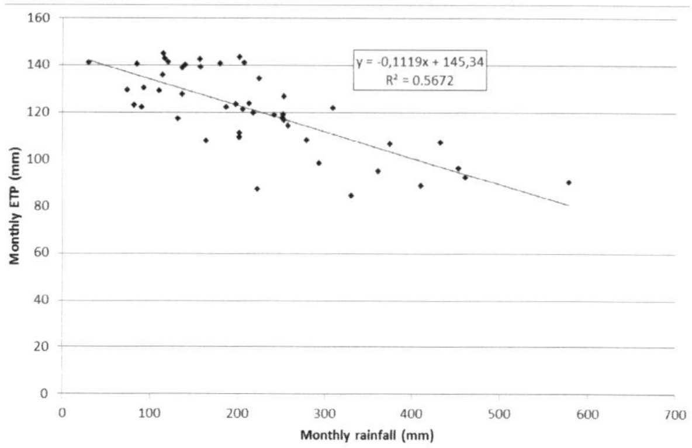


**Figure 1.** Example of regression for Brasilia, month of January.

### Results – relations between rainfall and ETP

Figure 1 shows an example of the resulting regressions, for the meteorological station of Brasilia (tropical Brasil central zone), for the month of January. It can be observed that, as expected, rainfall variability is in general much higher than that of ETP. In this case, while rainfall varied between 29 and 577 mm month$^{-1}$, ETP varied between 85 and 145 mm month$^{-1}$. The negative slope shown in this example was also obtained for all remaining regressions, emphasizing the inverse relationship between rainfall and ETP.

Although data dispersion is relatively high (also expressed in the low coefficient of determination $R^2$, it should be remembered that these regressions are an alternative for the use of long-term averages of ETP for hydrological simulation, i.e. when no time series are available.

This kind of regression is particularly useful during transition months between the dry and rain seasons in tropical regions (typically April/May and September/October in the Southern Hemisphere), which can be either very rainy or very dry, depending on the year. It is also useful in regions where the rain season presents periodical breaks, or periods of 10 to 15 days without rainfall. These periods, usually between December and March, can have very high ETP given the higher extraterrestrial radiation during summer months.

Only during very dry months, like July and August, ETP can have a higher variability, since rainfall is hardly larger than zero. In this case, the regressions are less useful, and the use of ETP averages seem to be more practical.

Figure 2 shows the spatialized values of the $R^2$ coefficient for the regression in two months of transition between rainy and dry seasons.

As can be seen, the higher values for the coefficient were obtained in the tropical zone. In the temperate and equatorial zones, the inverse relationship between rainfall and ETP is not so clear, perhaps because of a higher frequency of cloudy, but not necessarily rainy days.

**Table 8.** Monthly averages of $R^2$ statistics for the regressions in each climate zone.

| | Temperate | Tropical-central Brazil | Coastal north east | Semiarid north east | Equatorial | Month average |
|---|---|---|---|---|---|---|
| Jan | 0.26 | 0.51 | 0.54 | 0.45 | 0.29 | 0.41 |
| Feb | 0.29 | 0.46 | 0.51 | 0.43 | 0.25 | 0.37 |
| Mar | 0.19 | 0.41 | 0.33 | 0.35 | 0.20 | 0.32 |
| Apr | 0.31 | 0.29 | 0.37 | 0.43 | 0.21 | 0.29 |
| May | 0.18 | 0.16 | 0.41 | 0.45 | 0.27 | 0.23 |
| Jun | 0.08 | 0.19 | 0.34 | 0.27 | 0.20 | 0.18 |
| Jul | 0.17 | 0.15 | 0.38 | 0.32 | 0.22 | 0.19 |
| Aug | 0.11 | 0.15 | 0.32 | 0.17 | 0.18 | 0.15 |
| Sep | 0.26 | 0.28 | 0.30 | 0.14 | 0.13 | 0.23 |
| Oct | 0.28 | 0.33 | 0.38 | 0.23 | 0.11 | 0.26 |
| Nov | 0.31 | 0.33 | 0.43 | 0.21 | 0.18 | 0.28 |
| Dec | 0.27 | 0.31 | 0.30 | 0.25 | 0.27 | 0.28 |
| Zone average | 0.22 | 0.30 | 0.38 | 0.31 | 0.21 | |

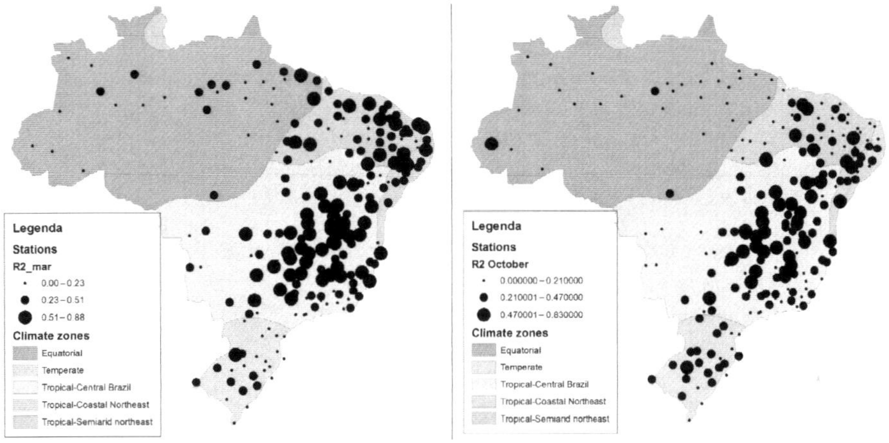


**Figure 2.** Spatialized values of $R^2$ statistic for transition months: March (left) and October (right).

Finally, Table 8 shows the average of $R^2$ statistics for each zone and month.

As mentioned, the winter months in the Southern Hemisphere (June to August) showed the lowest average regression performance, because of the low variability of rainfall during these months in most regions of Brazil.

## 5 Conclusions

This work is a first attempt to further explore the meteorological time series recently made available by INMET, based on a hydrological point of view. In the following sections, the main specific conclusions are presented.

### 5.1 Statistical correspondence between explaining meteorological variables

The statistical analysis performed over the meteorological variables explaining the evapotranspiration process supported the comprehension of the mechanisms behind the inverse relationship between rainfall and ETP. Incoming radiation, which is the main driver of ETP, is clearly higher during dry days than in rainy days. Relative humidity, which also strongly influences ETP, is clearly higher during rainy days. Both results are intuitive and are supported by the statistical analysis.

Mean air temperature, which is also positively correlated with ETP, is generally higher during dry days, in comparison to rainy days. The results also showed that minimum temperatures very often were higher during dry days, mostly in the tropics. This result shows that one very popular premise of hydrometeorology is not necessarily true for tropical regions, namely the belief that minimum temperatures should be lower during clear weather due to less heat retention in the lower atmosphere.

For wind velocity and atmospheric pressure, results were less clear. In most cases, null hypothesis of equal means could not be rejected, thus not allowing further conclusions. One singular fact that was observed is a tendency for stronger wind during dry days in coastal places, while in hilly locations wind was stronger in rainy days.

Thus, the inverse relationship between rainfall and ETP is explained mostly by the higher incoming radiation (or sunshine duration) and lower relative humidity in dry days, and by a lesser part, by higher mean air temperature in these conditions.

### 5.2 Regressions between rainfall and ETP

In general, the results of the regressions confirmed the inverse relationship between rainfall and ETP, since regression slopes were negative in most cases.

Although the numerical values of the coefficient of determination $R^2$ were low, it must be remembered that these regressions are an alternative to the use of ETP averages, i.e. monotonical ETP values, in order to take at least a share of temporal variability into account. Thus, while the obtained regressions explain around 50 % of ETP's variability, the usage of long-term averages would explain 0 %.

The obtained relations are particularly useful in tropical regions, during months of transition between rain and dry seasons, which can be either very wet or very dry, thus showing varying ETP.

These indirect estimates of ETP can be useful, for example, for extending ETP series into the past, in regions or periods where rainfall data is more abundant than meteorological data. As mentioned, the Brazilian hydropower network is planned and operated to supply energy demand, at a low risk, even in the case of a repetition of a critical hydrological period occurred in the 1950s, when hydrometeorological data was scarce nationwide (and even more so in low-populated areas like the amazon, where hydropower was expanded in recent years). More accurate ETP estimates would be helpful both for a better assessment of inflows during that period (given the need of proper rainfall-runoff modelling) and for a better estimation of evaporation losses.

Other potential areas of usage are reservoir design and reservoir yield estimation, where long-term ETP averages are still widely used, leading to an overestimation of reservoir yields. Irrigation management would also be an area for application, through a better estimation of crop water needs, based only on rainfall measurements and the regressions obtained. Finally, assessment of future hydrological conditions would also benefit, since ETP conditions could be easily estimated based only on rainfall predictions.

## 6 Data availability

Meteorological data used in this paper belongs to the Brazilian Meteorological Institute and is available (after registration) at http://www.inmet.gov.br/projetos/rede/pesquisa/inicio.php.

**Acknowledgements.** The authors thank the Brazilian National Institute of Meteorology (INMET) for making the hydrometerological time series available for academic purposes.

## References

Allen, R., Pereira, L., Raes, D., and Smith, M.: Guidelines for computing crop water requirements. Rome: FAO, 310 pp. (FAO-Irrigation and Drainage Paper, 56), 1998.

Brazilian Meteorological Institute: BDMEP: Meteorological Database for Research Purposes, available at: http://www.inmet.gov.br/projetos/rede/pesquisa/inicio.php, last access: December 2013.

Collischonn, W., Allasia, D., Silva, B., and Tucci, C. The MGB-IPH model for large-scale rainfall-runoff modelling, Hydrol. Sci. J., 52, 878–895, 2007.

Instituto Brasileiro de Geografia e Estatística: Map of Brazilian Climate Zones, available at: ftp://geoftp.ibge.gov.br/informacoes_ambientais/climatologia/mapas/brasil/clima.pdf, last access: December 2013.

Mudiare, O. J.: Influence of light rainfall and cloud cover on evapotranspiration demands, PhD Thesis, University of Saskatchewan, 1985.

Shuttleworth, W. J.: Terrestrial Hydrometeorology, Wiley-Blackwell, 448 pp., 2012.

Smith, M.: Report on the expert consultation on revision of FAO methodologies for crop water requirements, Technical Report, FAO/AGL, Rome, 1991.

Tucci, C.: Modelos Hidrológicos, UFRGS/ABRH, 678 pp., 1998 (in Portuguese).

Proc. IAHS, 374, 41–51, 2016
proc-iahs.net/374/41/2016/
doi:10.5194/piahs-374-41-2016


Open Access
PIAHS

# Crystal balls into the future: are global circulation and water balance models ready?

**Balázs M. Fekete**[1,2], **Giovanna Pisacane**[3], **and Dominik Wisser**[4]

[1]Department of Civil Engineering, CUNY CREST Institute, The City College of New York, New York, NY, USA
[2]Environmental Sciences Initiative, Advanced Science Research Center, City University of New York, New York, NY, USA
[3]CLIM Lab, ENEA – Italian National Agency for New Technologies, Energy and Sustainable Economic Development, Rome, Italy
[4]Center for Development Research (ZEF), University of Bonn, Bonn, Germany

*Correspondence to:* Balázs M. Fekete (bfekete@ccny.cuny.edu)

Published: 17 October 2016

**Abstract.** Variabilities and changes due to natural and anthropogenic causes in the water cycle always presented a challenge for water management planning. Practitioners traditionally coped with variabilities in the hydrological processes by assuming stationarity in the probability distributions and attempted to address non-stationarity by revising this probabilistic properties via continued hydro-climatological observations. Recently, this practice was questioned and more reliance on Global Circulation Models was put forward as an alternative for water management plannig.

This paper takes a brief assessment of the state of Global Circulation Models (GCM) and their applications by presenting case studies over Global, European and African domains accompanied by literature examples. Our paper demonstrates core deficiencies in GCM based water resources assessments and articulates the need for improved Earth system monitoring that is essential not only for water managers, but to aid the improvements of GCMs in the future.

## 1 Introduction

Climate and direct anthropogenic change altering the water cycle has called into question the traditional water management planning that is based on past records of water resources as a means to characterize the likelihood of extreme conditions affecting water availabilities (Milly et al., 2008). The suggested alternative is relying on complex water balance simulations driven by projected climate forcings from global circulation models (GCM).

Hydrologists recognized long ago that the stationarity assumption is often violated due to changes other than climate (e.g. land cover and land use change, engineering alteration of the river channel, constructions of reservoirs, etc.). Water managers were very much aware of the need to test the statistical homogeneity of past observations (Lins and Cohn, 2011) and adjust the long-term characterization of the water resources and extremes according to the anticipated changes in trends. The scientific challenge is if new techniques involving GCM simulations are indeed necessary or as Lins and Cohn (2011) expressed "humility may be more important than physics; a simple model with well-understood flaws may be preferable to a sophisticated model whose correspondence to reality is uncertain."

A growing number of institutions run GCMs and use dynamical downscaling of GCM with Regional Climate Models (GCM-RCM) to produce large sets of projections for both present climate and future scenarios. The resulting datasets are expected to thoroughly sample the climate system phase space, where each simulation corresponds to different trajectory, starting from different initial conditions and determined by different choices of model parameters and structural uncertainties. Coordinated efforts such as the Coupled Model Intercomparison Project (CMIP) offers a collection of stan-

Published by Copernicus Publications on behalf of the International Association of Hydrological Sciences.

dardized GCM simulations that are the backbone of the regularly revised assessment reports of the Intergovernmental Panel on Climate Change (IPCC) (Allen et al., 2014).

Unfortunately, GCMs have difficulties properly reproducing contemporary climate casting doubts on their abilities in projecting future changes in key climate variables (e.g. air temperature and precipitation) (Maslin and Austin, 2012; Stevens and Bony, 2013). The discrepancy between representing present day climate are often handled via bias corrections (Hempel et al., 2013) assuming that the projected changes themselves are either correct by magnitude (delta method) or proportionally (scaling method) relative to observed contemporary simulations. Bias correction breaks the integrity of the original GCM simulation and often violates conservation principles, neglects feedback mechanisms, and their time-invariance under climate change is largely questionable (Maurer and Hidalgo, 2008). Although a debate on their applicability and effectiveness has recently arisen (Ehret et al., 2012), bias corrections are still a common practice.

It is generally assumed (without real scientific basis beyond very few empirical evidence) that multi-model averages outperform individual model projections, as individual biases are expected to at least partly cancel out. This expectation might have merit for a large number of truly independent models, but in reality GCMs share lot of commonalities (Masson and Knutti, 2011) due to their genealogy. The similarities in GCMs are not surprising since they are meant to represent the same physical processes after all. Nevertheless, given the usually small number of independent models (Pennell and Reichler, 2011) and model sensitivity to parameter choice, it is often difficult to assess the reliability of model ensemble averages and unambiguously quantify the associated uncertainties. Moreover, averaging is liable to reduce the variability in climate change analyses, especially for spatially heterogeneous variables (e.g. precipitation), while the apparent independence of the predicted change from the model skill in representing present day conditions undermines the hypothesis that reduced model spread in present day projections directly implies greater confidence in future scenarios (Knutti et al., 2010).

In addition to the inherent biases from GCM, their coarse resolutions prevents them from capturing regional spatial variability that is essential for water management applications. The coarse resolution GCM projections are either downscaled "statistically" considering spatial climate variability from observed records or dynamically by performing higher resolution regional climate model simulations forced by coarse resolution GCM as boundary condition. Just like bias correction, statistical downscaling carried out on multiple variables inevitably breaks their integrity resulting in forcing data sets that are inconsistent with the plausible states of the climate variables. The validity of dynamic downscaling is disputed as a viable strategy (Pielke Sr. and Wilby, 2012) since the skills of the high resolution RCMs resolving fine scale processes are constrained by the erroneous GCM forcing. Since the main source of the GCM deficiencies originates exactly from their inability to represent the fine scale processes that RCMs intend to capture, the lack of feedbacks from RCM simulation into the GCM forcings puts a clear limit to the "improvement" from dynamic downscaling to the degree that some even question if RCMs have any added value (Kerr, 2013).

Impact assessment models designed for assessing the sensitivity of various economic sectors (e.g. water resources management, agriculture, human health, etc.) play a key role in translating the projected climatic changes into corresponding societal consequences. A growing number of hydrological models intended to support water managers and policy makers offer capabilities to assess water resources (Haddeland et al., 2011). Model inter-comparison exercises like Water and Global Change (WATCH) (Haddeland et al., 2011), Inter-Sectoral Impact Model Intercomparison Project (ISI-MIP) (Haddeland et al., 2014; Schewe et al., 2014) or Quantifying Projected Impacts under 2 °C Warming (IMPACT2C) shed lights on how these models in combination with state-of-the-art global circulation models perform.

Our paper presents a couple of examples applying a well established hydrological modeling framework based on the Water Balance Model introduced by Vörösmarty et al. (1989, 1998). While WBM itself was amongst the first in applying physical hydrological principles at continental and global scales its current implementation $WBM_{plus}$ serve more as a modeling platform than a particular water balance implementation. We regard $WBM_{plus}$ a modeling framework that can be configured on the fly to specific application both in terms of modeling domains and the complexity of simulated hydrological processes. $WBM_{plus}$ can be configured with complex land surface processes that are in par with some of the hydrological models designed to resolve the energy and water balances simultaneously (e.g. Variable Infiltration Capacity, [VIC] model, Liang et al., 1994; Nijssen et al., 2001) and enabled to represent direct anthropogenic activities such as irrigation or reservoir operations.

The presented applications span the global and regional domains and show the challenges in applying GCMs for multi-decadal water resources management planning. Our paper intends to present a couple of case studies highlighting the limitations in applying GCMs and GCM driven RCM climate forcings for long-term water resources projections.

## 2 Lessons from $WBM_{plus}$ simulations using GCM projected climate forcings

In the presented case study examples, $WBM_{plus}$ was configured with a parsimonious representation of the vertical water exchange between the land surface and the atmosphere using a temperature driven estimation of the potential evapotranspiration (Hamon, 1963; Federer et al., 1996) and rudimentary drying function (Vörösmarty et al., 1989). This configuration limits the climate variables needed to daily values of air temperature and precipitation and reduces the potential artifacts arising from the bias corrections.

WBM was criticized in the past for being too simplistic by neglecting the full energy balance in the vertical water exchange processes and lumping together various components of the evapotranspiration (evaporation from soil, from the canopy and transpiration through the stomates of the leaves). Our team resisted for years to apply the more elaborated water balance configuration in "production" experiments, where the uncertainties in forcing data and land cover parameterization appeared to outweigh the anticipated gain in more realistic representation of the hydrological processes. In recent years, a number of papers came to similar conclusion that added complexity does not necessary improve model performance (Perrin et al., 2001; van Griensven et al., 2006).

### 2.1 Global example

Impact model assessments rarely consider "raw" GCM climate forcings. Instead they either rely on a single GCM or an ensemble of multiple bias corrected GCMs as their starting points and expect the GCM community to characterize the discrepancies between GCMs. Fekete and Stakhiv (2013) carried out rudimentary comparison of three GCMs (CanESM2 from the Canadian Centre for Climate Modeling and Analysis, Canada; the MIROC5 model from the Atmosphere and Ocean Research Institute at the University of Tokyo, Japan; MRI-CGCM3 model from the Meteorological Research Institute, Japan) from the CMIP5 archive, which was at its early phase – at the time the Fekete and Stakhiv study was carried out – in assembling model result archive from the different modeling groups and only a few modeling team had completed all the model simulations (Fig. 1). The tested model scenarios were based on the most aggressive greenhouse gas emission trajectory following the RCP 8.5 representative concentration pathway.

Even without analyses (discussed in Fekete and Stakhiv, 2013), it is quite clear that the projected changes are well below the biases between runoff estimates based on "raw" GCM forcings versus observed climate data from the Climate Research Unit of East Anglia (New et al., 2000). Not only are the projected trajectories different from the different GCM projections but often the direction of their trends differs.

GCMs consistently project rising temperatures for the upcoming decades (although they have marked differences in the rate of change), but projected precipitation trends from different GCMs have a wide spread in the magnitude and direction of change. GCMs and their weather forecast model cousins are known to have deficiencies in representing cloud formation and precipitation processes for decades (WMO, 1975), and still remains their major shortcoming (Stevens and Bony, 2013).

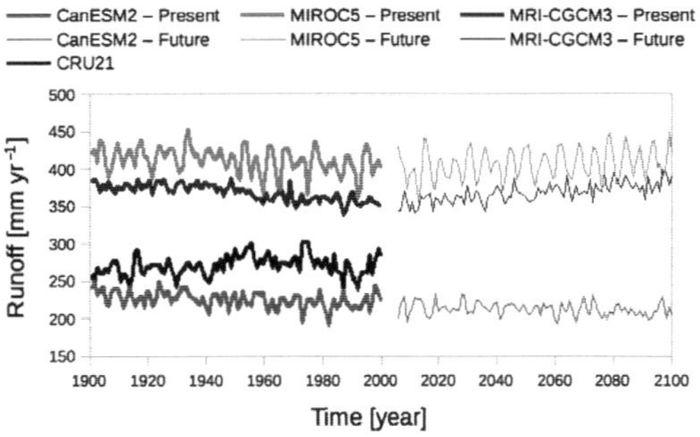


**Figure 1.** Annual estimates from GCM simulations under present and projected future conditions in contrast with observed climate forcings from the Climate Research Units, University of East Anglia.

The Inter-Sectorial Impact Model Intercomparison Project (ISI-MIP; (Warszawski et al., 2014) led by the Potsdam Institute for Climate Impact Research assembled an excellent compilation of GCM climate projections and applied consistent bias correction (Hempel et al., 2013) to five GCM/Earth System models: (1) HadGEM2-ES, Hadley Center, Met Office, United Kingdom (Jones et al., 2011); (2) IPSL-CM5A-LR, Climate Modelling Centre, Institut Pierre Simon Laplace, France (Dufresne et al., 2013); (3) MIROC-ESM-CHEM, Atmosphere and Ocean Research Institute, The University of Tokyo, Japan (Watanabe et al., 2011); (4) GFDL-ESM2M, Geophysical Fluid Dynamics Laboratory, NOAA, United States (Dunne et al., 2012) and (5) Norwegian Meteorological Institute, Norway (Bentsen et al., 2013) for four emission scenarios defined as resentative concentration pathawys (RCPs) and expressed in energy imbalances of 2.6, 4.5, 6.0 and 8.0 W $m^{-2}$ due to greenhouse gas effect.

Figure 2 shows the global terrestrial runoff estimates for contemporary and projected future climate conditions using the Water Balance Model ($WBM_{plus}$) (Vörösmarty et al., 1989, 1998; Wisser et al., 2010) global scale hydrological model. The bias correction removed some of the differences in the long-term mean under contemporary climate, but the seasonal variabilities according to the four models are still markedly different (Table 1, Fig. 2). The GFDL model in particular stands out with its rather hectic inter-annual variability in precipitation.

The projected linear trends in global air temperature (in °C per year), precipitation (in mm $yr^{-2}$) and WBMplus simulated runoff (in mm $yr^{-2}$) are summarized in Table 2. Only the air temperature has consistent rising trends with considerable spread (0.04–0.08 °C $yr^{-1}$ for the RCP8.5 scenario). The precipitation trends vary regionally with increasing and decreasing trends for individual continents, with a global average increase of 0.21–1.04 mm $yr^{-2}$ for RCP8.5) during the 2006–2099 period. The predicted runoff trends (–049–0.13 mm $yr^{-2}$ for RCP8.5) are the results of the competition between the rising air temperature (leading to more evapo-

**Table 1.** Contemporary (1950–2006) global average air temperature, precipitation and runoff (estimated using $WBM_{plus}$) based on five bias corrected GCM hind-casts.

| GCM | Air Temp [°C] | | Precipitation [$mm\,yr^{-1}$] | | Runoff [$mm\,yr^{-1}$] | |
|---|---|---|---|---|---|---|
| | Mean | SD | Mean | SD | Mean | SD |
| HadGEM2-ES | 13.25 | 0.37 | 875 | 21 | 344 | 15 |
| IPSL-CM5A-LR | 13.25 | 0.45 | 853 | 23 | 335 | 16 |
| MIROC-ESM-CHEM | 13.19 | 0.29 | 876 | 20 | 335 | 15 |
| GFDL-ESM2M | 13.22 | 0.36 | 858 | 44 | 349 | 33 |
| NorESM1M | 13.21 | 0.35 | 871 | 15 | 331 | 12 |

SD = Standard Deviation

**Table 2.** Linear regression slope of the projected air temperature, precipitation and WBMplus simulated runoff for the 2006–2099 period.

| | | | RCP2.6 | RCP5.4 | RCP6.0 | RCP8.5 |
|---|---|---|---|---|---|---|
| Air Temp. | [$°C\,yr^{-1}$] | HadGEM2-ES | 0.01 | 0.04 | 0.05 | 0.07 |
| | | IPSL-CM5A-LR | 0.01 | 0.03 | 0.04 | 0.07 |
| | | MIROC-ESM-CHEM | 0.02 | 0.04 | 0.05 | 0.08 |
| | | GFDL-ESM2M | 0.00 | 0.01 | 0.02 | 0.04 |
| | | NorESM1M | 0.01 | 0.03 | 0.03 | 0.05 |
| Precip. | [$mm\,yr^{-2}$] | HadGEM2-ES | 0.18 | 0.35 | 0.53 | 0.21 |
| | | IPSL-CM5A-LR | 0.13 | 0.59 | 0.39 | 0.96 |
| | | MIROC-ESM-CHEM | 0.32 | 0.54 | 0.76 | 1.04 |
| | | GFDL-ESM2M | 0.03 | 0.40 | −0.03 | 0.27 |
| | | NorESM1M | 0.32 | 0.34 | 0.47 | 0.71 |
| Runoff | [mm yr−2] | HadGEM2-ES | −0.02 | −0.10 | −0.02 | −0.49 |
| | | IPSL-CM5A-LR | −0.02 | 0.14 | −0.02 | 0.13 |
| | | MIROC-ESM-CHEM | 0.04 | −0.01 | 0.07 | −0.10 |
| | | GFDL-ESM2M | −0.01 | 0.20 | −0.21 | −0.08 |
| | | NorESM1M | 0.17 | −0.01 | 0.06 | −0.01 |

transpiration and hence less runoff) and more precipitation (that would lead to more runoff if there was no change in air temperature).

Perhaps, one of the most disturbing outcome of the ISI-MIP model intercomparison was the recognition that model uncertainties in the impact models were comparable to the uncertainties in the GCM forcings (Haddeland et al., 2011). While the true assessment of GCM uncertainties is probably better judged before bias corrections, the discrepancies amongst impact models (water balance models in the context of various applications such as water management, agricultural production, etc.) is disturbing given their relative simplicity compared climate models.

The large difference between global scale hydrological model simulations is not entirely new as the European Water and Global Change (EU-WATCH) program (a precursor to the ISI-MIP project) already came to the similar conclusions (Haddeland et al., 2011).

## 2.2 European example

IMPACT2C made use of the growing ensemble of high-resolution (∼ 12 km resolution) simulations from the Coordinated Downscaling Experiment – European Domain (Euro-CORDEX) and Med-CORDEX (Jacob et al., 2014). Climate projections for Europe are available through the CORDEX Climate Data Archive, while the bias corrected fields for hydrological impact studies are at present only available for internal use, and consist of a subset of five modeling GCM-RCM chains. A preliminary assessment of the robustness of the climate change signal for the RCP 4.5 scenario was performed on the original high resolution atmospheric temperature, precipitation and wind fields, following Tebaldi et al. (2011), although with more severe significance thresholds due to the small ensemble size (Fig. 3).

Robust signals could only be detected in temperature showing a 2 °C overall increase, a slightly weaker warming along the coasts of North-Western Europe in all seasons, and a more intense warming (up to +4 °C) in Northern and Eastern Europe in Winter and in Southern Europe in Summer (up

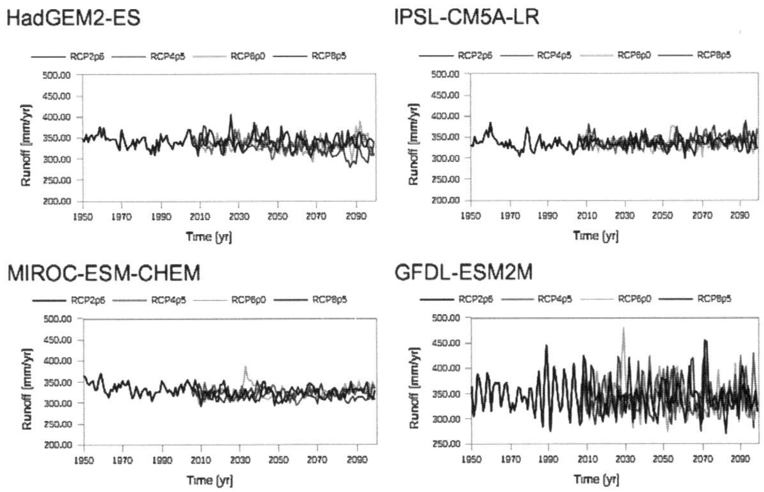


**Figure 2.** Global continental runoff estimates based on bias corrected present and projected future climate conditions using the ISI-MIP forcing data.

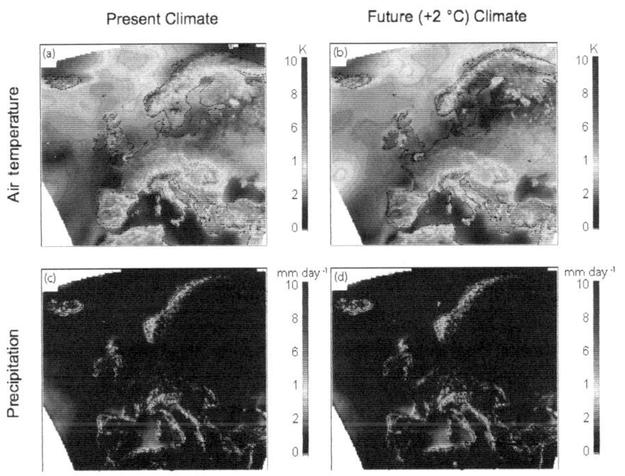


**Figure 3.** Ensemble spread (the difference between the minimum and the maximum from the ensemble members) of mean annual air temperature **(a, b)** and average precipitation **(c, d)** from uncorrected RCMs for present (1971–2000, **a, c**) and future ($+2\,°C$, **b, d**) Climate considering RCP 4.5 scenario.

to +3 °C). Similar patterns are detectable in daily minimum and maximum temperatures, especially in wintertime daily minima over Scandinavia. Daily precipitation does not show clear changes in most of Europe, but exhibits general increases over Central and Northern Europe in winter and only over Northern Europe in summer, while precipitation seems to decrease in Central/Southern Europe in summer. Extreme precipitation exhibits scattered robust increases across Europe in both seasons. Extreme winds are not found to robustly increase/decrease.

It has been documented, however, that a winter positive bias in surface temperature and precipitation rates can be induced in RCM simulations over Northern Europe through the prescribed large scale boundary conditions, which can lead to a too deep Icelandic low precipitation extending too far into the Nordic seas. Such feature also causes too low temperature and precipitation rates over Southern Europe. On the other hand, in summer warm and dry biases have been observed in RCMs simulations over Eastern Europe and to a lesser degree the Mediterranean, where too little simulated rainfall can dry out soil water reservoirs causing very high surface temperatures (Jacob et al., 2007). Therefore, there is definitely a need for thorough analyses of all the modeling chains before any conclusion can be reached (Jacob et al., 2014).

It is worth noting that the high resolution of the Med-CORDEX simulations might be expected to locally reduce the fraction of convective precipitation to total precipitation, therefore possibly reducing the summer warm and dry biases, thus improving both the spatial pattern and temporal evolution of precipitation. On the contrary, winter circulation being dominated by large scale features, it can mainly benefit from a finer representation of the flow-topography interactions in the domain interior (Rauscher et al., 2010).

Quite unfortunately, the bias correction procedures (in this case, quantile mapping) to the surface temperature and precipitation fields required the original RCM outputs to be upscaled from 12 to 25 km resolution, in order to match that of the E-OBS gridded dataset selected for reference (Haylock et al., 2008). In addition, when further examined, the E-OBS dataset exhibits an even lower effective resolution, in particular over complex topography, as a consequence of spatial interpolation, and clear underestimation of orographic precipitation and overestimation of surface temperature over mountainous areas, even when compared to lower resolution data (WFDEI).

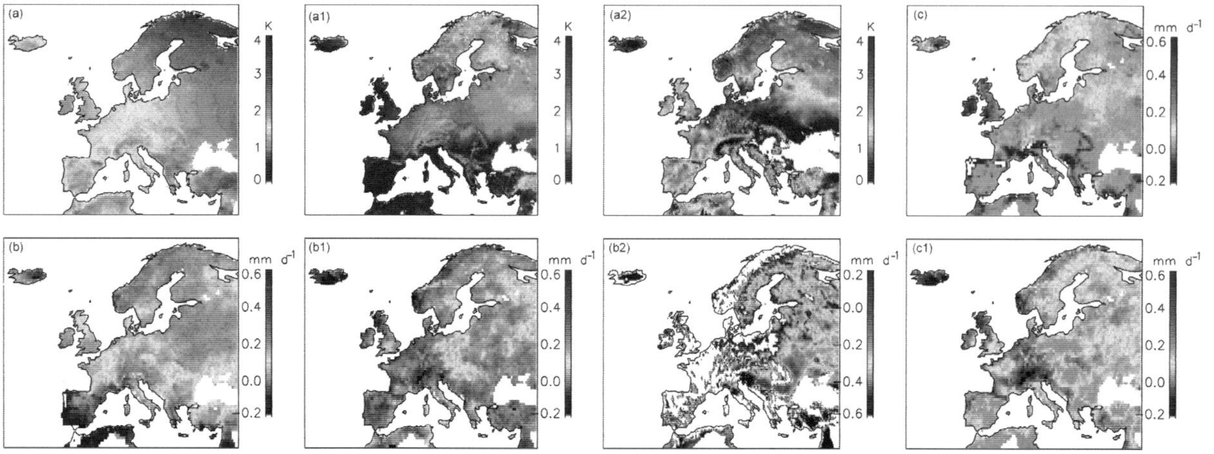


**Figure 4.** Predicted changes in mean annual temperature **(a)** average annual precipitation **(b)** and runoff **(c)** contrasted by their respective future climate bias-corrected spread (**a1**, **b1**, **c1** respectively) and present climate bias-correction patterns for temperature and precipitation only (**a2**, **b2** respectively).

When combined with the different representations of orography in the different RCMs, in vast areas this led to virtually correct altitude rather than intrinsic biases in the projected fields, as demonstrated by the comparison between the patterns of the uncorrected RCM ensemble spread in present and future climate for both precipitation and surface temperature, clearly carrying a stable topography signature, and between the correspondent average fields before and after bias correction (Fig. 4a2 and b2).

Such features are even more evident over specific regions, e.g. the Alps, over which individual model fields have been compared: single peaks can be identified and directly compared to high resolution geographic maps. The overall effect is a dramatic decrease and over-smoothing in precipitation projections and a similarly non-negligible increase in temperature at high altitudes, both severely inconsistent with the rest of modeled fields and, quite ironically, neutralizing just what is one of the major improvements expected from high resolution regional simulations.

Precipitation is the most critical input variable in hydrological modeling (Fekete et al., 2004; Biemans et al., 2009) and the main driver of river runoff. The corrected over-smoothed fields can therefore hardly be expected to produce accurate discharge projection when used to feed a hydrological model.

For four main European catchments discharging into the Mediterranean or the Black seas, $WBM_{plus}$ generally underestimates total discharge when forced with the bias corrected regional $P$ and $T$ fields. Figure 5 shows the resulting seasonal cycle of basin integrated $P-E$ and total runoff, the black horizontal line representing the average total $P^*-E$ over the climatological time window (1971–2005), where $P^*$ is the uncorrected precipitation. Evaporation is still derived from the corrected temperature field, as the correction applied to precipitation largely dominates the final adjustment. Quite notably, the uncorrected precipitation appears to give a better estimate of $P-E$ for the Danube, while it has comparable (but opposite) bias for the Po+Adige and the Ebro basins in contrast to the bias corrected IMPACT2C simulations. Such overestimation is mainly attributable to spurious summer precipitation peaks in one (Po+Adige case) or more (Ebro case) model realizations. The Rhone constitutes an evident exception, where the overestimation of precipitation is

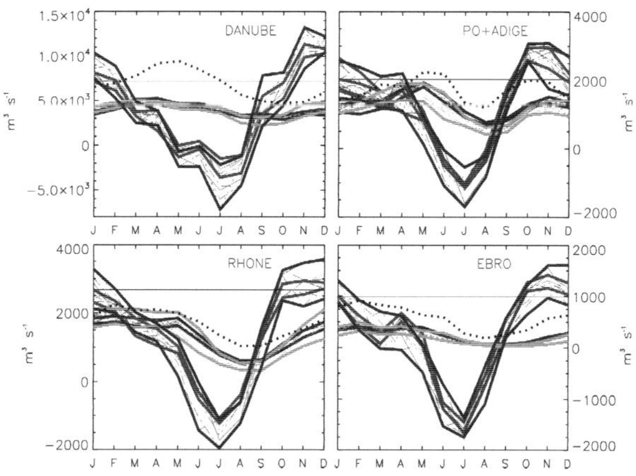


**Figure 5.** Basin-wide seasonal cycles of key water balance components ($P-E$: present – red, future $+2\,°C$ – blue, runoff: present – orange, future $+2\,°C$ – light blue). Thick lines outline the envelop of the five ensemble members, while thin lines correspond to single realizations. The black dotted line represent observed discharge. The black horizontal line represents average of the $P-E$ over the 1971–2005 period using uncorrected (supposedly biased) precipitation along with evapotranspiration estimates from the corrected temperature fields.

generally confined to the fall and winter months (except for one single ensemble member), when rainfall over the catchment is dominated by the large scale component rather than by local processes or interactions. Such bias is then probably attributable to systematic error propagation from the large scale GCM that the regional models are unable to correct, and that bias-correction techniques can only crudely mend in present day conditions.

By removing apparent systematic errors in the projected fields, bias correction effectively reduces model spread in control simulations, whereas it only partly succeeds in limiting noise in the projected water balance for the +2 °C scenario realizations, due to the variety of different model responses. Therefore there is no clear indication of a significant climate change signal, while the summer drying problem appears to be worsened.

Figure 4 allows extension of such conclusions to the continental scale hydrological balance. The +2 °C signal in the modelled precipitation climatological field appears as an increase over central eastern Europe and Scandinavia, and as a general decrease in southern Europe. On the other hand, evaporation likewise increases or decreases in the same areas, driven by rising temperatures and coupled to precipitation in wet regions via a soil-precipitation feedback amplification mechanism inducing more efficient extraction of moisture from the large scale atmospheric flow (Schar et al., 1999). It is elsewhere limited by water availability. A significant signal in total runoff can therefore only be detected over central eastern Europe and part of Scandinavia, where its magnitude is yet comparable to the relatively low local values, and numerical computations are possibly less accurate. It is difficult to quantify the role of upscaling, bias correction and ensemble averaging in determining the observed quasi-cancellation of signals in the atmospheric water balance, as all effectively act to decouple precipitation and soil moisture anomalies and to smooth out sub-grid variability.

The climate change signal in temperature alone also exhibits a definite meridional gradient, with the highest increments located over northern Europe, and southern countries experiencing lower though consistent warming. It is worth noting again, that both effects might originate in the reported model deficiencies, a hypothesis that is confirmed if the spatial patterns of both corrections to the data and model spread are considered. Under the assumption that model spread and correction magnitude together concur to give a rough estimate of uncertainties in model projections, if not as a disclosure of both model and data inadequacy, the climate change signal is in fact severely obscured by noise.

### 2.3 African example

For the African domain, the CORDEX community has produced a number of GCM-RCM combinations at a spatial resolution of 0.44° (~ 50 km). A total of 16 of such combinations are currently available. For West Africa, a consistent trend in temperature is visible (despite different magnitudes in the trends) in these models. The picture is less clear for precipitation owing to the difficulties of GCM in resolving the monsoon. The region has traditionally been one where climate models showed most disagreement (Giannini et al., 2008; Druyan, 2011). Despite some consistency in precipitation trends in the RCM ensemble, there are still large uncertainties in precipitation and it is not clear if wetter or drier conditions can be expected in the future (Fig. 6). Projected changes in annual precipitation from GCM indicate changes that are within the range of the observed variability of precipitation in the region (Fig. 7).

A change indicated by the majority of models is a delay in the onset of the rainy season together with a earlier cessation, leading to a total shortening of the wet period that lasts between 165 days in the South and only about 90 days in the Sahelian region in the North. With little change in precipitation this implies more precipitation falling on less days and generally more erratic rainfall.

Bias correcting the ambiguous climate model data in this region is likely not going to make the climate model for water balance modeling more meaningful. Despite the conceptual issues related to bias correction in general, the bigger problem is selecting the data sets on which the bias correction should be based on. For precipitation, for example, the observation network is very sparse and access to data is restricted by many national data sharing policies. Related to this, the satellite precipitation products (that rely on in-situ observations for calibration) in the region show large discrepancies that will translate through the bias correction chain and introduce further uncertainties.

## 3 Conclusions

Climate models are indispensable tools to understand atmospheric processes and the evolution of the Earth's climate regime, but they have clear shortcomings in providing climate projections for actual water management planning (Anagnostopoulos et al., 2010; Kundzewicz and Stakhiv, 2010). Hydrologist expressed concerns in the past about apparent inability of GCMs to reproduce contemporary climate conditions (Koutsoyiannis et al., 2008; Anagnostopoulos et al., 2010) with convincing fidelity. After decades of global circulation modeling research, the uncertainties in GCM projections are still increasing (Maslin and Austin, 2012).

In an anticipation that the computing power needed to enable GCM modelers to carry out computations at significantly higher spatial and temporal resolution (that are viewed as the key in improving GCM performance) is still decades away, some scientist are envisioning a possible alternative pathways by using supercomputers that are less accurate and don't necessary compute the same results from identical input data (Palmer, 2015). It is unclear, how such model can be

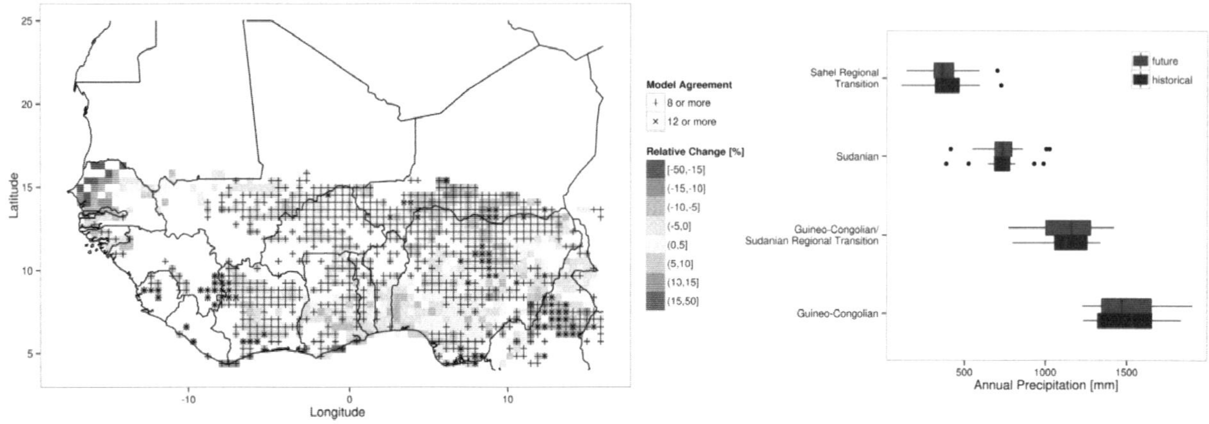


**Figure 6.** Relative change in ensemble median precipitation (left) and simulated absolute values of long term annual historical and future precipitation for 16 RCM/GCM combinations aggregated for 4 vegetation zones (right). "+" indicates an agreement of 8 or more models in the direction of change, "*" indicates an agreement of 12 or more models. Boxes extends from the 25th to 75th percentiles, whiskers indicate 1.5 times the interquantile range.

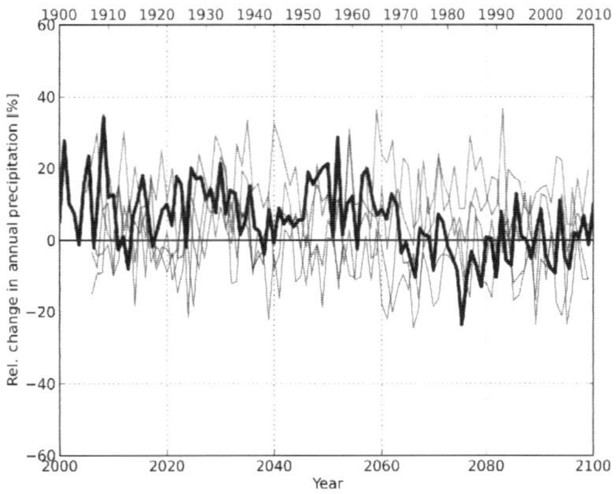


**Figure 7.** Projections of rel. changes in annual precipitation for West Africa, from 5 GCMs, randomly picked from ISI-MIP bias corrected data (grey). The blue line shows the observed variability in precipitation for the last 100 years (upper $x$ axis) from UDEL data (Willmott et al., 1994).

tested if the model simulation results could vary randomly, and how random perturbation from numerical inaccuracy can lead to sufficiently dense sample of the system phase space. Furthermore, such a system would violate perhaps the most important basis of sciences that relies on the reproducibility of the experiments carried out by other scientists.

Considering the incredible increase in computational power (which was always viewed as the major roadblock in improving GCM performance) during the last three decades (since the climate change agenda rose to its current prominence in geophysical research), one has to wonder if it is indeed the lack of computing power that prevents major breakthroughs or there are fundamental obstacles in our ability to predict the trajectories of the chaotic climate systems (Curry and Webster, 2011), and in our understanding and modeling of crucial processes (Held, 2005). Even if more detailed model simulations can improve model performance, the basic rule of computing "garbage in – garbage out" certainly will limit, how much improvement can be accomplished. Just like increased model complexity does not necessarily lead to better modeling skills processing more of the same poor quality data will only lead to more poor quality model results coming out.

The real improvements in our understanding of the Earth system processes will likely come only from better data that will need to come from improved Earth system monitoring both from in-situ and remote sensing sensors. A recent debate published in Science provided two distinct view about the role of in-situ monitoring (Fekete et al., 2015) and remote sensing (Famiglietti et al., 2015). While the right balance between putting sensors on the ground or in space is still up for scientific debate and ultimately might change over time as the various sensor technologies improve, but the debating papers were in full agreement in the need for both.

Given the demonstrated inefficiencies of both the Global Circulation Models and their operational cousins, the weather forecast models, in capturing the water cycle, better hydrological data are much needed. While remote sensing undoubtedly will play critical roles in monitoring precipitation and possibly soil moisture, river discharge, which is the most accurately monitored element of the hydrological cycle (Gutowski Jr. et al., 1997; Roads and Betts, 2000) will remain best monitored on the ground (Fekete et al., 2012). Data exchange of in-situ monitoring is often seen as an insurmountable obstacle, but surrendering to the difficulties in international data exchange amounts to nothing less than acknowledging that acting on larger goals such as combating climate change is impossible.

The biggest obstacle to more monitoring is the lack of financial resources. One would think that cost of operating high performance computing centers to support GCM modeling pales in comparison to the investment needed to maintain in-situ observing networks. In reality, it is not the case, as an example the Earth Simulator built in Japan (which was the fastest computer between 2002–2004) had a USD 700 million price tag and needed a full overhaul by 2009 (Fekete and Stakhiv, 2013). In contrast, operating 5000 river discharge gauges globally at USD 20 000 per year (United States Geological Survey's costs) would require USD 100 million annual spending (Fekete et al., 2012). Similarly, the price of the Earth Simulator's is comparable to the costs of two Earth observing NASA satellites (Anthes et al., 2007).

The only way the investments in GCM and impact model assessment capabilities can reach their full potential is if robust and reliable Earth observation can aid their development and calibration and permit sustained validations. "Historia est Magistra Vitae" – and history remains the life's teacher in our changing world.

## 4 Code and data availability

The data used in the presented research are all publicly available listed under the assets tab of the electronic version of the paper. The water balance model results are available from the ISIMIP project of the Potsdam Institute for Climate Impact Research (Warszavszki et al., 2014; Hempel et al., 2013). Their hosting data portal is part of the Earth System Grid Federation. The results from the European water balance model experiments are accessible at the ENSEMBLES project under theme RT3. The modeling and spatial analysis tools used in the present study are available on GitHub. Updated version of Willmott et al. (1994) was retrieved from Willtmott-Matsuura (2016).

The source code for the Water Balance/transport Model used in the presented studies are available on GitHub (Fekete, 2016).

**Acknowledgements.** The authors would like to acknowledge the Inter-Sectoral Impact Model Intercomparison Project, organized by the Potsdam Institute for Climate Impact Research that not only supplied the bias corrected global circulation model projections, but provided funding for the water balance model simulations. The European experiments were carried out under the IMPCAT2C project (Grant agreement#282746) funded by the European Union Seventh Framework Programme (FP7/2007–2013). Funding for the African example were provided by the German Ministry of Education and Research (BMBF) through the West African Science Service Center on Climate Change and Adapted Land Use (WASCAL, http://www.wascal.org).

## References

Allen, M. R., Barros, V. R., Broome, J., Cramer, W. P., Christ, R., Church, J. A., Clarke, L., Dasgupta, P., Dubash, N. K., Edenhofer, O., Elgizouli, I., Field, C. B., Forster, P. M., Friedlingstein, P., Fuglestvedt, J., Gomez-Echeverri, L., Hallegatte, S., Hegerl, G., Mark Howden, Jiang, K., Cisneros, B. J., Kattsov, V., Lee, H., Mach, K. J., Marotzke, J., Mastrandrea, M. D., Meyer, L., Minx, J., Mulugetta, Y., O'Brien, K., Oppenheimer, M., Pachauri, R. K., Pereira, J. J., Pichs-Madruga, R., Plattner, G.-K., Pörtner, H.-O., Power, S. B., Preston, B., Ravindranath, N. H., Reisinger, A., Riahi, K., Rusticucci, M., Scholes, R., Seyboth, K., Sokona, Y., Stavins, R., Stocker, T. F., Tschakert, P., van Vuuren, D. P., and van Ypersele, J.-P.: IPCC Fifth Assessment Synthesis Report, Tech. Rep. AR5 Synthesis Report, Intergovernmental Panel on Climate Change, 2014.

Anagnostopoulos, G. G., Koutsoyiannis, D., Christofides, A., Efstratiadis, A., and Mamassis, N.: A comparison of local and aggregated climate model outputs with observed data, Hydrol. Sci. J., 55, 1094–1110, doi:10.1080/02626667.2010.513518, 2010.

Anthes, R. A., Moore III, B., Anderson, J. G., Avery, S. K., Barron, E. J., Brown Jr., O. B., Cutter, S. L., DeFries, R. S., Gail, W. B., Hager, B. H., Hollingsworht, A., Janetos, A. C., Kelly, K. A., Lane, N. F., Lettenmaier, D. P., Marcus, B., Wahsington, W. M., Wilson, M. L., and Zoback, M. L.: Earth Science and Applications from Space: National Imperatives fro the Next Decade and beyond, National Academies Press, Washington DC, Washington, DC, 2007.

Bentsen, M., Bethke, I., Debernard, J. B., Iversen, T., Kirkevåg, A., Seland, Ø., Drange, H., Roelandt, C., Seierstad, I. A., Hoose, C., and Kristjánsson, J. E.: The Norwegian Earth System Model, NorESM1-M – Part 1: Description and basic evaluation of the physical climate, Geosci. Model Dev., 6, 687–720, doi:10.5194/gmd-6-687-2013, 2013.

Biemans, H., Hutjes, R. W. A., Kabat, P., Strengers, B. J., Gerten, D., and Rost, S.: Effects of precipitation uncertainty on discharge calculations for main river basins, J. Hydrometeorol., 10, 1011–1025, doi:10.1175/2008JHM1067.1, 2009.

European Network for Earth System Modeling (ENES): The Coordinated Regional Climate Downscaling Experiment (CORDEX) downscaled climate forcing data (https://verc.enes.org) which is part of the Earth System Grid Federation (http://esgf.llnl.gov), last access: August 2016.

Curry, J. A. and Webster, P. J.: Climate Science and the Uncertainty Monster, B. Am. Meteorol. Soc., 92, 1667–1682, doi:10.1175/2011BAMS3139.1, 2011.

Druyan, L. M.: Studies of 21st-century precipitation trends over West Africa, Int. J. Climatol., 31, 1415–1424, doi:10.1002/joc.2180, 2011.

Dufresne, J.-L., Foujols, M.-A., Denvil, S., Caubel, A., Marti, O., Aumont, O., Balkanski, Y., Bekki, S., Bellenger, H., Benshila, R., Bony, S., Bopp, L., Braconnot, P., Brockmann, P., Cadule, P., Cheruy, F., Codron, F., Cozic, A., Cugnet, D., de Noblet, N., Duvel, J.-P., Ethé, C., Fairhead, L., Fichefet, T., Flavoni, S., Friedlingstein, P., Grandpeix, J.-Y., Guez, L., Guilyardi, E., Hauglustaine, D., Hourdin, F., Idelkadi, A., Ghattas, J., Joussaume, S., Kageyama, M., Krinner, G., Labetoulle, S., Lahellec, A., Lefebvre, M.-P., Lefevre, F., Levy, C., Li, Z. X., Lloyd, J., Lott, F., Madec, G., Mancip, M., Marchand, M., Masson, S.,

Meurdesoif, Y., Mignot, J., Musat, I., Parouty, S., Polcher, J., Rio, C., Schulz, M., Swingedouw, D., Szopa, S., Talandier, C., Terray, P., Viovy, N., and Vuichard, N.: Climate change projections using the IPSL-CM5 Earth System Model: from CMIP3 to CMIP5, Clim. Dynam., 40, 2123–2165, doi:10.1007/s00382-012-1636-1, 2013.

Dunne, J. P., John, J. G., Adcroft, A. J., Griffies, S. M., Hallberg, R. W., Shevliakova, S., Stouffer, R. J., Cooke, W., Dunne, K. a., Harrison, Matthew J.Krasting, J. P., Malyshev, S. L., Milly, P. C. D., Phillips, P. J., Sentman, L. T., Samuels, B. L., Spelman, M. J., Winton, M., Wittenberg, A. T., and Zadeh, N.: GFDL's ESM2 Global Coupled Climate-Carbon Earth System Models. Part I: Physical Formulation and Baseline Simulation Characteristics, J. Climate, 25, 6646–6665, doi:10.1175/JCLI-D-11-00560.1, 2012.

Ehret, U., Zehe, E., Wulfmeyer, V., Warrach-Sagi, K., and Liebert, J.: HESS Opinions "Should we apply bias correction to global and regional climate model data?", Hydrol. Earth Syst. Sci., 16, 3391–3404, doi:10.5194/hess-16-3391-2012, 2012.

Famiglietti, J. S., Cazenave, A., Eicker, A., Reager, J. T., Rodell, M., and Velicogna, I.: Satellites prove the big picture, Science, 349, 684–685, 2015.

Federer, C. A., Vörösmarty, C. J., and Fekete, B. M.: Intercomparison of methods for calculating potential evaporation in regional and global water balance models, Water Resour. Res., 32, 2315–2321, 1996.

Fekete, B. M.: WBMplus, available at: https://github.com/bmfekete/WBMplus, last access: August 2016.

Fekete, B. M. and Stakhiv, E. Z.: Water management preparation strategies for adaptation to changing climate, in: Climatic Change and Global Warming of Inland Waters: Impacts and Mitigation for Ecosystems and Societies, edited by: Goldman, C. R., Kumagai, M., and Robarts, R. D., 415–427, Wiley-Blackwell, 2013.

Fekete, B. M., Vörösmarty, C. J., Roads, J. O., and Willmott, C. J.: Uncertainties in Precipitation and their Impacts on Runoff Estimates, J. Climatol., 17, 294–303, 2004.

Fekete, B. M., Looser, U., and Robarts, R. D.: Rationale for monitoring discharge on the ground, J. Hydrometeorol., 13, 1977–1986, 2012.

Fekete, B. M., Robarts, R. D., Kumagai, M., Nachtnebel, H.-P., Odada, E., and Zhuliov, A. V.: Time for in situ renaissance, Science, 349, 685–686, 2015.

Giannini, A., Biasutti, M., Held, I. M., and Sobel, A. H.: A global perspective on African climate, Climatic Change, 90, 359–383, doi:10.1007/s10584-008-9396-y, 2008.

ISIMIP: The global water balance model results from the ISIMIP fast track project are available from the ISIMIP project's (https://www.isimip.org) data portal (https://esg.pik-potsdam.de/projects/esgf-pik), last access: August 2016.

Gutowski Jr., W. J., Chen, Y., and Otles, Z.: Atmospheric water vapor transport in NCEP-NCAR reanalyses: compariosn with river discharge in the central United States, B. Am. Meteorol. Soc., 78, 1957–1969, 1997.

Haddeland, I., Clark, D. B., Franssen, W., Ludwig, F., Voss, F., Arnell, N. W., Bertrand, N., Best, M., Folwell, S., Gerten, D., Gomes, S. M., Gosling, S. N., Hagemann, S., Hanasaki, N., Harding, R., Heinke, J., Kabat, P., Koirala, S., Oki, T., Polcher, J., Stacke, T., Viterbo, P., Weedon, G. P., and Yeh, P.: Multimodel estimate of the global terrestrial water balance: setup and first results, J. Hydrometeorol., 12, 869–884, doi:10.1175/2011JHM1324.1, 2011.

Haddeland, I., Heinke, J., Biemans, H., Eisner, S., Florke, M., Hanasaki, N., Konzmann, M., Ludwig, F., Masaki, Y., Schewe, J., Stacke, T., Tessler, Z. D., Wada, Y., and Wisser, D.: Global water resources affected by human interventions and climate change, P. Natl. Acad. Sci., 111, 3251–3256, doi:10.1073/pnas.1222475110, 2014.

Hamon, W. R.: Computation of direct runoff amounts from storm rainfall, Int. Assoc. Sci. Hydrol. Publ., 63, 52–62, 1963.

Haylock, M. R., Hofstra, N., Klein Tank, A. M. G., Klok, E. J., Jones, P. D., and New, M.: A European daily high-resolution gridded data set of surface temperature and precipitation for 1950–2006, J. Geophys. Res.-Atmos., 113, D20119, doi:10.1029/2008JD010201, 2008.

Held, I. M.: The Gap between Simulation and Understanding in Climate Modeling, B. Am. Meteorol. Soc., 86, 1609–1614, doi:10.1175/BAMS-86-11-1609, 2005.

Hempel, S., Frieler, K., Warszawski, L., Schewe, J., and Piontek, F.: A trend-preserving bias correction – the ISI-MIP approach, Earth Syst. Dynam., 4, 219–236, doi:10.5194/esd-4-219-2013, 2013.

Hempel, S., Frieler, K., Warszawski, L., Schewe, J., and Piontek, F.: Bias corrected GCM input data for ISIMIP Fast Track, GFZ Data Services, available at: http://doi.org/10.5880/PIK.2016.001 (http://isimip.org), 2013.

Jacob, D., Bärring, L., Christensen, O. B. S., Christensen, J. H., De Castro, M., Déqué, M., Giorgi, F., Hagemann, S., Hirschi, M., Jones, R., Kjellström, E., Lenderink, G., Rockel, B., Sánchez, E., Schär, C., Seneviratne, S. I., Somot, S., Van Ulden, A., and Van Den Hurk, B.: An inter-comparison of regional climate models for Europe: Model performance in present-day climate, Climatic Change, 81, 31–52, doi:10.1007/s10584-006-9213-4, 2007.

Jacob, D., Petersen, J., Eggert, B., Alias, A., Christensen, O. B. S., Bouwer, L. M., Braun, A., Colette, A., Déqué, M., Georgievski, G., Georgopoulou, E., Gobiet, A., Menut, L., Nikulin, G., Haensler, A., Hempelmann, N., Jones, C., Keuler, K., Kovats, S., Kröner, N., Kotlarski, S., Kriegsmann, A., Martin, E., van Meijgaard, E., Moseley, C., Pfeifer, S., Preuschmann, S., Radermacher, C., Radtke, K., Rechid, D., Rounsevell, M., Samuelsson, P., Somot, S., Soussana, J. F., Teichmann, C., Valentini, R., Vautard, R., Weber, B., and Yiou, P.: EURO-CORDEX: New high-resolution climate change projections for European impact research, Reg. Environ. Change, 14, 563–578, doi:10.1007/s10113-013-0499-2, 2014.

Jones, C. D., Hughes, J. K., Bellouin, N., Hardiman, S. C., Jones, G. S., Knight, J., Liddicoat, S., O'Connor, F. M., Andres, R. J., Bell, C., Boo, K.-O., Bozzo, A., Butchart, N., Cadule, P., Corbin, K. D., Doutriaux-Boucher, M., Friedlingstein, P., Gornall, J., Gray, L., Halloran, P. R., Hurtt, G., Ingram, W. J., Lamarque, J.-F., Law, R. M., Meinshausen, M., Osprey, S., Palin, E. J., Parsons Chini, L., Raddatz, T., Sanderson, M. G., Sellar, A. A., Schurer, A., Valdes, P., Wood, N., Woodward, S., Yoshioka, M., and Zerroukat, M.: The HadGEM2-ES implementation of CMIP5 centennial simulations, Geosci. Model Dev., 4, 543–570, doi:10.5194/gmd-4-543-2011, 2011.

Kerr, R. A.: Forecasting Regional Climate Change Flunks Its First Test, Science, 339, p. 638, 2013.

Knutti, R., Furrer, R., Tebaldi, C., Cermak, J., and Meehl, G. A.: Challenges in combining projections from multiple climate models, J. Climate, 23, 2739–2758, doi:10.1175/2009JCLI3361.1, 2010.

Koutsoyiannis, D., Efstratiadis, A., Mamassis, N., and Christofides, A.: On the credibility of climate predictions, Hydrol. Sci. J., 53, 671–684, doi:10.1623/hysj.53.4.671, 2008.

Kundzewicz, Z. W. and Stakhiv, E. Z.: Are climate models "ready for prime time" in water resources management applications, or is more research needed?, Hydrol. Sci. J., 55, 1085–1089, doi:10.1080/02626667.2010.513211, 2010.

Liang, X., Lettenmaier, D. P., Wood, E. F., and Burges, S. J.: A Simple hydrologically Based Model of Land Surface Water and Energy Fluxes for GSMs, J. Geophys. Res., 99, 14415–14428, 1994.

Lins, H. F. and Cohn, T. A.: Stationarity: Wanted Dead or Alive?, J. Am. Water Resour. Assoc., 47, 475–480, 2011.

Maslin, M. and Austin, P.: Climate models at their limit?, Nature, 486, 183–184, 2012.

Masson, D. and Knutti, R.: Climate model genealogy, Geophys. Res. Lett., 38, 1–4, doi:10.1029/2011GL046864, 2011.

Maurer, E. P. and Hidalgo, H. G.: Utility of daily vs. monthly large-scale climate data: an intercomparison of two statistical downscaling methods, Hydrol. Earth Syst. Sci., 12, 551–563, doi:10.5194/hess-12-551-2008, 2008.

Milly, P. C. D., Betancourt, J., Falkenmark, M., Hirsch, R. M., Kundzewicz, Z. W., Lettenmaier, D. P., and Stouffer, R. J.: Stationarity is Dead: Whither Water Management?, Science, 319, 573–574, 2008.

New, M., Hulme, M., and Jones, P. D.: Representing Twentieth Century Space-time Climate Variability: II. Development of 1901-1996 monthly grids of terrestrial surface, J. Climatol., 13, 2217–2238, 2000.

Nijssen, B., O'Donnell, G. M., Lettenmaier, D. P., Lohmann, D., and Wood, E. F.: Predicting the discharge of global rivers, J. Climate, 14, 3307–3323, 2001.

Palmer, T. N.: Build imprecise supercomputers, Nature, 526, 32–33, doi:10.1038/526032a, 2015.

Pennell, C. and Reichler, T.: On the effective number of climate models, J. Climate, 24, 2358–2367, doi:10.1175/2010JCLI3814.1, 2011.

Perrin, C., Michel, C., and Andreassian, V.: Does a large number of parameters enhance model performance? Comparative assessment of common catchment model structures on 429 catchments, J. Hydrol., 242, 275–301, 2001.

Pielke Sr., R. A. and Wilby, R. L.: Regional Climate Downscaling: What's the Point?, Eos Transactions American Geophysical Union, 93, 52–53, 2012.

Rauscher, S. a., Coppola, E., Piani, C., and Giorgi, F.: Resolution effects on regional climate model simulations of seasonal precipitation over Europe, Clim. Dynam., 35, 685–711, doi:10.1007/s00382-009-0607-7, 2010.

Roads, J. O. and Betts, A.: NCEP-NCAR and ECMWF Reanalysis Surface Water and Energy Budgets for the Mississippi River Basin, J. Hydrometeorol., 1, 88–94, doi:10.1175/1525-7541(2000)001<0088:NNAERS>2.0.CO;2, 2000.

Schar, C., Luthi, D., Beyerle, U., and Heise, E.: The soil-precipitation feedback: A process study with a regional climate model, J. Climate, 12, 722–741, 1999.

Schewe, J., Heinke, J., Gerten, D., Haddeland, I., Arnell, N. W., Clark, D. B., Dankers, R., Eisner, S., Fekete, B. M., de Jesus Colon Gonzalez, F., Gosling, S. N., Kim, H., Liu, X., Masaki, Y., Portmann, F. T., Satoh, Y., Stacke, T., Tang, Q., Wada, Y., Wisser, D., Albrecht, T., Frieler, K., Piontek, F., Warszawski, L., and Kabat, P.: Multimodel assessment of water scarcity under climate change, P. Natl. Acad. Sci., 111, 3245–3250, doi:10.1073/pnas.1222460110, 2014.

Stevens, B. and Bony, S.: What Are Climate Models Missing?, Science, 340, 1053–1054, doi:10.1126/science.1237554, 2013.

Tebaldi, C., Arblaster, J. M., and Knutti, R.: Mapping model agreement on future climate projections, Geophys. Res. Lett., 38, 1–5, doi:10.1029/2011GL049863, 2011.

van der Linden, P.: The results of the European water balance model experiments are hosted at the ENSEMBLES project (http://ensembles-eu.metoffice.com) under theme RT3 (http://ensemblesrt3.dmi.dk), last access: August 2016.

van Griensven, a., Meixner, T., Grunwald, S., Bishop, T., Diluzio, M., and Srinivasan, R.: A global sensitivity analysis tool for the parameters of multi-variable catchment models, J. Hydrol., 324, 10–23, doi:10.1016/j.jhydrol.2005.09.008, 2006.

Vörösmarty, C. J., Moore III, B., Grace, A. L., Gildea, M., Melillo, J. M., Peterson, B. J., Rastetter, E. B., and Steudler, P. A.: Continental scale models of water balance and fluvial transport: An application to South America, Global Biogeochem. Cy., 3, 241–265, 1989.

Vörösmarty, C. J., Federer, C. A., and Schloss, A. L.: Potential evaporation functions compared on US watersheds: Possible implications for global-scale water balance and terrestrial ecosystem modeling, J. Hydrol., 207, 147–169, 1998.

Warszawski, L., Frieler, K., Huber, V., Piontek, F., Serdeczny, O., and Schewe, J.: The Inter-Sectoral Impact Model Intercomparison Project (ISI-MIP): Project framework, P. Natl. Acad. Sci., 111, 3228–3232, doi:10.1073/pnas.1312330110, 2014.

Watanabe, S., Hajima, T., Sudo, K., Nagashima, T., Takemura, T., Okajima, H., Nozawa, T., Kawase, H., Abe, M., Yokohata, T., Ise, T., Sato, H., Kato, E., Takata, K., Emori, S., and Kawamiya, M.: MIROC-ESM 2010: model description and basic results of CMIP5-20c3m experiments, Geosci. Model Dev., 4, 845–872, doi:10.5194/gmd-4-845-2011, 2011.

Willmott, C. J., Robeson, S. M., and Feddema, J. J.: Estimating continental and terrestrial precipitation averages from rain-gauge networks, Int. J. Climatol., 14, 403–414, 1994.

Willmott, C. J.: Matsuura Climate Data (air temperature and precipitation) is available from the "Willmott, Matsuura and Collaborators' Global Climate Resource Pages" http://climate.geog.udel.edu/~climate, last access: August 2016.

Wisser, D., Fekete, B. M., Vörösmarty, C. J., and Schumann, A. H.: Reconstructing 20th century global hydrography: a contribution to the Global Terrestrial Network-Hydrology (GTN-H), Hydrol. Earth Syst. Sci., 14, 1–24, doi:10.5194/hess-14-1-2010, 2010.

WMO: The physical basis of climate and climate modelling, Tech. Rep. 16, World Meteorological Organization, Geneva, Switzerland, available at: http://library.wmo.int/pmb_ged/garp-ps_16.pdf (last access: August 2016), 1975.

Proc. IAHS, 374, 53–62, 2016
proc-iahs.net/374/53/2016/
doi:10.5194/piahs-374-53-2016


Open Access
PIAHS

# Impact of climate forcing uncertainty and human water use on global and continental water balance components

**Hannes Müller Schmied[1,2], Linda Adam[1], Stephanie Eisner[3], Gabriel Fink[3], Martina Flörke[3], Hyungjun Kim[4], Taikan Oki[4], Felix Theodor Portmann[1], Robert Reinecke[1], Claudia Riedel[1], Qi Song[1], Jing Zhang[1], and Petra Döll[1]**

[1]Institute of Physical Geography, Goethe University Frankfurt, Frankfurt, Germany
[2]Senckenberg Biodiversity and Climate Research Centre (BiK-F), Frankfurt, Germany
[3]Center for Environmental Systems Research (CESR), University of Kassel, Kassel, Germany
[4]Institute of Industrial Science, The University of Tokyo, Tokyo, Japan

*Correspondence to:* Hannes Müller Schmied (hannes.mueller.schmied@em.uni-frankfurt.de)

Published: 17 October 2016

**Abstract.** The assessment of water balance components using global hydrological models is subject to climate forcing uncertainty as well as to an increasing intensity of human water use within the 20th century. The uncertainty of five state-of-the-art climate forcings and the resulting range of cell runoff that is simulated by the global hydrological model WaterGAP is presented. On the global land surface, about 62 % of precipitation evapotranspires, whereas 38 % discharges into oceans and inland sinks. During 1971–2000, evapotranspiration due to human water use amounted to almost 1 % of precipitation, while this anthropogenic water flow increased by a factor of approximately 5 between 1901 and 2010. Deviation of estimated global discharge from the ensemble mean due to climate forcing uncertainty is approximately 4 %. Precipitation uncertainty is the most important reason for the uncertainty of discharge and evapotranspiration, followed by shortwave downward radiation. At continental levels, deviations of water balance components due to uncertain climate forcing are higher, with the highest discharge deviations occurring for river discharge in Africa (−6 to 11 % from the ensemble mean). Uncertain climate forcings also affect the estimation of irrigation water use and thus the estimated human impact of river discharge. The uncertainty range of global irrigation water consumption amounts to approximately 50 % of the global sum of water consumption in the other water use sector.

## 1 Introduction

The interest in global-scale water resources assessments has increased in the last two decades. There has been an increasing number of publications in this field (Web of Science, topic "global scale" AND "water resources", 1981–1990: 0 entries; 1991–2000: 6 entries; 2001–2010: 64 entries; 2011–2015: 85 entries), and a number of global hydrological models (GHMs) have been developed (Bierkens, 2015). The UN and other international organizations require global-scale information on water resources and their use, e.g. UNESCO's World Water Assessment Programme (www.unesco.org/water/wwap) or the Transboundary Waters Assessment Programme (TWAP, http://www.geftwap.org/twap-project), which can only be provided by modelling approaches due to a lack of observations with global coverage. Such model-based assessments require meteorological variables as climate forcing input. Currently, a number of state-of-the-art global-scale climate forcings are available that are all based on weather models and differ in terms of methodology including the underlying reanalysis and in terms of observation data used for bias correction. Different climate forcings result in large differences in simulated water fluxes and states as has already been shown by Biemans et al. (2009) for precipitation uncertainty and by Müller Schmied et al. (2014) regarding the uncertainty caused by two climate forcings that differ with respect to other climate variables. Analyses of the impact of different climate forcings are currently the focus of

Published by Copernicus Publications on behalf of the International Association of Hydrological Sciences.

**Table 1.** Summary of climate forcing characteristics used in this study. Abbreviations: precipitation – $P$, temperature – $T$, shortwave downward radiation – SWD, longwave downward radiation – LWD.

| Name | Time span | Basis | Bias correction | Reference |
|---|---|---|---|---|
| GSWP3 | 1901–2010 | 20th Century Reanalysis using NCEP atmosphere land model | GPCC v6 ($P$), and undercatch correction (Hirabayashi et al., 2008) CRU TS3.21 (other variables) | http://hydro.iis.u-tokyo.ac.jp/GSWP3 |
| PGFv2 | 1901–2012 | NCEP-NCAR reanalysis | CRU TS3.21 ($P$, $T$), no precipitation undercatch correction, U Maryland, CRU TS3.21 cloud cover (SWD), U Maryland (LWD) | Updated version of Sheffield et al. (2006), information based on personal communication with J. Sheffield (2015) |
| WFD | 1901–2001 | ERA-40 reanalysis | GPCCv4 ($P$), undercatch correction using Adam and Lettenmaier (2003), CRU TS 2.1 cloud cover (SWD), CRU TS 2.1 temperature ($T$) | Weedon et al. (2010) |
| WFD_WFDEI | 1901–2010 | WFD 1901–1978, WFDEI (based on ERA-Interim reanalysis) afterwards | GPCC v5 (v6 for 2010) ($P$), undercatch correction using Adam and Lettenmaier (2003), CRU TS 3.1/3.21 cloud cover (SWD), CRU TS 3.1/3.21 temperature ($T$) | Weedon et al. (2014) |
| WFDEI_hom | 1901–2010 | As WFD_WFDEI, but WFD homogenized using a multiplicative approach for SWD and LWD and additive approach for $T$ | Homogenization: Haddeland et al. (2012), Müller Schmied et al. (2016) | |

model intercomparison studies such as the Inter-Sectoral Impact Model Intercomparison Project (ISIMIP) in its phase 2a, where (among other sectors) several global and regional water models are driven by four state-of-the-art climate forcings and compared to historical observations of, for example, discharge and actual evapotranspiration. In particular, the propagation of climate forcing uncertainty at multiple scales (grid-cell level, continental, global) is one topic to be addressed in ISIMIP2a.

Humans increasingly influence the water cycle through water abstractions (Oki and Kanae 2006), in particular for irrigation (e.g. Siebert et al., 2015) but also for other purposes like thermal power plant cooling, manufacturing, livestock production and domestic sectors (Flörke et al., 2013). Quantification of sectoral water abstractions and consumptive water use (also called water consumption, the amount of the abstracted water that evapotranspires during human water use or is incorporated in products), and in particular of the source of water, is highly uncertain due to lack of data (Döll et al., 2016). In some regions, irrigation by groundwater leads to groundwater depletion problems (Döll et al., 2014a; Wada, 2016), and it has been estimated that in groundwater depletion areas, farmers irrigate with only 70 % of the optimal amount of water (Döll et al., 2014a).

Given the large uncertainties, we aim to answer the following research questions by using the Water Global Assessment and Prognosis (WaterGAP) GHM in its version 2.2 (ISIMIP2a):

1. How does climate forcing affect computed runoff at the grid-cell level?
2. How does climate forcing uncertainty and human water use affect long-term average water balance components (including human water use) on global and continental scales?

In Sect. 2 we briefly present the model and climate forcings used in this study. Results are presented and discussed in Sect. 3. The paper ends with a conclusion (Sect. 4), where we answer the research questions, followed by an outlook.

## 2 Data and methods

The global water availability and water use model WaterGAP (Alcamo et al., 2003; Döll et al., 2003; Müller Schmied et al., 2014) was applied using version WaterGAP 2.2 (ISIMIP2a). The main model characteristics of version 2.2 are described in Müller Schmied et al. (2014), and the differences to the ISIMIP2a version are described in Müller Schmied et

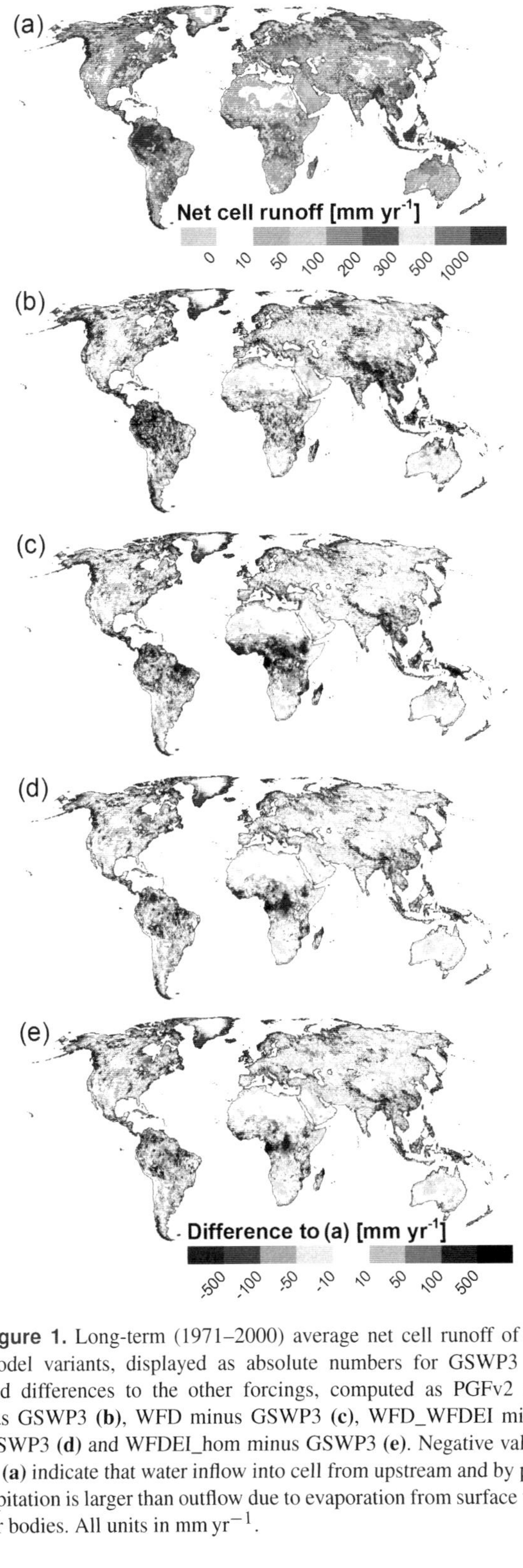


**Figure 1.** Long-term (1971–2000) average net cell runoff of the model variants, displayed as absolute numbers for GSWP3 **(a)** and differences to the other forcings, computed as PGFv2 minus GSWP3 **(b)**, WFD minus GSWP3 **(c)**, WFD_WFDEI minus GSWP3 **(d)** and WFDEI_hom minus GSWP3 **(e)**. Negative values in **(a)** indicate that water inflow into cell from upstream and by precipitation is larger than outflow due to evaporation from surface water bodies. All units in mm yr$^{-1}$.

al. (2016). WaterGAP has a spatial resolution of $0.5 \times 0.5°$ (i.e. $55 \times 55$ km at the Equator) and consists of five water use models that are linked through the Ground Water Surface Water Use submodule with the WaterGAP global hydrology model (WGHM). Within WGHM, water storage changes in several compartments and freshwater fluxes are modelled at a daily time step. WGHM is calibrated to match long-term average discharge at 1319 observation points (from GRDC database) within 1 % deviation by adjusting one to three parameters (calibration details in Müller Schmied et al., 2014).

Four state-of-the-art climate forcings provided by the Inter-Sectoral Impact Model Intercomparison Project (ISIMIP) in its current phase 2a (https://www.isimip.org/about/#simulation-rounds-isimip2a) plus a fifth homogenized forcing were used to force WaterGAP. Table 1 summarizes the main characteristics of all five climate forcing datasets. For a detailed description, the reader is referred to Müller Schmied et al. (2016). The names of the model runs are similar to the names of the climate forcings.

## 3 Results and discussion

### 3.1 Global runoff at grid-cell level

Net cell runoff (computed as outflow minus inflow of each grid cell) differs considerably between the different climate forcing datasets (Fig. 1). This can be attributed to large differences between climate forcings at grid-cell level, in particular with respect to precipitation. Different observational datasets are used to bias-correct *P* (PGFv2 based on CRU and the others based on different versions of GPCC). This results in large differences (in both directions) at the regional scale for South America and South East Asia. Obviously, the unequal *P* gauging networks underlying the observational datasets and/or varying regionalization approaches lead to the large differences. In addition, PGFv2 is not corrected for undercatch of solid precipitation (J. Sheffield, personal communication, 2015) while all other datasets are. As undercatch correction (e.g. Adam and Lettenmaier, 2003) leads to the highest *P* increases in northern (snow-dominated) latitudes, *P* (and consequently net cell runoff) is lower for PGFv2 (red areas in the northern latitudes in Fig. 1b).

The main reason for the large discrepancies between GSWP3 and WFD in equatorial regions (Fig. 1c) is attributable to systematically smaller SWD (Fig. A3c) in energy-limited areas for the WFD dataset. This effect is lessened in the combined WFD_WFDEI dataset (Fig. 1d) and even more in the homogenized forcing WFDEI_hom (Fig. 1e), as WFDEI shows systematically higher SWD than WFD. The higher SWD in parts of Asia, western Africa and Australia (Fig. A3d, e) does not influence net cell runoff significantly because these regions are water-limited: evaporation and runoff are mainly controlled by precipitation and not by available energy. In many regions where SWD is in-

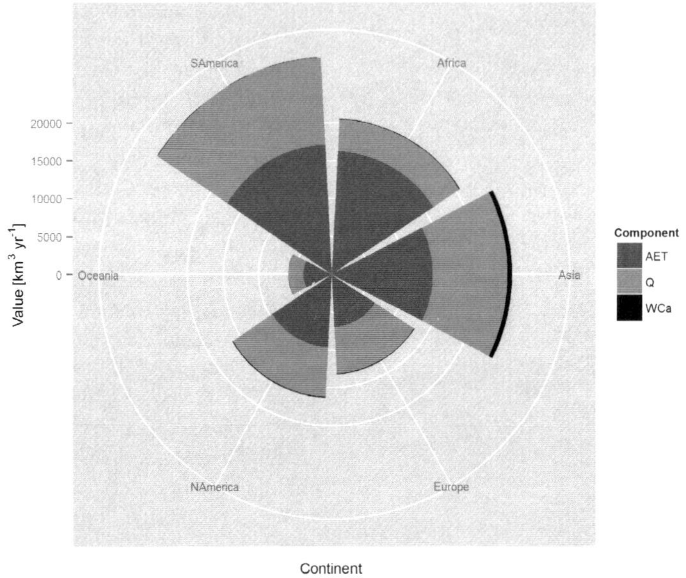


**Figure 2.** Continental water balance components 1971–2000 (km$^3$ yr$^{-1}$); ensemble mean of the five climate forcings. Abbreviations: actual evapotranspiration (AET), discharge into oceans and inland sinks ($Q$), actual water consumption (WCa).

**Table 2.** Global water balance components for land area (except Antarctica and Greenland) in % of precipitation (row 1) for the five model variants and 1971–2000. Cells representing inland sinks were excluded but discharge into inland sinks was included.

| No. | Component | GSWP3 | PGFv2 | WFD | WFDEI_hom | WFD_WFDEI |
|---|---|---|---|---|---|---|
| 1 | Precipitation $P$ (km$^3$ yr$^{-1}$) | 109 631 | 103 525 | 110 690 | 111 050 | 111 050 |
| 2 | Actual evapotranspiration AET[a] | 62.0 | 61.3 | 61.1 | 63.0 | 62.0 |
| 3 | Discharge into oceans and inland sinks $Q$[b] | 37.1 | 37.8 | 38.1 | 36.2 | 37.2 |
| 4 | Water consumption (actual) WCa | 0.9 | 0.9 | 0.8 | 0.9 | 0.8 |
| 5 | Change of total water storage $\mathrm{d}S\,/\,\mathrm{d}t$[c] | −0.01 | −0.03 | −0.02 | −0.02 | −0.07 |

[a] AET does not include evapotranspiration caused by human water use, i.e. actual water consumption WCa. [b] Taking into account anthropogenic water use.
[c] Total water storage (TWS) of 31 December 2000 minus TWS of 31 December 1970, divided by the number of 30 years.

creased, LWD is decreased (and vice versa), which reduces the effect on net radiation. $T$ effects model results via the equation for potential evapotranspiration (Priestley and Taylor, 1972), via snow dynamics as well as the leaf area index model that affects canopy evaporation (details in Müller Schmied et al., 2014). As $T$ differs only little between the forcing datasets (Fig. A2), effects of $T$ differences on simulated net cell runoff are expected to be relatively small.

### 3.2 Continental water balance components

Figure 2 displays the continental-scale partitioning of precipitation into actual evapotranspiration AET, river discharge $Q$, and human water consumption WCa. South America and Africa have nearly the same absolute amount of AET, but values for $Q$ differ strongly. As a consequence of extensive irrigated agriculture especially in India and China (Siebert et al., 2015), the highest water consumption occurs in Asia, where 2.5 % of precipitation is evapotranspired, mainly due to irrigation. For the other continents, water consumption plays – relative to the other water balance components – only a marginal role. The lowest runoff coefficient ($Q/P$) is found in Africa (0.21), whereas runoff coefficients vary between 0.34 (Oceania) and 0.47 (Europe) for the other continents. Hence, differences in $P$ result in higher relative uncertainties of estimated water resources for Africa. The deviation from the mean continental value for $Q$ among the climate forcings is between −5.9 (calculated as min $Q$ / mean $Q$) and 10.9 % (max $Q$ / mean $Q$) for Africa, whereas for all other continents deviations are lower (−5.4 and −2.2 as minimum $Q$, 2.5–5.0 % as maximum $Q$).

### 3.3 Global water balance components

Compared to the continental-scale deviation of water balance components, the impact of climate forcing uncertainty

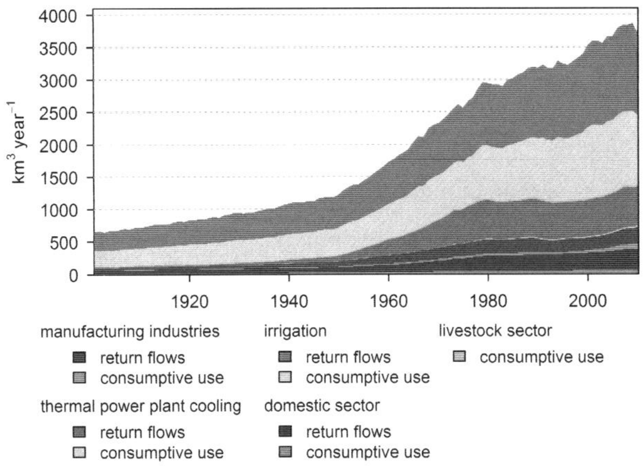


**Figure 3.** Development of water abstractions (sum of return flows and consumptive use) and water consumption (the amount of water that is evapotranspired or incorporated in products, light colours) of the five water use sectors considered in WaterGAP for 1901–2010. Values for irrigation (modelled with 70 % of demand in grid cells with groundwater depletion) are averaged across the five climate forcings; other sectors are modelled independently of climate forcing and taken from Flörke et al. (2013).

is levelling out at the global scale to a certain degree (Table 2). Global runoff coefficients vary between 0.362 and 0.381, and deviation of global $Q$ from the ensemble mean is −3.8 to 3.7 %. Actual evapotranspiration is estimated to range between 61.1 and 63.0 % of global $P$ (Table 2). The lowest value for AET (and highest value for $Q$) is computed when using WFD climate forcing. Here, average global SWD is 15 W $m^{-2}$ lower compared to the other forcings (see also Müller Schmied et al., 2016, their Fig. 1). In absolute numbers, differences in AET and $Q$ resulting from the five climate forcings are considerable. For example, global discharge values range from 39 200 to 42 200 $km^3$ $yr^{-1}$: the uncertainty range is equal to thrice the total water consumption (∼ 930 $km^3$ $yr^{-1}$).

On the global scale, sectoral water uses have strongly increased since 1901 (Fig. 3). Whereas overall water abstractions (consumptive use) are about 650 (260) $km^3$ $yr^{-1}$ in the year 1901, values are about 5 times higher with 3700 (1250) $km^3$ $yr^{-1}$ in 2010. In contrast to Müller Schmied et al. (2016, their Fig. 1), where water consumption of each climate forcing is presented using different time step aggregations, Fig. 3 shows the proportion of potential (if water were available without limitation) consumptive water use components (light colours) and the amount of return flows (dark colours). The most important water use sector regarding both abstraction and consumption is the irrigation sector. The sum of potential water consumption of all water use sectors (except irrigation) throughout the period 1971–2000 is 112 $km^3$ $yr^{-1}$, whereas the sums of potential irrigation water consumption vary between 834 and 894 $km^3$ $yr^{-1}$ depending on the climate forcings. Together with the other potential water uses (manufacturing, cooling of thermal power plants, domestic and livestock sector), the demand of consumptive water uses ranges from 946 to 1006 $km^3$ $yr^{-1}$. Due to limited water availability to satisfy the demand, actual water consumption (WCa) ranges between 915 (WFD) and 960 (PGFv2) $km^3$ $yr^{-1}$ (all numbers 1971–2000). Hence, water availability reduces the impact of climate forcing uncertainty when modelling water use demand. The uncertainty range of estimated global irrigation water consumption due to the climate forcing is therefore about 50 % of the sum of all the other water use sectors.

## 4 Conclusions

Within this study, the WaterGAP 2.2 (ISIMIP2a) model was used to assess water balance components on grid-cell, continental and global scale as well as the development of human water use on the global scale. The research questions can be answered as follows:

1. How does climate forcing affect computed runoff at the grid-cell level?

On the grid-cell level, the effect of climate forcing uncertainty on computed runoff is very large. In particular, usage of different observation-based products to bias-correct reanalysis data affects the spatial distribution of runoff. Furthermore, undercatch correction (or the lack thereof) of $P$ leads to differences in model estimates. Whereas $T$ uncertainty does not lead to clearly visible spatial differences in computed runoff, SWD uncertainty was found to have a large impact in energy-limited regions like tropical Africa. For water-limited areas, this is not the case.

2. How does climate forcing uncertainty and human water use affect long-term average water balance components on global and continental scales?

Climate forcing uncertainty is high (Figs. A1–A4), and most important are differences in $P$ and SWD. At the continental scale, these uncertainties lead to large differences in calculated water balance components, in particular in regions with high $P$ uncertainty and low runoff coefficient (e.g. Africa). Global-scale values vary less in relative terms as deviations even out with spatial aggregation. The uncertainty range of estimated global irrigation water consumption due to uncertain climate forcing is around 50 % of the water consumption in the other water use sectors.

Multi-model hydrological assessments as done by the ISIMIP initiative for both historical periods (e.g. Haddeland et al., 2011) and future scenarios (e.g. Schewe et al., 2014) will help to relate the uncertainties of water balance components at different scales of aggregation that are caused by different climate forcings to uncertainties due to the hydrological models themselves. To constrain both types of uncertainty, model calibration not only of mean annual river discharge (as done for the WaterGAP model) but also of remote-sensing-based data like total water storage variations from GRACE (Eicker et al., 2014; Döll et al., 2014b, 2016) is promising, but collection and sharing of in situ data remains crucial (Fekete et al., 2015).

## 5 Data availability

The WaterGAP output will become freely available to the public within the framework of the ISIMIP project phase 2a, but it is not yet clarified where the data will be hosted (please check https://www.isimip.org/outputdata/ for updates). The homogenized climate forcing WFDEI_hom is not included within the ISI-MIP project. All model outputs used in this study are available on request from the corresponding author.

## Appendix A

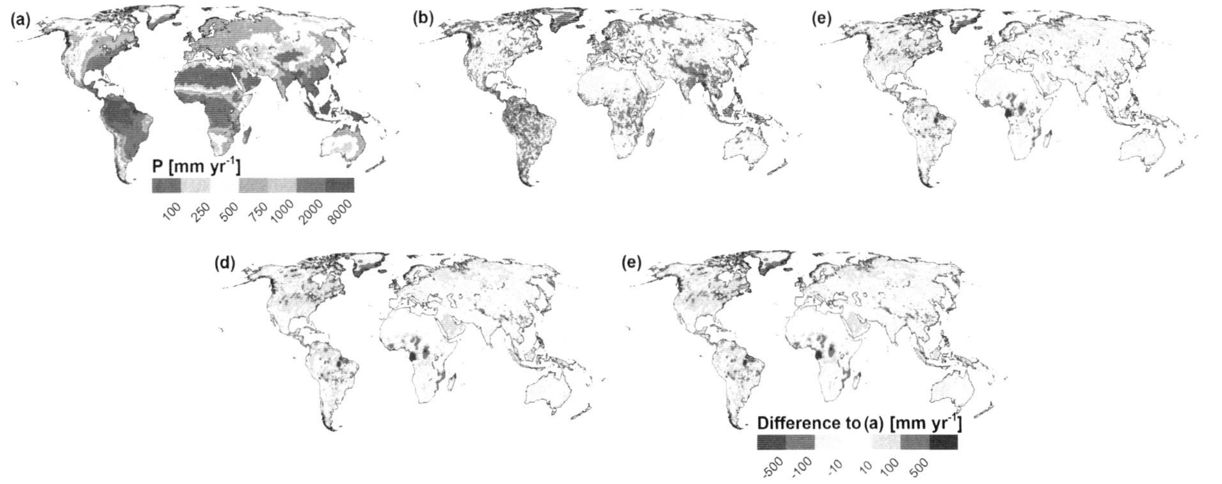


**Figure A1.** Long-term (1971–2000) average precipitation of the model variants, displayed as absolute number for GSWP3 (**a**) and differences to the other forcings, computed as PGFv2 minus GSWP3 (**b**), WFD minus GSWP3 (**c**), WFD_WFDEI minus GSWP3 (**d**), and WFDEI_hom minus GSWP3 (**e**). All units in mm yr$^{-1}$.

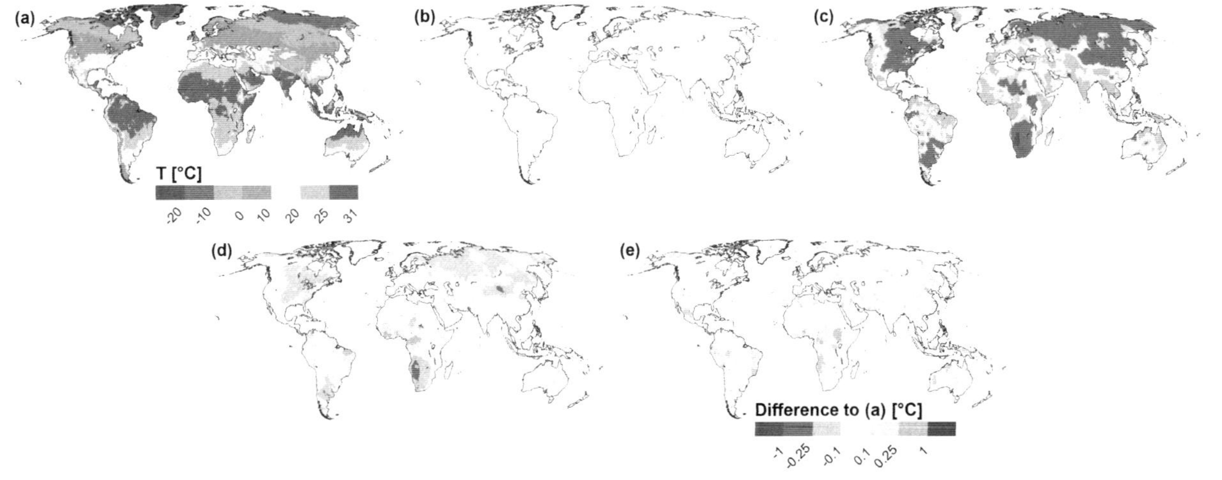


**Figure A2.** Long-term (1971–2000) average temperature of the model variants, displayed as absolute number for GSWP3 (**a**) and differences to the other forcings, computed as PGFv2 minus GSWP3 (**b**), WFD minus GSWP3 (**c**), WFD_WFDEI minus GSWP3 (**d**), and WFDEI_hom minus GSWP3 (**e**). All units in °C.

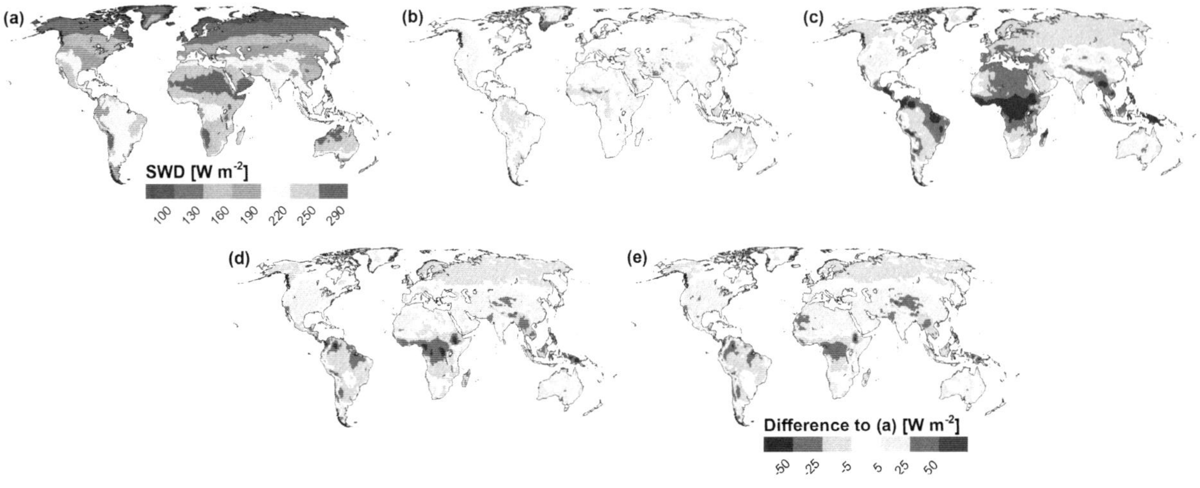


**Figure A3.** Long-term (1971–2000) average shortwave downward radiation of the model variants, displayed as absolute number for GSWP3 (**a**) and differences to the other forcings, computed as PGFv2 minus GSWP3 (**b**), WFD minus GSWP3 (**c**), WFD_WFDEI minus GSWP3 (**d**), and WFDEI_hom minus GSWP3 (**e**). All units in W $m^{-2}$.

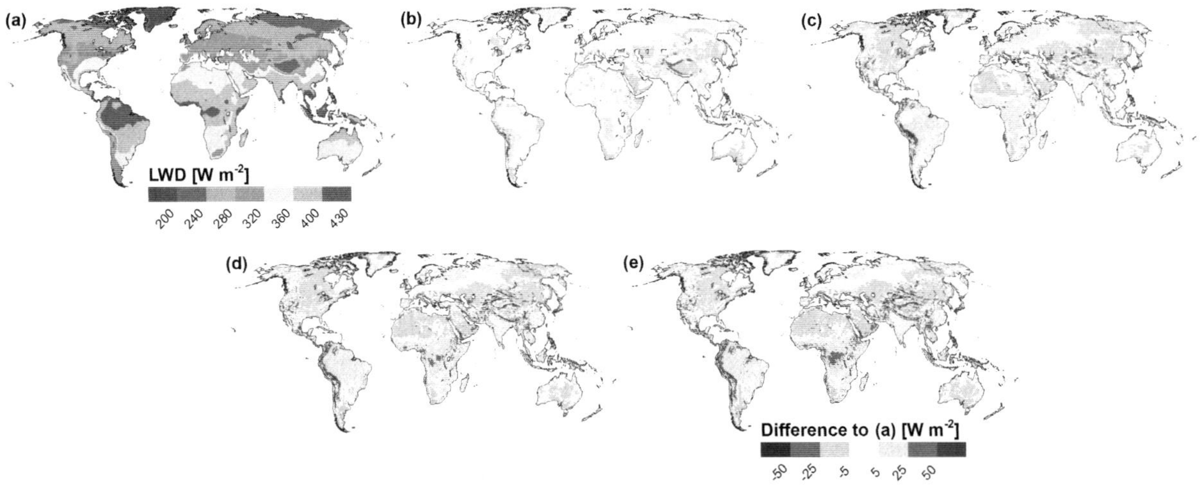


**Figure A4.** Long-term (1971–2000) average longwave downward radiation of the model variants, displayed as absolute number for GSWP3 (**a**) and differences to the other forcings, computed as PGFv2 minus GSWP3 (**b**), WFD minus GSWP3 (**c**), WFD_WFDEI minus GSWP3 (**d**), and WFDEI_hom minus GSWP3 (**e**). All units in W $m^{-2}$.

**Acknowledgements.** The authors thank the Global Runoff Data Centre (GRDC, http://grdc.bafg.de), 56068 Koblenz, Germany, for providing the discharge data used in this study for calibrating the model. We are also grateful to the ISI-MIP coordination team as well as the leaders of the water sector (Simon Gosling and Rutger Dankers) for providing the climate forcings and the support. Furthermore, we thank Wolfgang Grabs for organizing the international conference "Water Resources Assessment & Seasonal Prediction" (13–16 October in Koblenz, Germany), where some content of this paper was presented.

## References

Adam, J. C. and Lettenmaier, D. P.: Adjustment of global gridded precipitation for systematic bias, J. Geophys. Res., 108, 4257, doi:10.1029/2002JD002499, 2003.

Alcamo, J., Döll, P., Henrichs, T., Kaspar, F., Lehner, B., Rösch, T., and Siebert, S.: Development and testing of the WaterGAP 2 global model of water use and availability, Hydrol. Sci. J., 48, 317–337, doi:10.1623/hysj.48.3.317.45290, 2003.

Biemans, H., Hutjes, R. W. A., Kabat, P., Strengers, B. J., Gerten, D., and Rost, S.: Effects of precipitation uncertainty on discharge calculations for main river basins, J. Hydrometeorol., 10, 1011–1025, doi:10.1175/2008JHM1067.1, 2009.

Bierkens, M. F. P.: Global hydrology 2015: state, trends and directions, Water Resour. Res., 51, 4923–4947, doi:10.1002/2015WR017173, 2015.

Döll, P., Kaspar, F., and Lehner, B.: A global hydrological model for deriving water availability indicators: model tuning and validation, J. Hydrol., 270, 105–134, doi:10.1016/S0022-1694(02)00283-4, 2003.

Döll, P., Müller Schmied, H., Schuh, C., Portmann, F. T., and Eicker, A.: Global-scale assessment of groundwater depletion and related groundwater abstractions: Combining hydrological modeling with information from well observations and GRACE satellites, Water Resour. Res., 50, 5698–5720, 2014a.

Döll, P., Fritsche, M., Eicker, A., and Müller Schmied, H.: Seasonal water storage variations as impacted by water abstractions: Comparing the output of a global hydrological model with GRACE and GPS Observations, Surv. Geophys, 35, 1311–1331, doi:10.1007/s10712-014-9282-2, 2014b.

Döll, P., Douville, H., Güntner, A., Müller Schmied, H., and Wada, Y.: Modelling freshwater resources at the global scale: Challenges and prospects, Surv. Geophys., 37, 195–221, doi:10.1007/s10712-015-9347-x, 2016.

Eicker, A., Schumacher, M., Kusche, J., Döll, P., and Müller Schmied, H.: Calibration/data assimilation approach for integrating GRACE data into the WaterGAP Global Hydrology Model (WGHM) using an Ensemble Kalman Filter: First results, Surv. Geophys., 35, 1285–1309, doi:10.1007/s10712-014-9309-8, 2014.

Fekete, B. M., Robarts, R. D., Kumagai, M., Nachtnebel, H.-P., Odada, E., and Zhulidov, A. V.: Time for in situ renaissance, Science, 349, 685–686, doi:10.1126/science.aac7358, 2015.

Flörke, M., Kynast, E., Bärlund, I., Eisner, S., Wimmer, F., and Alcamo, J.: Domestic and industrial water uses of the past 60 years as a mirror of socio-economic development: A global simulation study, Global Environ. Chang., 23, 144–156, doi:10.1016/j.gloenvcha.2012.10.018, 2013.

Haddeland, I., Clark, D. B., Franssen, W., Ludwig, F., Voß, F., Arnell, N. W., Bertrand, N., Best, M., Folwell, S., Gerten, D., Gomes, S., Gosling, S. N., Hagemann, S., Hanasaki, N., Harding, R., Heinke, J., Kabat, P., Koirala, S., Oki, T., Polcher, J., Stacke, T., Viterbo, P., Weedon, G. P., and Yeh., P.: Multi-model estimate of the global terrestrial water balance: Setup and first results, J. Hydrometeorol., 12, 869–884, doi:10.1175/2011JHM1324.1, 2011.

Haddeland, I., Heinke, J., Voß, F., Eisner, S., Chen, C., Hagemann, S., and Ludwig, F.: Effects of climate model radiation, humidity and wind estimates on hydrological simulations, Hydrol. Earth Syst. Sci., 16, 305–318, doi:10.5194/hess-16-305-2012, 2012.

Hirabayashi, Y., Kanae, S., Motoya, K., Masuda, K., and Döll, P.: A 59-year (1948–2006) global meteorological forcing data set for land surface models. Part II: Global snowfall estimation, Hydrol. Res. Lett., 2, 65–69, doi:10.3178/hrl.2.65, 2008.

Müller Schmied, H., Eisner, S., Franz, D., Wattenbach, M., Portmann, F. T., Flörke, M., and Döll, P.: Sensitivity of simulated global-scale freshwater fluxes and storages to input data, hydrological model structure, human water use and calibration, Hydrol. Earth Syst. Sci., 18, 3511—3538, doi:10.5194/hess-18-3511-2014, 2014.

Müller Schmied, H., Adam, L., Eisner, S., Fink, G., Flörke, M., Kim, H., Oki, T., Portmann, F. T., Reinecke, R., Riedel, C., Song, Q., Zhang, J., and Döll, P.: Variations of global and continental water balance components as impacted by climate forcing uncertainty and human water use, Hydrol. Earth Syst. Sci., 20, 2877–2898, doi:10.5194/hess-20-2877-2016, 2016.

Oki, T. and Kanae, S.: Global hydrological cycles and world water resources, Science, 313, 1068–1072, doi:10.1126/science.1128845, 2006.

Priestley, C. H. B. and Taylor, R. J.: Assessment of surface heat flux and evaporation using large-scale parameters, Mon. Weather Rev., 100, 81–92, doi:10.1175/1520-0493(1972)100<0081:OTAOSH>2.3.CO;2, 1972.

Schewe, J., Heinke, J., Gerten, D., Haddeland, I., Arnell, N. W., Clark, D. B., Dankers, R., Eisner, S., Fekete, B. M., Colón-González, F. J., Gosling, S. N., Kim, H., Liu, X., Masaki, Y., Portmann, F. T., Satoh, Y., Stacke, T., Tang, Q., Wada, Y., Wisser, D., Albrecht, T., Frieler, K., Piontek, F., Warszawski, L., and Kabat, P.: Multimodel assessment of water scarcity under climate change, P. Natl. Acad. Sci. USA, 111, 3245–3250, doi:10.1073/pnas.1222460110, 2014.

Sheffield, J., Goteti, G., and Wood, E. F.: Development of a 50-year high-resolution global dataset of meteorological forcings for land surface modeling, J. Climate, 19, 3088–3111, doi:10.1175/JCLI3790.1, 2006.

Siebert, S., Kummu, M., Porkka, M., Döll, P., Ramankutty, N., and Scanlon, B. R.: A global data set of the extent of irrigated land from 1900 to 2005, Hydrol. Earth Syst. Sci., 19, 1521–1545, doi:10.5194/hess-19-1521-2015, 2015.

Wada, Y.: Modeling groundwater depletion at regional and global scales: Present state and future prospects, Surv. Geophys., 37, 419–451, doi:10.1007/s10712-015-9347-x, 2016.

Weedon, G. P., Gomes, S., Viterbo, P., Österle, H., Adam, J. C., Bellouin, N., Boucher, O., and Best, M.: The WATCH Forcing Data: a meteorological forcing dataset for land surface- and hydrological models, Watch Techn. Rep., 22, 41 pp., available at: http://www.eu-watch.org/media/default.aspx/emma/org/10376311/

WATCH+Technical+Report+Number+22+The+WATCH+forcing+data+1958-2001+A+meteorological+forcing+dataset+for+land+surface-+and+hydrological-models.pdf (last access 29 July 2016), 2010.

Weedon, G. P., Balsamo, G., Bellouin, N., Gomes, S., Best, M. J., and Viterbo, P.: The WFDEI meteorological forcing data set: WATCH Forcing Data methodology applied to ERA-Interim reanalysis data, Water Resour. Res., 50, 7505–7514, doi:10.1002/2014WR015638, 2014.

Proc. IAHS, 374, 63–67, 2016
proc-iahs.net/374/63/2016/
doi:10.5194/piahs-374-63-2016


Open Access
PIAHS

# Impact of climate change and anthropogenic pressure on the water resources of India: challenges in management

**K. Shadananan Nair**

Nansen Environmental Research Centre (India), Kochi-682016, Kerala, India

*Correspondence to:* K. Shadananan Nair (nair59@yahoo.com)

Published: 17 October 2016

**Abstract.** Freshwater resources of India are getting fast degraded and depleted from the changing climate and pressure of fast rising population. Changing intensity and seasonality of rainfall affect quantity and quality of water. Most of the rivers are polluted far above safety limits from the untreated domestic, industrial and agricultural effluents. Changes in the intensity, frequency and tracks of storms salinate coastal aquifers. Aquifers are also under the threat from rising sea level. Groundwater in urban limits and industrial zones are far beyond safety limits. Large-scale destruction of wetlands for industries and residential complexes has affected the quality of surface and groundwater resources in most parts of India. Measures to maintain food security and the new developments schemes such as river linking will further deteriorate the water resources. Falling water availability leads to serious health issues and various socio-economic issues. India needs urgent and appropriate adaptation strategies in the water sector.

## 1 Introduction

Demand in water has been escalating in India with fast rising population. Population has already crossed 1.2 Billion and it is expected to stabilize only by the year 2050 at 1.5 Billion. Increasing domestic, agricultural, and industrial water requirements are serious challenges for India as the economy is still not very strong to implement large development projects and to find adaptive mechanisms for changing climate conditions. Indian economy and life of more than half of the population have been traditionally linked to agriculture or agriculture-related industries. Monsoons, the source of water for India undergoes wide inert-annual variation associated with global anomalies. Wide disparity in rainfall is reflected in local climate and water resources. As a result, vast areas of interior India are arid and semi-arid. Most of these regions are rich in agriculture where another agricultural revolution is possible, if some more water is available for irrigation. Any extremes in climate leading to further water shortage in such regions will disrupt all measures to attain food security. Available resources are getting fast deteriorated and depleted. Groundwater level in the country is receding fast, as a result of changing rainfall seasonality, land use changes and overexploitation. Proper assessment and careful management of water resources are utmost important. India is going to face a very serious water scarcity in two or three decades, unless new resources are identified and effective conservation and management measures are not resorted to in its major river basins (Fig. 1). Two-third of India's available freshwater is lost due to evaporation and runoff into the Sea, while there exists a freshwater crisis in many parts at different times of a year (UN Newsletter, 1999). Study of the World Watch Institute predicts that the Ganges and other rivers vital for farming in heavily populated India will run dry for a part or all of the dry season, 30 years from now (Science update, 1997).

Renewable water resources of India are about 4 % of the global availability (IWRS, 1998). Today, around 83 % of the population comes under the drinking water supply schemes. Only 35 % of the cultivable land is irrigated, while 70 % of the runoff wastefully joins the Sea. Water deliveries in India rarely correspond in quantity and timing of the true requirements (Postel, 1989). The annual precipitation of nearly 4000 $Bm^3$ contributes to the formation of twelve major river basins and eight other basins formed by the combinations of medium and minor basins. The major and medium river basins contribute over 90 % of the total runoff in the country.

The Indo-Gangetic planes have enormous amount of water, but the current method of utilization is not appropriate (Serageldin, 1998). Of the total water potential of 1869 $Bm^3$, only 1122 $Bm^3$ can be put to beneficial use. Out of this, 690 $Bm^3$ is surface water and 432 $Bm^3$ is replenishable groundwater (Ministry of Water Resources, 1998). There will be a considerable gap between the water need and availability in various basins in two or three decades from now. An in-depth study based on water balance and of the expected changes in water resources become necessary to formulate measures to meet the challenges in future.

## 2 Analysis and results

Characteristics of rainfall over India have been analysed using the data provided by the India Meteorological Department. Changes in the seasonality of rainfall have been studied using the method suggested by Walsh and Lawler (1981). Changes in per capita water availability in the major river basins during the period 2010–2030 have been estimated, considering the decadal population growth and changes in rainfall and temperature patterns adopted by Government of India (MoEF, 2010), For this, changes in hydrological parameters have been computed using the water balance method developed by Thornthwaite (1948) and modified by Thornthwaite and Mather (1955) that has been modified to suit the local conditions.

### Rainfall and water availability

In general, monsoons and the local weather systems produce a mean annual rainfall of around 108 cm over India. Effect of orography results in up to 400 cm rainfall in the extreme northeast region and in the southwest coast. Isolated regions in the northeast get more than 1000 cm and in the west coast, up to 700 cm. Rainfall is less than 20 cm in the north-western drylands. The southwest monsoon (June–September) is the principal rainy season for the country. During the north-east monsoon period (October–December), the chief areas of rainfall is confined to the southern tip of the peninsula, the east coast and isolated areas in the north-east and the extreme north. Major part of interior India is semi arid, but rich agricultural lands and if some more water becomes available here by efficient water management, it may help a lot in solving future food crisis.

It is the temporal distribution of the rainfall, more than the total amount that affects the water availability of a region. A less seasonal (more distribution through months) rainfall permits more soil moisture retention and increases water availability in dry months. Though the rainfall is heavy in the west coast hill regions, most of it is highly seasonal and confined to two or three months. Water from the heavy rainfall flow fast wastefully into the ocean through the steep hill slopes, before it could be harnessed. In general, rainfall is less seasonal in the southern tip of Peninsular India, as this region gets rainfall from both monsoons and local systems. Similarly, it is less in the northern parts also, due to the winter rainfall associated with the passage of Western Disturbances. Effect of the northeast monsoon and depressions and cyclones in winter produce a less seasonal pattern in the east coast. Rainfall is becoming more and more seasonal in almost all parts of India. The seasonality of rainfall is always reflected in water resources.

The per capita water availability (from surface water surplus, after meeting requirements for evapotranspiration and soil moisture recharge) in the basins ranges from 17 730 $m^3$ in the Brahmaputra to 269 $m^3$ in Sabarmati (Table 1). Even today, availability on annual basis is less than requirement in many of the basins, especially in the interior peninsula. When the entire basin is considered, it is satisfactory only for the Brahmaputra. The study reveals that by the year 2030, at the current rate of growth of population and with the predicted increase in global temperature, the availability will be drastically reduced in all the cases. However, situation will be hopeful in the northeast Brahmaputra region. In almost all parts of India, seasonal water deficiencies show an increasing trend and the surpluses show a decreasing trend. According to WMO, rainfall increase may be considerable in the north-west India, but, it may be very low in the south and north, where the increase in precipitation may not be able to compensate for the reduction in soil moisture due to rise in temperature (Bhalme, 1997). Northeast India has high values of per capita availability. But, the very high growth rate of population in these regions will reduce the availability up to one-fifth of the present level by 2030. Seasonality of rainfall is reflected in the water surpluses. If the rainfall is highly seasonal, there can be a seasonal surplus though the amount is small (needs for evapotranspiration and soil moisture recharge are temporarily met here). Therefore, change in seasonality may lead to more seasonal water crisis. In India, population increase is a more serious issue compared to the effect of global warming on water resources.

## 3 Management issues

The study points out that there will be a considerable shortage of water in all river basins of India by the year 2030. Government is planning to meet the estimated freshwater requirements from exploitable surface and ground water resources. In fact, any further need will have to be met with new resources to be explored or better management. Study of the World Watch Institute predicts that the Ganges and other rivers vital for farming in heavily populated India will run dry for a part or all of the dry season, 30 years from now (Science update, 1997). Increased urbanization in India has stressed water use and water supply infrastructure beyond capacity. One-third of the urban population and one-twelfth of the total population live in 23 metropolitan cities and most of them are near major rivers. Growth of mega cities adds

**Table 1.** Water availability in the major river basins.

| River basins | Per capita water availability ($m^3$) (Year 2010) | Per capita by year 2030 ($m^3$) |
|---|---|---|
| Indus | 1250 | 534 |
| Ganaga | 1150 | 490 |
| Brahmaputra | 17 730 | 6252 |
| Narmada | 1148 | 434 |
| Tapi | 721 | 504 |
| Brahmani & Baitarni | 2112 | 1147 |
| Mahanadi | 2107 | 1243 |
| Godavari | 1695 | 670 |
| Krishna | 1014 | 393 |
| Pennar | 514 | 255 |
| Cauvery | 605 | 320 |
| Sabarmati | 269 | 107 |

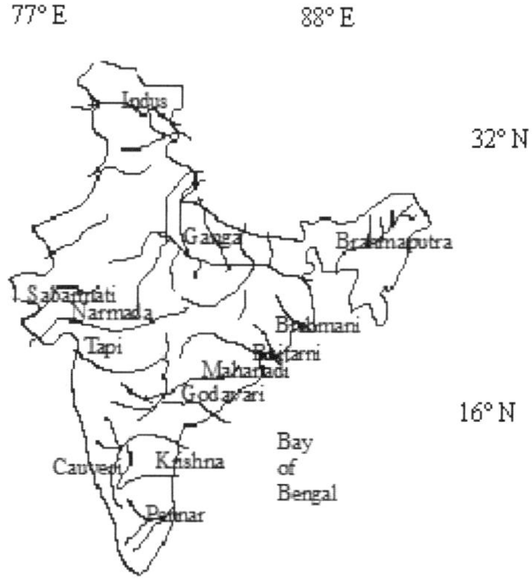


**Figure 1.** India – Major river systems.

more stress on water. Mega-cities have to cope with intense competition from agriculture and industry to provide their expanding populations, especially the urban poor, with adequate water supply. Mega-cities also have to address the increasing pollution of their freshwater sources from growing volumes of urban waste and the increasing environmental risks from over-abstraction of groundwater resources, inadequate drainage and floods. If the present rate of migration to cities continues, not in far future, water planning in cities will become complicated with no easy solutions. Water consumption by the thousands travelling daily to the cities for job is not considered in the urban water planning.

Only a portion of the water surplus in the basins is available for human use due to deterioration of water resources, especially during dry months when need is more. Increased demands in agricultural, industrial and domestic sectors lead to considerable imbalances in the quantity and quality of water resources. Extremes and variability in climate are likely to worsen the existing situation by further limiting the water availability. Potential changes in temperature and precipitation pattern may adversely affect soil moisture condition, annual runoff, and ground water recharge. Despite extensive efforts to improve the use and management of water resources, water demands continue to rise, contamination degrades water quality, and natural hazards, such as floods and droughts, disrupt human activities and cause extensive human suffering and economic losses. Population growth and urbanization are among the key factors underlying the enormous growth in the demand of water and the increase of environmental degradation.

Changing climate pattern is of serious concern for India. IPCC (1998) predicts an increased recession of Himalayan glaciers and increasing danger from glacial lake outburst floods in an increased temperatures and increased seasonal variability in precipitation. Reduction in average flow of snow-fed rivers and an increase in peak flows and sediment yield would have major impacts on hydropower generation, urban water supply, and agriculture. Almost all parts of India are vulnerable to floods or droughts. In an altered global climate, change in such extremes can be catastrophic. Any change in the strength or seasonality of monsoons would lead to soil moisture deficiency or floods in the basins and affect water availability and runoff rates. There can be tremendous increase in the transport of sediment load of the Ganges and Brahamaputra, which already carry an extraordinarily heavy sediment load and have a high, though irregular, rate of downstream deposition. The Ganges-Brahmaputra delta is one of the world's most densely populated areas, and the effects of climate change could cause serious drainage and sedimentation problems, in addition to bank erosion and land loss. These impacts clearly would have immense socio-economic costs.

Extreme water conditions and quality issues are major problems in water resources management. The magnitude and severity of these problems varies from one basin to another. Increasing demand is characterized by increasing competition for available water. The social, environmental and political issues related to the rivers are worsening. There exists dispute over water sharing, either national or international in almost all the basins. In India, domestic water disputes are more severe and difficult to resolve compared to international, because of various political and social reasons. Water-related environmental problems are becoming transboundary in scope as local pollution problems spread across borders due to the pressure of population growth, increased fertilizer and pesticide use, more industries and inadequate pollution controls. When the availability decreases, existing disputes may worsen or new one may start in any basins. Even the updated national water policy (MWR, 2012) of India doesn't provide any clear suggestion for conflict resolution.

Water related health issues and economic issues are also worsening. Water-borne and vector-borne diseases spread into new areas. In two or three decades, one-third of the country will be prone to malaria. Large investments will be required to provide safe water to the growing population and for the measures for adaptation and impact mitigation. Price of reliable water is escalating. It becomes unaffordable to millions of poor people, especially in urban limits.

Shortage of surface water will increase dependency on groundwater. More water withdrawal for agriculture may lead to increased desertification, land degradation and loss of soil fertility. Substantial reduction in the area under agriculture and extensive damage to structures will not only affect the economy but also will create problems associated with shift in population to reliable basin areas. Even now, overextraction of groundwater in some basin areas has resulted in reduction of summer runoff in rivers and lowering of water level in nearby open water bodies.

Increased population and increasing demand in the agricultural, industrial, and hydropower sectors will put additional stress on water resources. Agriculture is the predominant water user in India and the gap between the need in the sector and availability will be widening, unless viable alternatives are not found. In addition to these, limited water management capacity, fragmented organizational structures, and inadequate water planning, management, and conservation are among the contributing factors to water crisis. Situation seriously underlines the need for taking up integrated plans for water conservation and utilization at basin level to meet the increasing demands of irrigation, water harvesting, human and livestock consumption, expanding industry, hydro-electric power generation, recreation, navigation and other uses.

What India needs is an appropriate and frequently updated water policy and a strong political will to implement it. Policy development should involve climate change impact on socio-economic conditions in different zones. Water management should be brought under a central control instead of the various departments making delays in decisions. Legal and institutional mechanisms are to be improved to cope with the need of the time. Even constitutional amendments become necessary to solve water problems in large basins. At present, water is a state matter and the central government has only limited role in its control. Therefore, disputes over water among states continue to be a major hazard in water related development activities.

It is true that lack of government funds have haltered progress of many projects in India. An option is private sector participation. But, private sector financing should be seen to be complementary and in no way a substitute for the state's responsibility in providing basic necessities for all. Private sector involvement is a failure in India's basic requirements sector. Any development scheme cannot succeed without giving due care to the millions living well below poverty line. Water should continue to be provided free to the extreme poor and the marginal farmers. Also, social situation in India may not permit pricing easily. Awareness in conservation and management is more important.

The fundamental right of access to safe water and sanitation with environmental protection is to be ensured. But, strict control is to be made to avoid overuse and wasteful use. Satisfactory water allocation with consensus among different users is a key factor in the development of society and in maintaining harmony. Involvement of non-governmental organizations and charitable organizations are better performed in India than govt in providing basic necessities like water. People's participation should be an integral part of all aspects of water resources management.

Water has to be used in an environmentally sustainable manner in order to maximize its economic and social benefits. It should not be used faster than it is replenished, nor should it be polluted. Addressing water problems requires an inter-sectoral approach that recognizes the interlinkages – for example between land and water, agriculture and water, technology and water, health and water – that affect water management. No single mechanism or approach will be enough. However, reformation in agricultural sector is urgently needed in maximum and efficient use of water and in minimising input of pollutants. Policy packages using a mutually reinforcing mix of institutional and policy reform, and legal, economic and management instruments will be needed. Strong political decision by the state to implement the rules and regulations is most important. Traditional methods of water collection and conservation, which are environment-friendly and less expensive may be promoted. Modernisation of hydroelectric systems is necessary to get more power and to reserve more water. Water conservation should be started at the domestic level. In heavy rainfall regions, roof top water collection alone can meet the water requirements for cooking and drinking. Proper enforcement of the environmental protection laws can effectively control the degradation of water resources.

## 4 Conclusions

Rainfall over India is becoming more seasonal and this is reflected in groundwater recharge and summer water availability. Per capita availability of freshwater in all major river basins is fast decreasing, as a result of changes in climate and rise in population. Most of the basins will experience water stress by the year 2030. Available resources are getting fast depleted and degraded. Falling water availability leads to several socio-economic issues such as disputes over sharing, migration, pricing of water. National policies do not address the issues seriously. Measures for adaptation often fail because of various social, economic and political reasons. There are possibilities to overcome the growing water crisis through better conservation and management. For this India need a better policies and better institutional mechanism to implement the policy guidelines.

## 5 Data availability

Data of rainfall and temperature available from the India Meteorological Department (IMD) on payment, Website of IMD: http://www.imd.gov.in/pages/services_climate.php Data of population available from the Census Department, Government of India: http://www.censusindia.gov.in/2011census/population_enumeration.html Information on climate projections available from the Ministry of Environment, Forest and Climate Change, Government of India, Website: http://www.moef.nic.in/downloads/public-information/fin-rpt-incca.pdf.

## References

Bhalme, H.: Agriculture, food and climate, WCASP, 42, 5–12, 1997.

Government of India: Population Enumeration Data (Final Population), available at: http://www.censusindia.gov.in/2011census/population_enumeration.html, last access: May 2016.

India Meteorological Department (IMD): Climatological Services, available at: http://www.imd.gov.in/pages/services_climate.php, last access: September 2015.

IPCC: The regional impacts of climate change: an assessment of vulnerability, Cambridge University Press, 1998.

IWRS: Five Decades of Water resources Development in India, IWRA Report, New Delhi, 1998.

Lonergan, S.: Climate warming and India, in: Measuring the Impact of Climate Change on Indian Agriculture, edited by: Dinar, A., World Bank Technical Paper No. 402, 1998.

Ministry of Water Resources: Water Resources Development in India 1947–1997, Govt. of India Publication, New Delhi, 1998.

Ministry of Environment & Forests: Climate Change and India: A $4 \times 4$ Assessment, available at: http://www.moef.nic.in/downloads/public-information/fin-rpt-incca.pdf, last access: September 2015.

MoEF (Ministry of Environment and Forests): Climate Change and India: A 4X4 Assessment – A sectoral and regional analysis for 2030s, Indian Network for Climate Change Assessment, Government of India, 2010.

MWR (Ministry of Water Resources): Revised Draft National Water Policy, p. 11, Govt. of India, 2012.

Postel, S.: Water for Agriculture Facing the Limits, World Watch Paper 93, The World Resources Institute, USA, 1989.

Science Update: Water Scarcity seen in future, Publication of the US Consulate, 3–4, 1997.

Serageldin, I.: Water in the 21st Century: A Dialogue, Water Policy, 1, 29–137, 1998.

Thornthwaite, C. J.: An Approach Towards the Rational Classification of Climate, Geogr. Rev., 38, 1948.

Thornthwaite, C. J. and Mather, J. R.: The Water Balance, Publ.inClimat. Lab. Of Climat., 8, 1955.

UN Newsletter: UN and Water, 54, 3–8, 1999.

Walsh, R. P. D. and Lawler, D. M.: Rainfall Seasonality: Description, Spatial Patterns and Change Through Time, Weather, 36, 201–208, 1981.

Proc. IAHS, 374, 69–73, 2016
proc-iahs.net/374/69/2016/
doi:10.5194/piahs-374-69-2016


# Integrated water resource assessment for the Adelaide region, South Australia

**James W. Cox**[1,3], **Michele Akeroyd**[2], **and Danielle P. Oliver**[2,4]

[1]South Australian Research and Development Institute, Adelaide, 5064, Australia
[2]Goyder Institute for Water Research, Adelaide, 5001, Australia
[3]The University of Adelaide, Adelaide, 5001, Australia
[4]CSIRO Land and Water, Adelaide, 5064, Australia

*Correspondence to:* James W. Cox (jim.cox@sa.gov.au)

Published: 17 October 2016

**Abstract.** South Australia is the driest state in the driest inhabited country in the world, Australia. Consequently, water is one of South Australia's highest priorities. Focus on water research and sources of water in the state became more critical during the Millenium drought that occurred between 1997 and 2011. In response to increased concern about water sources the South Australian government established The Goyder Institute for Water Research – a partnership between the South Australian State Government, the Commonwealth Scientific and Industrial Research Organisation (CSIRO), Flinders University, University of Adelaide and University of South Australia. The Goyder Institute undertakes cutting-edge science to inform the development of innovative integrated water management strategies to ensure South Australia's ongoing water security and enhance the South Australian Government's capacity to develop and deliver science-based policy solutions in water management. This paper focuses on the integrated water resource assessment of the northern Adelaide region, including the key research investments in water and climate, and how this information is being utilised by decision makers in the region.

## 1 Introduction

Historically, South Australia has relied on three rain-dependent sources of water – the River Murray, Mt. Lofty Ranges catchment and groundwater. However, like much of the southern regions of the continent, many areas of South Australia have experienced a decline in surface water flows and groundwater levels over the past decade compared to long term averages (CSIRO, 2007). This has resulted in an increased threat to the security of water supplies for regional communities, industry and the environment. In the future the pressure on these sources of water is expected to increase with a decrease in average annual rainfall in the northern Adelaide region of approximately 8–14 %. At the same time, South Australia's population is expected to grow from the current 1.67 million people to reach 1.79 million by 2016 and 2.28 million by 2036 (Government of South Australia, 2015), increasing the demand on water supplies.

To combat this threat, the State established The Goyder Institute for Water Research – a partnership between the South Australian State Government, the Commonwealth Scientific and Industrial Research Organisation (CSIRO), Flinders University, University of Adelaide and University of South Australia. The Goyder Institute undertakes cutting-edge science to inform the development of innovative integrated water management strategies to ensure South Australia's ongoing water security and enhance the South Australian Government's capacity to develop and deliver science-based policy solutions in water management.

There are a number of key research investments that have been targeted to inform integrated water management in State Government water planning and policy development. This new science will be presented in the context of the government policy and development drivers, with a focus on several key research issues.

Published by Copernicus Publications on behalf of the International Association of Hydrological Sciences.

## 2 Background

### 2.1 Northern Adelaide Plains region

Horticultural production in the Virginia area of the Northern Adelaide Plains (NAP) (Fig. 1) has been developed utilising a good supply of groundwater for irrigation (Stevens, 2004). However, the long-term supply of this water has been threatened due to total pumping volumes exceeding the total recharge (Anon, 1995). This has led to the utilisation of reclaimed water (reclaimed sewage effluent) as a water source for agriculture in the region. The Virginia Triangle is now one of the largest Class A water reclamation and reuse schemes in the world (Stevens, 2004). The South Australian Government would like to see the agricultural production area extended as shown in Fig. 2. The extension of agricultural production in this area would require identification of future water sources and an understanding of the quality and quantity of these sources.

### 2.2 Water resources research

There are a number of key research investments that have been targeted to inform integrated water management in State Government water planning and policy development. This new science will be presented in the context of the government policy and development drivers, with a focus on the following key research areas.

#### 2.2.1 Adelaide Plains groundwater

This project has built upon existing knowledge to provide a thorough assessment of the groundwater resources beneath Adelaide, the capital city of South Australia, and the impacts of current and future extraction and climate change. It is the first study to provide an integrated assessment of the groundwater resources of the entire Adelaide metropolitan region.

#### 2.2.2 Optimal water resources mix

A key product of this project is the development of methods for determining trade-offs between the multiple objectives of water security, economic efficiency and environmental benefits of water supply options that are consistent with the city's social values. The trade-offs analysis methodology developed also provided a framework that could be applied to other cities/regions to inform the development of total water cycle management plans.

#### 2.2.3 Managed aquifer recharge and stormwater use options

This project assessed a range of potential uses of stormwater including via managed aquifer recharge at various locations around Australia and overseas. The project addressed

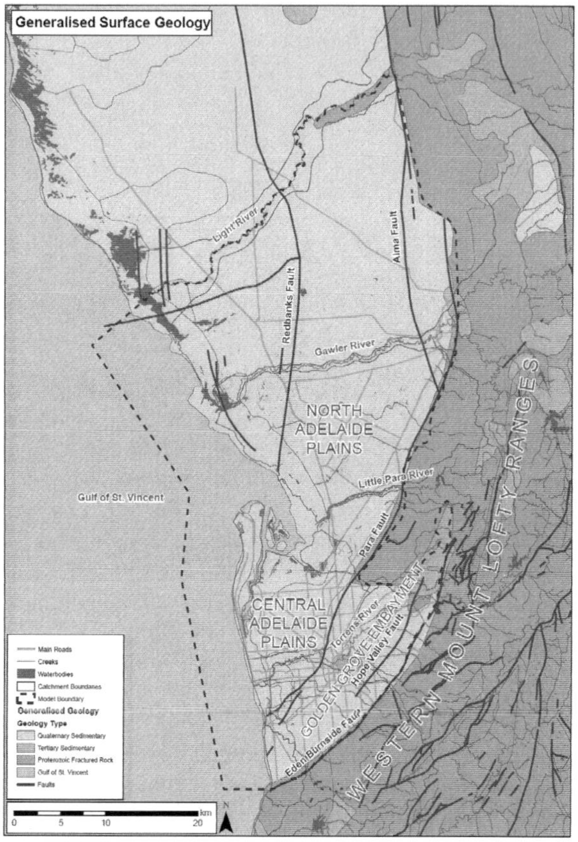


Figure 1. Map indicating the Northern Adelaide Plains region (from Bresciani et al., 2015a).

the health risk assessment of uses for public open space irrigation, third pipe non-potable supplies to households and industry and for drinking water supplies. For several of those uses prototype risk management plans were developed. The economics of some specific options were also considered taking account of environmental impacts, and surveys of public acceptance of the more novel options that were considered.

#### 2.2.4 Mt. Lofty Ranges catchment

The Mt. Lofty Ranges catchments are the source of surface water for Adelaide reservoirs. This project built on existing knowledge, management and monitoring systems in the Mt. Lofty Ranges to understand environmental water needs, patterns of flow and water quality within the catchment. To support decision making, a tool has been developed that brings together all the information needed to make robust decisions on water allocations and assess the potential outcomes of alternative management options. The research has also involved collecting important hydro-ecological data and water quality assessments that support development of water allocation plans for the region, including installing monitoring equipment in high priority regions to measure water flow and quality and the subsequent ecological response from fish, vegetation and macro-invertebrates.

### 2.2.5 SA Climate Ready

SA Climate Ready is the most comprehensive set of down-scaled climate projections data ever available in South Australia. The data can be used in future planning decisions. For example, to ensure that planned future infrastructure such as roads, bridges, farm dams and mine tailings dams, are designed to take into account; the impacts of climate change enable water resource and catchment managers to assess the security of future water supplies and protect water supplies for all water users; and anticipate changes in extreme heat and fire risk to inform planning for South Australia's emergency, health and social services sectors.

### 2.2.6 Stormwater Interventions – linking catchment actions to coastal water quality

The government of South Australia is developing an Integrated Urban Water Management Plan (IUWMP) for Greater Adelaide that will embrace all available water sources including stormwater and wastewater. Currently, investment in stormwater and wastewater infrastructure is made or guided by state and local government agencies of South Australia. This does not currently include an evaluation of the relative effectiveness and efficiency of the investment (separately or in combination) in achieving goals such as water supply security, flood management and improved receiving water quality. In addition, there is a lack of understanding on catchment-scale implications of water sensitive urban design (WSUD) interim targets. While previous work has partially addressed some of the knowledge gaps more information is required to address the most efficient and cost effective way to manage stormwater and wastewater discharges to improve coastal water quality. This research has brought together the available information and tools to enable the assessment of stormwater interventions in the catchment to coastal water quality and other benefits.

These research investments are being drawn together in the peri-urban area north of Adelaide. This is a growing region for urban development, industry and agriculture. All require access to reliable, fit-for-purpose water supplies. The Goyder Institute is leading a program to coordinate the existing knowledge, including the latest science from the Goyder Institute's research portfolio and the specific water infrastructure projects undertaken by local authorities and state government. This coordination will bring together the best available assessment on the available quantity and quality of the northern Adelaide water resources including groundwater, surface water, stormwater and wastewater recycling, desalination, inter-basin transfers and projected demand to deliver an integrated water resource assessment now and into the future to support sustainable development. Outcomes from two of the Goyder Institute projects are presented in relation to the development of the NAP region.

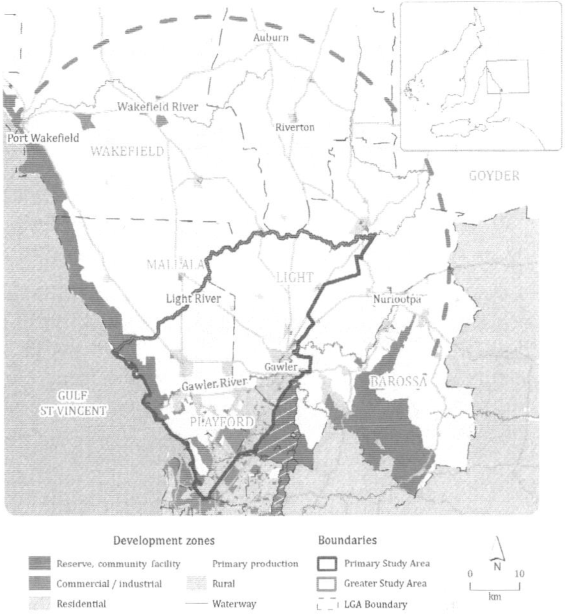


**Figure 2.** Proposed development region in Northern Adelaide Plains, South Australia.

## 3 Application of findings to Northern Adelaide Plains

Outcomes from two projects as applied to the NAP region are presented below.

### 3.1 Adelaide Plains groundwater

The largest reserves of fresh groundwater in the Adelaide Plains are contained in the Tertiary T1 and T2 aquifers which extend from the Eden-Burnside Fault and Para Fault in the east up to an unknown distance offshore in the west (Gerges, 1999). These groundwater resources have been utilised since the start of the 20th century and continue to act as a vital water resource for Adelaide's industry, agriculture and horticulture, and recreational facilities. With growing pressures such as increasing groundwater demand and a changing climate, the longevity of these resources is a critical concern, which calls for effective groundwater management. In this context, a robust understanding of groundwater flow mechanisms is required.

A modelling platform was developed by RPS Aquaterra (Georgiou et al., 2011) for the South Australian Government Department of Environment Water and Natural Resources (DEWNR) and has recently been used to help manage the

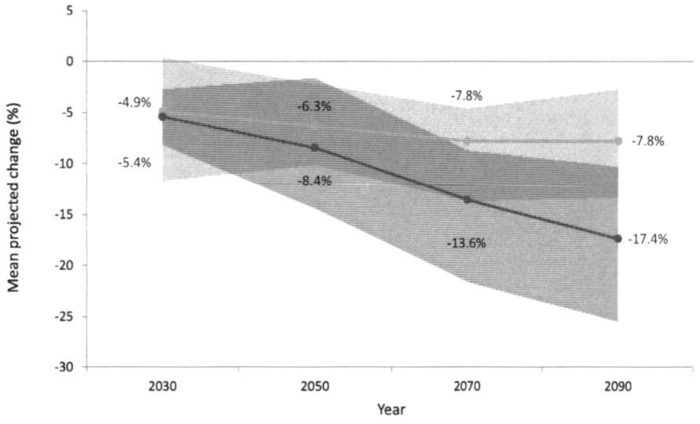


**Figure 3.** Predicted changes in average annual rainfall in the Adelaide Mt. Lofty Ranges region (from Goyder Institute for Water Research, 2015).

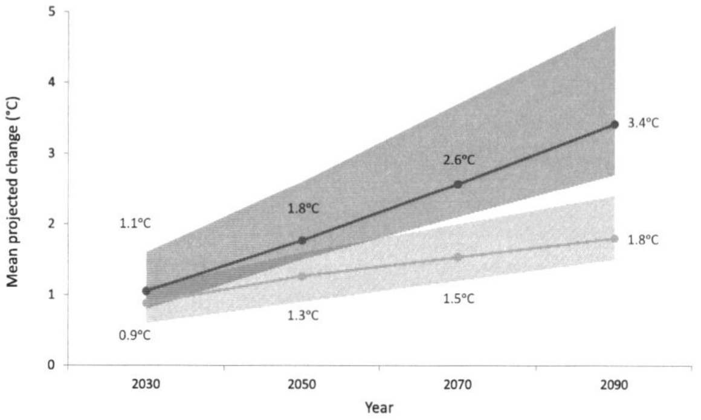


**Figure 4.** Predicted changes in average annual maximum temperature in the Adelaide Mt. Lofty Ranges region (from Goyder Institute for Water Research, 2015).

groundwater resources of the Adelaide Plains. However, the model showed some significant contradictions with the accepted conceptualization of the groundwater flow mechanisms in the region. Namely, diffuse recharge largely dominated the water budget (Georgiou et al., 2011). In contrast, diffuse recharge across the Adelaide Plains is believed to be insignificant, with recharge being mainly the result of infiltration from surface water features (Miles, 1952; Shepherd, 1975; Gerges, 1999). The model also showed groundwater flowing from the plains towards the hills along a significant portion of the eastern model boundary, especially in the NAP when groundwater should globally flow from the hills towards the plains. Furthermore, the model did not benefit from a number of modern techniques available for calibration.

This research showed that both river leakage flow and lateral flow contribute to groundwater flow mechanisms. Simulations also showed the observed widespread decline of potentiometric levels indicates that total pumping of groundwater is presently occurring at a rate that largely exceeds natural inflows, resulting in a strong decline in storage in the groundwater system and increased cross-formational flow. Furthermore, despite the fact that abstraction is largest from the T1 and T2 aquifers, most of the storage depletion occurs the Quaternary units which have relatively high groundwater salinity. Consequently, there is an increased risk in salinity in the Tertiary aquifers (Bresciani et al., 2015b), which could severely limit the crops able to be grown in the area.

The development of the groundwater flow and transport modelling platform will allow the groundwater supplies to be optimally managed to ensure there is not over-extraction and to minimise the salinisation of the water source.

### 3.2 SA Climate Ready

One of the key drivers affecting future water availability and hence development in South Australia will be climate change, which will place new pressures on water use and threaten supply. In 2011, the Goyder Institute commenced a project to develop downscaled climate change projections for South Australia, that is, possible future climates generated at a local scale. There are over 40 global climate models (GCMs) used by the Intergovernmental Panel on Climate Change (IPCC) for generating climate projections. The Goyder Institute identified a subset of these GCMs based on their ability to reproduce the State's climate drivers such as the Indian Ocean Dipole and the El Niño-Southern Oscillation (ENSO). This sub-set of GCMs were then used to generate climate projections at a local scale – providing the best available science to ensure that South Australia is Climate Ready. SA Climate Ready is the most comprehensive set of downscaled climate projections data ever available in South Australia. Data is available for six climate variables (rainfall, temperature maximum, temperature minimum, areal potential evapotranspiration, solar radiation, vapour pressure deficit), using two emission scenarios (intermediate and high "representative concentration pathways") through to 2100.

Regional downscaling for the Adelaide and Mt. Lofty Ranges region includes the NAP area that is planned for future expansion. Climate modelling suggests that average annual rainfall in the region could decline by up to 7.8 to 17.4 % by the end of the 21st century (Fig. 3). In addition, average annual maximum temperatures have been modelled to increase by 1.8 to 3.4 °C (Fig. 4). The greatest decline in average annual future rainfall is projected in spring which coincides with warming in spring (Goyder Institute for Water Research, 2015). This projected drier and warmer spring potentially may have implications with regard current agricultural practices in the NAP where horticultural and broadacre crops rely on spring rains to maximise yields. The type of crops grown and the growing season may have to be reassessed based on projected climate changes in the region. In addition alternative sources of water for crop production may need to be identified in the event of reduction in surface water.

## 4 Conclusions

The findings from projects from the Goyder Institute for Water Research are providing valuable data and models for planning future development in the NAP region. In addition the research will allow better integration of current available water sources and projections of the effect of increasing demands on these sources.

## 5 Data availability

The Goyder Institute projects have generated substantial data sets. Generally the datasets have been stored with the research partner that was responsible for the project work. However, these data sets are described on the Goyder Institute for Water Research page at the Australian National Data Service (ANDS) http://researchdata.ands.org.au/goyder-institute-water-research/267198/ (Piantadosi et al., 2015; Banks et al., 2015). Further contact details for accessing the data or contacting the key researchers involved are provided at this site. Model Metadata Templates have also been developed for a number of models modified or developed within Goyder Institute projects. These Templates are available at http://goyderinstitute.org/index.php?id=67 (Charles, 2014a, b; Bresciani, 2015).

**Acknowledgements.** The authors thank their colleagues from the Goyder Institute for Water Research partners, namely CSIRO, University of Adelaide, University of South Australia, Flinders University and the South Australian Government for the research outcomes that have contributed to this paper.

## References

Anon: Virginia Pipeline Committee, Bolivar-Virginia Reuse Project, Roger Stokes and Associates, 1995.

Banks, E., Batelaan, O., Batlle Aguilar, J., Kozuskanich, J. C., and Post, V. E.: Adelaide Plains Groundwater Model 2015: Assessment of Adelaide Plains Groundwater Resources, available at: https://researchdata.ands.org.au/adelaide-plains-groundwater-groundwater-resources/638490/?refer_q=, 2015.

Bresciani, E.: Adelaide Plains Groundwater Model 2015, available at: http://www.goyderinstitute.org/uploads/Adelaide_Plains_Groundwater_Model_Template_v5.pdf, 2015.

Bresciani, E., Batelaan, O., Banks, E. W., Barnett, S. R., Batlle-Aguilar, J., Cook, P. G., Costar, A., Cranswick, R. H., Doherty, J., Green, G., Kozuskanich, J., Partington, D., Pool, M., Post, V. E. A., Simmons, C. T., Smerdon, B. D., Smith, S. D., Turnadge, C., Villeneuve, S., Werner, A. D., White, N. and Xie, Y.: Assessment of Adelaide Plains Groundwater Resources: Summary Report, Goyder Institute for Water Research Technical Report Series No. 15/31, Adelaide, South Australia, available at: http://www.goyderinstitute.org/index.php?id=20 (last access: August 2016), 2015a.

Bresciani, E., Partington, D., and Xie, Y.: Adelaide Plains groundwater model, available at: http://www.goyderinstitute.org/uploads/Adelaide_Plains_Groundwater_Model_Template_v5.pdf (last access: August 2016), 2015b.

Charles, S.: An implementation of a multivariate daily weather generator, available at: http://www.goyderinstitute.org/uploads/ClimateChange_Multivariate_Daily_Weather_Generator.pdf, 2014a.

Charles, S.: MVNHMM Multi-Variate Non-homogeneous Hidden Markov Model, available at: http://www.goyderinstitute.org/uploads/ClimateChange_MVNHMM.pdf, 2014b.

CSIRO: Climate change in Australia, Technical Report, available at: http://www.climatechangeinaustralia.gov.au (last access: August 2016), 2007.

Georgiou, J., Stadter, M., and Purczel, C.: Adelaide Plains groundwater flow and solute transport model, available at: http://www.goyderinstitute.org/uploads/Adelaide_Plains_Groundwater_Model_Template_v5.pdf (last access: August 2016), 2011.

Gerges, N. Z.: The Geology and Hydrogeology of the Adelaide Metropolitan Area Volume 1 and Volume 2, PhD thesis, Flinders University, South Australia, 1999.

Government of South Australia, Department of Planning and Local Government: Population Projections for South Australia and Statistical Divisions, 2011-41, September, available at: http://www.dpti.sa.gov.au/__data/assets/pdf_file/0007/177937/Population_Projections_SA_SD_2011_41_Report.pdf (last access: August 2016), 2015.

Goyder Institute for Water Research: SA Climate Ready Regional Summaries, Adelaide Mount Lofty Ranges, available at: http://www.goyderinstitute.org/uploads/Fact%20Sheets/SA%20Climate%20Ready%20Regional%20Summary%20-%20AMLR.pdf (last access: August 2016), 2015.

Piantadosi, J., Beecham, S., Kamruzzaman, M., Boland, J., and Fleming, N.: Development of an agreed set of climate projections for South Australia, available at: https://researchdata.ands.org.au/development-an-agreed-south-australia/625589/?refer_q=, 2015.

Miles, K. R.: Geology and underground water resources of the Adelaide Plains area, Department of Mines: Adelaide, 1952.

Shepherd, R. G.: Northern Adelaide Plains groundwater study, Stage II, 1968–74, Department of Mines South Australia, Report Book 75/38, 1975.

Stevens, D.: Sustainable use of recycled water for horticultural irrigation on the Northern Adelaide Plains, Horticulture Australia Ltd Technical Report Project VG97081, available at: http://ausveg.com.au/intranet/technical-insights/docs/VG97081.pdf (last access: August 2016), 2004.

Proc. IAHS, 374, 75–77, 2016
proc-iahs.net/374/75/2016/
doi:10.5194/piahs-374-75-2016


Open Access
PIAHS

# Water resources of the Russian rivers and their changes

**Mikhail Georgievsky**

State Hydrological Institute, Saint-Petersburg, 199053, Russia

*Correspondence to:* Mikhail Georgievsky (mgeorgievsky@hotmail.com)

Published: 17 October 2016

**Abstract.** This paper is a review of the recent investigations related to changes in river water resources of the Russian Federation performed at the State Hydrological Institute (SHI). The long-term water resources fluctuations and their changes during the last decades including the significant changes in seasonal runoff are presented. The preliminary estimates of the possible future changes are also given.

## 1 Introduction

Water resources of any country are a very important factor of its social and economic development. Water availability and deficits influence the rate and efficiency of water use. Russia is one of the most prosperous countries of the world with respect to renewable water resources. The country's renewable water resources amounted to 4270 $km^3$ $year^{-1}$ until the mid-1970s (Shiklomanov and Georgievsky, 2007), which represented about 11 % of the global resources.

This paper summarizes the results of a study of changes in renewable water resources carried out by the SHI specialists. It should be emphasized that an assessment of the Russian water resources is a very complex task since there are about 2.5 million streams in the country while the hydrological network consists of only about 3100 observational stations at the moment. Therefore, the role of scientific hydrology, especially that of the studies of runoff formation processes for different physiographic regions and different sizes of watersheds, is very important. It is essential for generalization of the existing hydrological data and assessment of water resources for rivers and areas, where stationary hydrological observations are not available.

## 2 River water resources of Russia and their long-term dynamics

According to the SHI estimates (Shiklomanov, 2008), the average annual water resources of the Russian Federation amount to 4324 $km^3$ $year^{-1}$ including 4118 $km^3$ $year^{-1}$ formed within the country and 206 $km^3$ $year^{-1}$ coming from the neighbouring states.

Distribution of the renewable water resources within Russia is very uneven. Their predominant part belongs to basins of the Arctic (70 % of river runoff) and the Pacific Oceans (20 %). There are only about 10 % of the total river runoff resources in the Atlantic Ocean, the Black and the Caspian Sea basins where population density is the highest in the country.

The long-term dynamics of the Russian renewable water resources is presented in Fig. 1. As can be seen, rising temperatures have been accompanied by increase in renewable water resources since the late 1970s–early 1980s.

There were only three years (1982, 1987 and 1992) with water resources insignificantly below the normal (1936–1980) during the past three decades. It is considered at SHI that there have been significant changes in water regime of the Russian rivers as a result of climate impacts since the early 1980s. Therefore, annual normal flows are basically defined over the period from the beginning of observations up to 1980 (conditionally stationary regime). The maximum volume of water resources observed previously in 1974 (4706 $km^3$) was exceeded in 1990, 1997, 1999, 2002 and 2007 (see Fig. 1). The maximum water resources (4875 $km^3$) for the entire observation period were observed in 2007.

The most significant increase of annual runoff occurred in the major rivers of the Arctic Ocean. The water resources of the Pechora, Yenisei and Lena rivers in 2001–2011 exceeded the normal by 11–13 %.

The total increase of the Russian water resources for 1981–2012 amounted to 211 $km^3$ $year^{-1}$ on average, which is 5.0 % higher than the increase for 1930-1980. The water

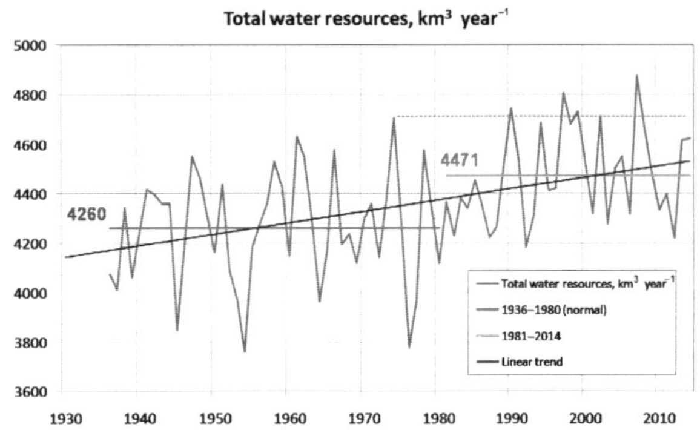


**Figure 1.** The long-term dynamics of renewable water resources of the Russian Federation.

resources increase was representative for all federal districts of Russia. The largest absolute increase was observed in the Siberian (54.7 $km^3$ $year^{-1}$), Far Eastern (50.8 $km^3$ $year^{-1}$) and North-western (46.5 $km^3$ $year^{-1}$) Federal Districts, the smallest one in the Central (9.9 $km^3$ $year^{-1}$) and Ural (11.9 $km^3$ $year^{-1}$) Federal Districts. The largest relative runoff increase was observed in the Volga (13.5 %), Southern (including the North Caucasus) (8.8 %), Central (7.9 %) and North-western (7.7 %) Federal Districts, the lowest one in the Ural (2.0 %) and Far Eastern (2.7 %) Federal Districts.

Thus, the data presented show that ongoing warming in Russia over the last decades has been accompanied by increase in water resources in relation to the previous long-term period (1936–1980).

## 3 Changes in seasonal runoff

The results of a comprehensive statistical analysis of seasonal and monthly runoff of rivers with a natural water regime led to the conclusion that the ongoing climate changes had a strong impact on the regime of winter and summer-autumn river flows.

The water content ($km^3$ $year^{-1}$) of the majority of rivers over the largest part of the European Russia (hereinafter EPR) increased by 50–120 %. In addition, there are positive trends of winter runoff with significantly increased runoff variability. Such significant increase in winter flow observed over the last several decades demonstrates that the conditions for runoff formation have changed significantly as a result of ongoing global warming. This conclusion is also confirmed by the analysis of observational data of the ROSHYDROMET observation network (the water balance and wetland observation stations), as well as by physical and mathematical modelling of migration and infiltration of moisture in freezing and thawing soils (Gurevich, 2009; Kalyuzhnyi et al., 2012; Kalyuzhnyi and Lavrov, 2012; Zhuravin et al., 2012).

It was found out that the winter air temperature increase led to:

- a decrease in soil freezing depth with increase of its draining properties;
- an increase of number and duration of winter thaws during which both snowmelt and meltwater outflow from the snowpack occur;
- ground water recharge and surface runoff formation.

As a result, the river runoff during winter period increased and snow water equivalent before melting period was reduced, which created the conditions for decreased spring flood runoff.

In addition to the direct influence of thaws on the winter runoff, the processes of winter moisture migration and moisture accumulation in a frozen zone make a substantial contribution to the runoff increase during the cold period. The winter air temperature increase coupled with the decrease in soil freezing depth delay the permafrost processes in a catchment which promotes the growth of the winter runoff. Higher air temperature also lead to a decrease of runoff withdrawals from a streamflow required for ice formation and increase the stream hydraulic conductivity, which is also important factor in increasing of low flows during the cold season (Gurevich, 2009; Lavrov and Kalyuzhnyi, 2012).

The increase of summer-autumn runoff in 1978–2012 was observed over most of the Russian regions. These anomalies are most clearly manifested in rivers of the southern part of the forest zone and in the forest-steppe and steppe zones of the EPR. In these regions, the summer-autumn runoff during 1978–2012 was higher in comparison with the previous period (1946–1977) by 40–70 % on average with increased runoff variability. Over the Asian part of Russia (APR), the winter runoff increase (up to 40 %) occurred in the Lena river and the Northeastern rivers (Olenek, Yana and Indigirka).

The spring flood runoff changes in various regions of the country are diverse. In the rivers of the Baltic Sea basin, the Volga (except for northern and north-eastern parts of the river), Don, Dnieper and Upper Ob river basins, the spring runoff was reduced by 10–30 % in the recent decades. The dispersion of spring runoff decreased overmost parts of these rivers. The spring flood flow reduction was due to a decrease in snow water equivalent by the beginning of the snowmelt period as a result of the increase in the number and duration of thaws during the cold season. The spring flood runoff increase was observed in the Kama basin. This applies for most of the EPR's rivers which belong to the Arctic Ocean basin, in the Lena, Yana and Indigirka river basins and over the northern part of the Ob and Yenisei basins.

## 4 Conclusions and possible future changes

Summarizing the results and analysing the long-term river water recourses fluctuations in Russia the following conclusions can be formulated. The main feature of the modern water regime changes in most parts of the country is due to a significant increase in water content during low-flow periods, especially during the winter months in the past 30 years. There are positive trends of increase in winter and summer-autumn low-flow for most rivers considered within the major regions of Russia. The observed "synchronization" of low flow changes (especially winter runoff) for large country areas and the scale of these changes are extraordinary and have no analogues in the 20th century. The low flow growth caused an increase in water resources during 1980–2010 even in river basins where the decrease of spring flood runoff took place. The analysis of observational data over the last hundred years allow us to conclude that such a situation happened for the first time since previously all significant low and high water phases were determined primarily by the magnitude of spring flood runoff.

A precise forecast of future water resources in several decades is required for the purposes of water supply planning as well as for the justification of major water economic practices for river runoff control, its intra basin and territorial re-distribution. Taking into account that climate change is a reality, a methodology of these assessments should be based on results of an analysis of regional river runoff changes and data on future climate scenarios generated by the Coupled Atmosphere-Ocean General Circulation Models.

In accordance with the assessments made at SHI (Georgievsky and Shalygin, 2012; Shiklomanov et al., 2008), there is no reason to expect any significant changes in water resources of the main Russian rivers as a result of climate warming in the coming decades. Over most parts of the country an insignificant increase in annual runoff within its natural variability (within 5 %) will probably be observed.

However, the most problematic regions in terms of water availability were identified based on the assessments. These are the Don basin in the EPR and the upper part of the Ob basin in the APR. There is a possibility of water resources decrease in these basins as a result of climate change which must be taken into account in advance in water supply planning.

Regarding possible future changes in seasonal river runoff it is expected that the identified tendencies based on the observational data will remain. There will be an increase in winter runoff which will require adaptation efforts including revision of water reservoirs management system.

## 5 Data availability

The data relating to the water resources of the Russian Federation are taken from the annual illustrated guides "Water Cadastre of the Russian Federation. Use and quality of surface and ground water resources" which have been published (in Russian) by ROSDYFROMET in cooperation with ROSVODRESURSY and ROSNEDRA since 1981. These guides are publically accessible and present the results of the assessments of water resources of the Russian Federation, its federal districts as well as the largest and most important river basins from economic and ecological perspectives. The data relating to Sect. 3 "Changes in seasonal runoff" are individual data of the SHI specialists cited in this article.

**The Supplement related to this article is available online at doi:10.5194/piahs-93-75-2016-supplement.**

**Acknowledgements.** The author is grateful to all colleagues of the State Hydrological Institute whose materials and publications were used in this article.

## References

Georgievsky, V. Yu. and Shalygin, A. L.: Hydrological regime and water resources, in: Methods for Assessing Climate Change Impacts on Physical and Biological Systems, Chapter 2, Moscow: Rosgidromet, 53–86, 2012.

Gurevich, E. V.: Influence of air temperature on the river runoff in winter (the Aldan river catchment case study), Russian Meteorology and Hydrology, 34, 628–633, 2009.

Kalyuzhnyi, I. L. and Lavrov, S. A.: Basic physical processes and regularities of winter and spring river runoff formation under climate warming conditions, Russ. Meteorol. Hydrol., 37, p. 44, 2012.

Kalyuzhnyi, I. L., Lavrov, S. A., and Romanyuk, K. D.: Variations in Water Regime of Bogs in Northern and Northwestern Russia under the Effect of Climatic Factors, Water Resources, 39, 13–22, 2012.

Lavrov, S. A. and Kalyuzhnyi, I. L.: Physical Processes and Regulation Concerning the Volga River Basin Rivers Winter and Spring Runoff Formation in the Conditions of Climate Change, Water Sector of Russia, 4, 74–84, 2012.

Shiklomanov, I. A. and Georgievsky, V. Yu.: Current and future river runoff changes in Russia under the influence of climatic factors, in: Inland waters in a changing climate, St. Petersburg, Nauka, 20–32, 2007.

Shiklomanov, I. A., Georgievsky, V. Yu., Shalygin, A. L., Georgievsky, M. V., Golovanov, O. F., and Shiklomanov, A. I.: Water resources: predictive estimates, in: Water resources of Russia and their use, edited by: Shiklomanov, I. A., Chapter 11.2, 442–464, 2008.

Zhuravin, S. A., Kaliuzhny, I. L, Lavrov, S. A., Rakcheyeva, E. A., and Gaponov, I. N.: Hydrological response to climate variability in European mixed forest zone (Podmoskovnaya WBS case study), in: Studies of Hydrological Processes in Research Basins: Current Challenges and Prospects, 14th Biennial Conference ERB 17–20 September 2012, St. Petersburg, Russia, Book of Abstracts, 19–22, 2012.

Shiklomanov, I. A. (Ed.): Water resources of Russia and their use, St. Petersburg, State Hydrological Institute, 600 pp., 2008.

Proc. IAHS, 374, 79–84, 2016
proc-iahs.net/374/79/2016/
doi:10.5194/piahs-374-79-2016


Open Access
PIAHS

# Water resources assessment and prediction in China

**Wang Guangsheng, Dai Ning, Yang Jianqing, and Wang Jinxing**
Bureau of Hydrology of MWR, Baiguang Road, Beijing, 100053, China
*Correspondence to:* Wang Guangsheng (gshwang@mwr.gov.cn)

Published: 17 October 2016

**Abstract.** Water resources assessment in China, can be classified into three groups: (i) comprehensive water resources assessment, (ii) annual water resources assessment, and (iii) industrial project water resources assessment. Comprehensive water resources assessment is the conventional assessment where the frequency distribution of water resources in basins or provincial regions are analyzed. For the annual water resources assessment, water resources of the last year in basins or provincial regions are usually assessed. For the industrial project water resources assessment, the water resources situation before the construction of industrial project has to be assessed. To address the climate and environmental changes, hydrological and statistical models are widely applied for studies on assessing water resources changes. For the water resources prediction in China usually the monthly runoff prediction is used. In most low flow seasons, the flow recession curve is commonly used as prediction method. In the humid regions, the rainfall-runoff ensemble prediction (ESP) has been widely applied for the monthly runoff prediction. The conditional probability method for the monthly runoff prediction was also applied to assess next month runoff probability under a fixed initial condition.

## 1 Introduction

Water is the most vital natural resource on the earth that cannot be substituted. It is necessary for human lives and for maintaining the environment on which all creatures rely. Although water is a renewable resource, which can be replenished, the total volume of available water in a definitive area for a definitive period is limited. The volume of water resources in China, either per capita or per hectare farmland, is less than the world average, and it is unevenly distributed in time and space. In pace with the social and economic progress and population growth, the human water demand increases every year. As a result, water shortage has occurred in many regions, conflicts between water supply and demand are becoming increasingly serious. Water resources assessment and prediction tries to assess normal and probability distributions of available water resources, and to predict the volume of available water in future. It does not only concern industry, agriculture and the aquatic environment, but it also deals with people's daily life. Therefore,water resources assessment and prediction are highly important to society.

## 2 General situation of water resources assessment

Based upon the basin' area, there are 10 first class water resources regions in China, namely, the Yangtze River, Yellow River, Huaihe River, Haihe River, Liaohe River, Songhua River, Northwest Rivers, Southwest Rivers, Pearl River and Southeast Rivers. These water resources regions are the basic unit of water resources management. The first water resources assessment in China was carried out in the 1980s for the above-mentioned water resources regions, and these comply with the provincial administrative regions to meet the need of administrative management.

Water resources in China are monitored through hydrological stations; the monitoring elements include, e.g., precipitation, surface water level, groundwater level, discharge. Most of the time series of hydrological observations started in the 1950s.

The water resources assessment index includes precipitation, water quantity and quality. In detail:

- Precipitation: recorded by standard rain-gauges
- Surface water: formed by local precipitation, inflow and outflow of the region

Published by Copernicus Publications on behalf of the International Association of Hydrological Sciences.

- Groundwater: formed by local precipitation
- Total volume of water resources: surface water plus groundwater
- Water storage: reservoirs, shallow groundwater storage in the plains
- Development and utilization of water resources: water supply, water use, water consumption, wastewater discharge
- Water quality

The water resources assessment aims to evaluate the volume of natural available water. It often means to assess the runoff generated from precipitation. Nowadays, observed runoff series are often disturbed by water diverting, water withdrawal and reservoir operation. Therefore, the difference between observed and the natural water flow is rather significant. Therefore, to assess the volume of natural water resources, the data series need to be naturalized. For the naturalization, water withdrawal by industry and agricultural irrigation, diverted water of cross-basins and dam regulation have to be considered by water balance modeling. Currently, water use of many small irrigation projects has not been monitored yet. Especially, volume of groundwater irrigation, is generally estimated by using water quota according to the irrigation area and kinds of crops, thus the required accuracy often cannot be satisfied.

Water resources assessment in China can be classified into three groups: (i) comprehensive water resources assessment, (ii) annual water resources assessment, and (iii) industrial project water resources assessment.

Ad (i): for the comprehensive water resources assessment, the annual normal distribution and probability distribution of water resources are analyzed and assessed at the scales of basins and provincial regions. Generally, the Pearson type III distribution function is used for the probability analysis. Two comprehensive water resources assessment have been carried out in China. The first assessment was carried out in 1980 with the data series from 1956–1979; and the second assessment was in 2003 with the data series from 1956–2000.

Ad (ii): for the annual water resources assessment, the water resources of the last year are assessed in basins and provincial regions.Their results areusually published in the China Water Resources Bulletin, the Gazette of River Sediment in China, and the Groundwater Bulletin.

Ad (iii): for the industrial project water resources assessment, the water resources situation before the constructing of industry projects has to be assessed, including rationality of water abstraction and water use, available water resources (quantity and quality), reliability of water supply, and effect of wastewater discharge into water environment.

In recent years, the increasing human activity and climate change, change the status of water resources.They also cause non-consistent water resources data series (the past situation deviates from the future situation, i.e. non-stationarity). The basic principle of water resources assessment is to use the historic data series to assess future water resources, the historic data series is expected to represent the water resources situation in the future. Thus, problems have emerged, as the environment and climate have changed. To solve this problem, studies are undertaken. One effective solution is hydrological modeling (Zhang and Wang, 2007), where the change of water resources as response to precipitation and temperature change was studied. The hydrological model Xinanjiang (Zhao, 1992), SWAT and VIC are applied in the studies. Another approach is statistics modeling, which combines the changing trend in runoff series with a probability model (Xie et al., 2009). It is suitable for basins whose landscape has been changed by human activity. The trend in runoff change is brought into regular pattern year by year in the model. The water resources situation is predicted under the assumption that the human activity continues.

## 3 Water resources prediction

In operational water resources management, long-term runoff predictions are frequently made since it is important in supporting dam operation, river flow dispatch, cross-basin water diversion, especially during a drought. As water resources become more important and water resources management is enforced, long-term runoff predictions are playing a more important role. For this kind of prediction, the probability model was widely used, and methods based on flow recession model have been developed for low flow conditions, while the ensemble prediction (ESP) also has been put into use in recent years. Usually, these predictions have the mean monthly discharge as the predicting target. Similar as with water resources assessment, time series that are used to calibrate the water resource prediction model need to be naturalized. Water withdrawal by industry and agriculture irrigation, cross-basin water transfer and dam regulation have to be considered through water balance modeling.

### 3.1 Low flow period prediction model

Low flow period usually are characterized by lack of rainfall, and hence runoff mainly is the base flow from water storage in a basin. In the northeast and northwest mountainous areas, runoff in spring also comes from snow melting. Because of the shortage of runoff in low flow period, the prediction becomes vital to irrigation and water supply management. Base on the recession model $Q_t = Q_0 e^{-\beta t}$, many correlation diagrams and simple equations can be further developed. One of the most simple and preferred method by engineering hydrologist is described below.

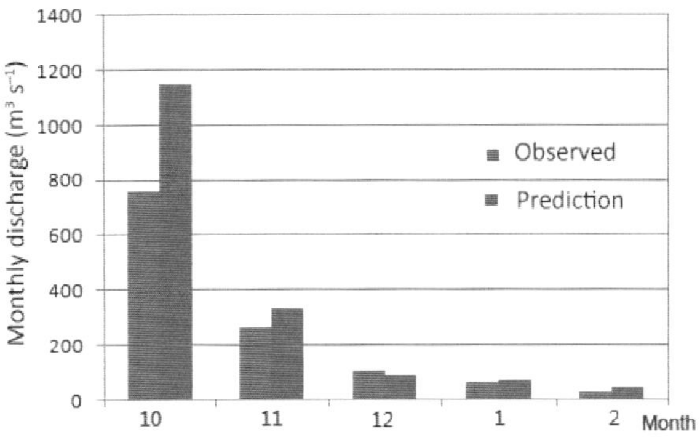


**Figure 1.** Predicted and observed discharge from October 2003 to February 2004 at the Jiangqiao station in the Nenjiang river.

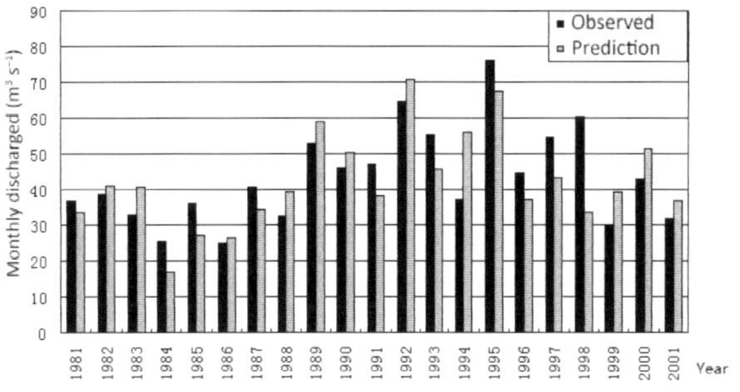


**Figure 2.** Predicted and observed monthly discharge in March at the Jiangqiao station.

The mean monthly discharge can be derived from the following recession model:

$$\overline{Q_{\mathrm{m}}} = \frac{1}{nd}\sum_{i=1}^{nd} Q_{t0}\left(e^{-\beta}\right)^{i} = Q_{t0}\left[\frac{1}{nd}\sum_{i=1}^{nd}\left(e^{-\beta}\right)^{i}\right] \quad (1)$$

where, $\overline{Q_{\mathrm{m}}}$ is mean monthly discharge, $Q_{t0}$ is initial time discharge; $nd$ is number of days in a month. The above equation can be rewritten as follows:

$$\overline{Q_{m1}} = K_{m1} Q_{t0}$$

$$\text{where, } K_{m1} = \left[\frac{1}{nd}\sum_{i=1}^{nd}\left(e^{-\beta}\right)^{i}\right] \quad (2)$$

Equation (2) is used to predict next month runoff. The monthly runoff for the low flow period also can be predicted one step forward:

$$\overline{Q_{mi}} = K_{mi} Q_{t0} \quad (3)$$

where, $mi$ is month $i$.

In practice, the mean monthly discharge at the end of rainy season or even within the whole low flow period can be predicted with reliable results. Since the base flow from groundwater varies for different $Q_{t0}$, the coefficient of percolation that is represented by $\beta$ in Eq. (1) very likely is different. Therefore, it is recommended to look for a representative value or relatively stable value of $Q_{t0h}$ for the same time period and its corresponding mean monthly discharge $\overline{Q_{mih}}$ from historical data series, to allow to fix the $K_{mi}$ as follows:

$$K_{mi} = \frac{\overline{Q_{mih}}}{Q_{t0h}} \quad (4)$$

Figure 1 gives the predicted monthly discharge at the Jiangqiao station in the Nenjiang river in the Songhua river basin in northeast China from October 2003 to February 2004. The basin area is 162 600 km$^2$, and the daily discharge of 30 September 2003 was take as the initial discharge $Q_{t0}$.

In the snow season, snow melting hydrological models and multiple regression equations are used to predict the monthly runoff. Due to its simplicity, in principle, and ease of use, multiple regression is more preferred by engineering hydrologists, where the monthly snowfall, temperature, initial base flow are the main variables in the equation. For example, the monthly discharge in March (Mar) at the Jiangqiao station can be predicted using the following regression equation:

$$\overline{Q_{m3}} = 1.1894\overline{Q_{m2}} + 0.45894 P_{10m-2m} + 3.28631\overline{T_{m3}} \quad (5)$$

where, $\overline{Q_{m3}}$ is the mean monthly discharge in March (Mar); $\overline{Q_{m2}}$ is the mean monthly discharge in February (Feb); $P_{10m-3m}$ is the basin average snowfall from October to February; $\overline{T_{m3}}$ is the mean temperature in March (Mar). The predicted monthly discharge in March for the years 1981–2001 is shown in Fig. 2.

### 3.2 The rainfall runoff ensemble prediction

Since last decade,the rainfall runoff ensemble prediction (ESP) has been studied and monthly runoff prediction has been carried out in humid regions. Usually, in humid regions, hydrological models can produce reliable forecasts and the initial soil moisture condition has a determinative effect on runoff generation. In arid regions, soil moisture generally is stable, remaining at a very low level, and runoff generation is less affected by initial soil moisture. Generally speaking, future runoff is mainly determined by future rainfall, where, future precipitation is rather a random term, while the initial soil moisture can be modeled reliably and affect future runoff in humid regions. Thus, the rainfall runoff ESP is more valuable in humid regions than in arid regions.

The Xinanjiang model is often used in rainfall runoff ESP in China. It is usually operated with a time interval of 1 day. Most hydrological models are used for flood forecasting, while long-term ESP should meet the need of water resources management since low flow forecast accuracy is especially important for water resources management. However, in this case, it is likely that model parameters need to be refitted for low flow forecasting. In general, precipitation series used for

**Table 1.** Observed and predicted monthly discharge from July to October 2007 at the Danjiangkou reservoir.

| Month | Observed ($m^3\,s^{-1}$) | Prediction ($m^3\,s^{-1}$) | | | | | |
|---|---|---|---|---|---|---|---|
| | | Expectation | 10 % | 25 % | 50 % | 75 % | 90 % |
| 7 | 4408.2 | 2056.5 | 3169.1 | 2613.1 | 1994.8 | 1376.9 | 821.0 |
| 8 | 2567.2 | 1992.6 | 3394.9 | 2703.1 | 1932.9 | 1164.1 | 471.2 |
| 9 | 1381.6 | 1969.5 | 3761.2 | 2893.9 | 1930.2 | 966.8 | 99.2 |
| 10 | 565.6 | 672.4 | 2265.9 | 1344.8 | 659.0 | 265.8 | 96.0 |

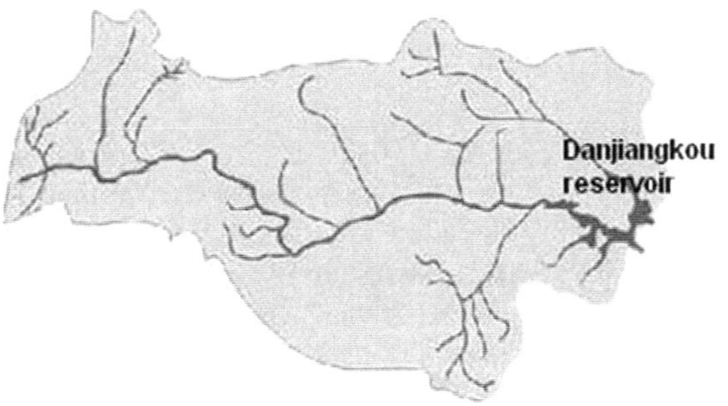


**Figure 3.** Basin area and river network of the Danjiangkou reservoir.

ESP requires time series exceeding 30 years. Practically, 1-week forecasted precipitation can be considered to be valuable for water resources management and flood mitigation. To improve the prediction, precipitation of the first week in each historical monthly precipitation series used in ESP should be replaced by the forecasted precipitation. Thus, the mathematical expectation value and probability distribution of the monthly discharge can be reliably predicted. The Pearson III, normal and log-normal probability distributions can be used for discharge probability analysis.

In China, application of the long-term rainfall runoff ESP is still in a development phase. A number of research studies have been carried out, but have been limited to part of southern China rivers. The most significant case study is for the Danjiangkou reservoir.

The Danjiangkou reservoir is located upstream of the Han river in the Yangtze river basin in south China. The basin area is about 95 200 $km^2$, annual precipitation varies between 800–1100 mm, and the mean annual discharge is 1200 $m^3\,s^{-1}$. The reservoir is located in a humid, and mountainous area. The basin area, river network and location of the reservoir are shown in Fig. 3. The Danjiangkou reservoir is an important water resource in the area. The special significance of the reservoir is that it is the source of water for the south-to-north water transfer project, which diverts water from south to north China.

The Xinanjiang model has been used in rainfall runoff modeling of the Danjiangkou reservoir, where, 30 years historical precipitation data series are available, and the Pearson III probability distribution function was used for the determination of the runoff probability. Since the reservoir is in a large humid basin and initial conditions substantially affect next month runoff, precipitation predictions of the first week are also used in each historical monthly precipitation series. The predicted (i.e. probabilities) and observed mean monthly discharge from July to October 2007 are shown in Table 1.

In operational ESP, the correlation of initial soil moisture and river base flow with next monthly river discharge are analyzed, and the higher correlation the more valuable the ESP. The ESP takes the same initial soil moisture, base flow, and input historical precipitation series to the hydrological model. Obviously, the coefficient of variation (CV) of the probability distribution of predicted monthly discharge should be less than the coefficient of variation (CV) of the probability distribution of the historical monthly discharge. Otherwise, the rainfall runoff model would not adequately simulate the real situation or the initial condition does not affect the next month runoff.

In probability computing, a few historical severe storms, whose return period significantly exceed the length of ESP historical precipitation series, ought to be specially treated. A better way is to analyze their return period with longer hydrological records, instead within the ESP historical precipitation series. Otherwise, the same frequency would be likely to predict a higher result.

### 3.3 Conditional probability

For large river basin,whose area is at the scale of hundred thousand square kilometers, their upstream area usually is brumal mountainous, and the density of rainfall gauge network cannot satisfy the need of rainfall runoff models. Then, conditional probability models are commonly used for such kind of predictions as the model does not require precipitation input. The conditional probability method (Wang et al., 2008) has been used for the monthly discharge prediction since late 1990s, originally for spring season prediction for irrigation in the Songhua river basin in northeast China. Conditional probability means assessment of the probability distribution of future runoff under a fixed initial condition.It is recommended to use initial conditions to reduce the coefficient of variation (CV, mean quadratic error) of the proba-

**Table 2.** Correlation coefficient of the discharge of the end day of the current month and monthly discharge of the next month for major sections in large river basins.

| | Songhua river basin | | | Yellow river basin | | | Huaihe river basin | Yangtze river basin | | | Pearl river basin |
|---|---|---|---|---|---|---|---|---|---|---|---|
| Month | Jiangqiao | Haerbin | Jianmusi | Tongguan | Hishiguan | Huaxian | Lutaizi | Cuntan | Xiangtan | Waizhou | Wuzhou |
| 1 | 0.97 | 0.75 | 0.93 | 0.54 | 0.96 | 0.94 | 0.64 | 0.87 | 0.84 | 0.83 | 0.83 |
| 2 | 0.95 | 0.92 | 0.96 | 0.72 | 0.93 | 0.90 | 0.63 | 0.84 | 0.58 | 0.59 | 0.62 |
| 3 | 0.78 | 0.84 | 0.95 | 0.67 | 0.88 | 0.85 | 0.57 | 0.70 | 0.63 | 0.73 | 0.90 |
| 4 | 0.70 | 0.68 | 0.67 | 0.11 | 0.54 | 0.56 | 0.54 | 0.56 | 0.21 | 0.46 | 0.50 |
| 5 | 0.69 | 0.79 | 0.90 | 0.82 | 0.43 | 0.42 | 0.80 | 0.24 | 0.45 | 0.48 | 0.48 |
| 6 | 0.84 | 0.85 | 0.71 | 0.89 | 0.50 | 0.25 | 0.55 | 0.31 | 0.32 | 0.40 | 0.56 |
| 7 | 0.78 | 0.83 | 0.74 | 0.63 | 0.16 | 0.35 | 0.57 | 0.43 | 0.59 | 0.68 | 0.58 |
| 8 | 0.82 | 0.85 | 0.80 | 0.68 | 0.47 | 0.19 | 0.67 | 0.38 | 0.59 | 0.30 | 0.32 |
| 9 | 0.76 | 0.90 | 0.93 | 0.86 | 0.63 | 0.30 | 0.79 | 0.49 | 0.78 | 0.86 | 0.78 |
| 10 | 0.85 | 0.90 | 0.93 | 0.90 | 0.61 | 0.89 | 0.74 | 0.55 | 0.60 | 0.62 | 0.47 |
| 11 | 0.97 | 0.96 | 0.95 | 0.95 | 0.88 | 0.81 | 0.62 | 0.78 | 0.56 | 0.62 | 0.83 |
| 12 | 0.96 | 0.90 | 0.70 | 0.79 | 0.98 | 0.92 | 0.91 | 0.79 | 0.74 | 0.82 | 0.82 |

bility distribution of the prediction monthly runoff, and thus reliability of the prediction can be improved.

Discharge of the end day of the current month is taken as the initial condition to predict next monthly discharge. The value of the conditional probability method is determined by the correlation between the initial condition and next month runoff. Generally, large basins have longer memories, with good correlations between initial condition and future runoff. The correlation coefficient of the discharge of the end day of the current month and the monthly discharge of the next month at selected river sections in major basins are shown in Table 2. The drainage areas of those sections are over ten thousands or hundred thousand square kilometers. The table shows that correlation of most sections in the low flow season are obviously strong, and even better than that in summer rainy season in the Songhua river and part of the Yellow river (northern region).

Two dimensional Log normal probability distribution functions are used for the conditional probability study. This function has been used for hydrological probability analysis already for a long time. It is a two dimensional analytic function:

$$f(x,y)=\frac{1}{2\pi\sigma_x\sigma_y\sqrt{1-\rho^2}}e^{-\frac{1}{2(1-\rho^2)}\left[\frac{(\ln x-a_x)^2}{2\sigma_x^2}-2\rho\frac{(\ln x-a_x)(\ln y-a_y)}{\sigma_x\sigma_y}+\frac{(\ln y-a_y)^2}{2\sigma_y^2}\right]} \tag{6}$$

where, $\rho$ is correlation coefficient between $\ln(x)$ and $\ln(y)$; $a_x$ is mean $\ln(x)$; $\sigma_x$ is mean quadratic error of $\ln(x)$; $a_y$ is mean $\ln(y)$; $\sigma_y$ is mean quadratic error of $\ln(y)$.

**Table 3.** Observed and predicted monthly discharge ($m^3\,s^{-1}$) of July 1999 at the Haerbin station.

| observed | Expectation | 10 % | 25 % | 50 % | 75 % | 90 % |
|---|---|---|---|---|---|---|
| 1160 | 1680 | 2460 | 2000 | 1680 | 1370 | 1170 |

Base on Eq. (6), the conditional Log normal probability distribution function $f(y)$ at $x=x^*$ can be derived:

$$f\left(y|x=x^*\right)=\frac{1}{\sigma_y\sqrt{2\pi(1-\rho^2)}}e^{-\frac{1}{2\sigma_y^2(1-\rho^2)}\left[\ln y-a_y-\rho\frac{\sigma_y}{\sigma_x}(\ln x^*-a_x)\right]^2} \tag{7}$$

Equation (7) is still a Log normal distribution function. The mathematical expectation value of $\ln(y)$ is $a_y+\rho\frac{\sigma_y}{\sigma_x}(\ln x^*-a_x)$. The mean quadratic error of $\ln(y)$ is $\sigma_y\sqrt{(1-\rho^2)}$ where, if $\rho>0$, correlation exists between $x$ and $y$, and the mean quadratic error $<\sigma_y$; and if $\rho=0$, then the mean quadratic error $=\sigma_y$. The higher the correlation between $x$ and $y$, the smaller the mean quadratic error of probability distribution of $y$ is. The Log normal distribution of $x$ and $y$ should be formulated, then $a_x$, $\sigma_x$, $a_y$, $\sigma_y$ can be obtained, the correlation coefficient $\rho$ can be computed, and parameters of conditional distribution of $y$ can be fixed.

Figure 4 gives the cumulative Log normal distribution of the discharge on 30 June, at the Haerbin station in Songhua river in northeast China, whose basin area is 557 200 $km^2$. Figure 5 provides the comparison of conditional and non-conditional cumulative log normal distributions of the monthly discharge of July, at the Haerbin station, where, the mean quadratic error of conditional distribution is less than that of the non-conditional distribution. Observed and predicted monthly discharge of July 1999, at the Haerbin station are shown in Table 3.

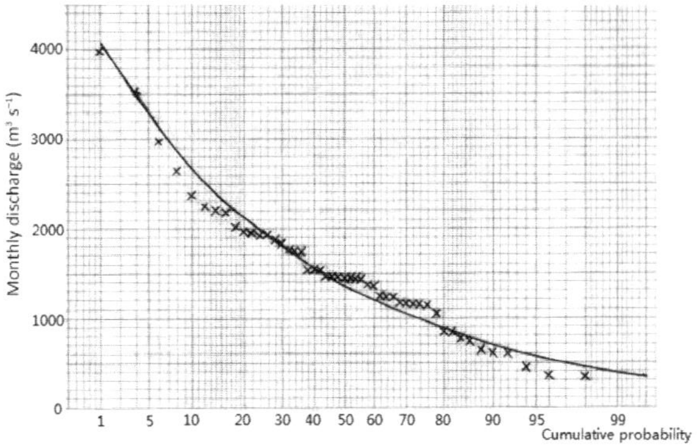


**Figure 4.** Predicted distribution of the discharge on 30 June at the Haerbin station using a conditional Log normal probability distribution.

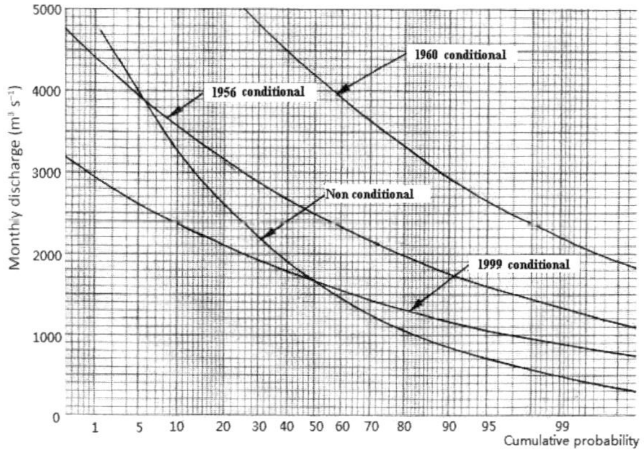


**Figure 5.** Conditional probability distribution of the monthly discharge of July for several years and the non-conditional distribution at the Haerbin station.

Conditional probability models to predict monthly runoff is applied mainly to large rivers. From the late 1990s, the model is used for water resources prediction in the Songhua and Laohe river basins in northeast China. In the 2005 and 2006 low flow seasons, severe low water occurred in the Pearl river, and sea water intruded the downstream river. To safeguard fresh water supply and to protect the aquatic environment, dam operation was optimized to release water to counteract the salty water tide. A conditional probability model was used to predict inflow of the Upper Pearl river at the Wuzhou station, which provided firm support to the decision-making.

## 4 Conclusions

Water resources assessment and prediction have provided strong support to water resources management and drought relief. As a follow-up of the implementation of a strict water resources management policy in China, water resources assessment and prediction will receive more attention in future.

Human activity, incl. climate change substantially changed the water resources status, and meanwhile bring non consistency in the past data series (non-stationarity) that are used for assessment of water resources and calibration of prediction models. This means that the historical data cannot represent future situation adequately. To make reliable water resources assessment and prediction, these changes impose serious challenges to the models. When climate changes, extreme weather events increase, such as severe floods and drought, inter annual variability of runoff increases, and within one year variability of runoff increases. Moreover, climate change in mountainous regions in north China causes decrease of snow melt runoff significantly, and water resources in the region become scarce in spring. All these issues need to be paid attention to and further studied for water resources assessment and prediction.

## 5 Data availability

Research data mainly refer information of river flow, can be accessed on web site of Ministry of Water Resources (http: //xxfb.hydroinfo.gov.cn/ssIndex.html.

**Acknowledgements.** The authors wish to thank Henny A. J. van Lanen from the Wageningen University for his comments, revision, and English smooth to improve the paper.

## References

Wang, G., Ning, F., Xiao, F., and Jiang, T.: Applied Hydrological Forecasting Method, China Water and Power Press Beijing China, 2008 (in Chinese).

Xie, P., Chen, G., Lei, H., Han, S., Xia, J., and Zhu, Y.: Water resources assessment method under changing environment, Science Press, 2009 (in Chinese).

Zhang, J. and Wang, G.: Impact of climate change on hydrology and water resources, Science Press, 2007 (in Chinese).

Zhao, R.: The Xinanjiang model applied in China, J. Hydrol., 135, 371–381, 1992.

Proc. IAHS, 374, 85–91, 2016
proc-iahs.net/374/85/2016/
doi:10.5194/piahs-374-85-2016


Open Access
PIAHS

# A tentative discussion on the monitoring of water resources in China

**Yang Jianqing, Dai Ning, Wu Mengying, and Wang Guangsheng**

Bureau of Hydrology of Ministry of Water Resources, Baiguang Road, Beijing, 100053, China

*Correspondence to:* Yang Jianqing (jianqing.yang@mwr.gov.cn)

Published: 17 October 2016

**Abstract.** With the rapid economy development and social civilization progress, the Chinese Government also is improving ecological environmental conditions. More efforts have been made to solve water problems through the implementation of stringent water resources management, as a key government policy on water. Thus, monitoring of water resources has been strengthened, being a main component of the hydrological work in recent years. Compared with routine hydrological monitoring, water resources monitoring pays more attention to the quantity and quality variations of regional waters, to reflect the status of water in river basins and administrative regions. In this paper, the overall layout of the hydrometric network in China is presented, monitoring efforts of the natural water cycle and water consumptions are analyzed, methodologies of water resources monitoring, which are commonly applied in the country, are summed up. Taking the hydrometric network planning on interprovincial boundary waterbodies as example, a summary of the planning at interprovincial boundary river sections is presented. The planning can meet the need of water resources management of administrative divisions. It can also improve the overall water resources monitoring for the country.

## 1 Introduction

Water is not only an irreplaceable natural resource, but also an important ecological and environmental element. According to the second National Water Resources Assessment in China, the total amount of water resources of the country was about 2841 billon $m^3$ in early 21st Century, which ranks the 6th place in the world. However, the amount of available waters in China per capita is about 2200 $m^3$, only 28 % of the world average, and the volume of water per unit land area is 83 % of the world average. Moreover, China's water resources are spatially unevenly distributed. Where in southern region, land area accounts for 36 % of the country, population accounts for 40 %, and the Gross Domestic Product (GDP) is about 56 %, with 81 % of total water resources. While in northern region, the land area is about 64 %, population is about 46 %, and GDP accounts for 44 %, but the total water resources is only 19 % of the country. The regional distribution of water resources in the country according to the China Water Resources Bulletin is shown in Fig. 1.

With the impact of climate change and human activities such as urbanization in the country, water scarcity due to changes of surface conditions in the north and water pollution in the south have not been effectively curbed. Contradiction between water supply and water demand is yet outstanding, and water utilization efficiency is still poor in comparison with those in developed countries.

In recent years, the Chinese Government devoted to promote the country's ecological and environmental condition by implementing stringent water management as a key policy on water. Three "red lines", i.e., (i) the control of total amount of water exploited, (ii) efficiency of water use, and (iii) water pollutions, have been set up in response to the increasing water problems across the country (Wang et al., 2015). To cope with the contradiction of temporally and spatially unevenly distributed available water and regional water consumptions, the Chinese Government also proposed the arduous task of inter-basin water transfer plans to realize regional water resources allocation. Above all, monitoring of water resources is the basis for the management of water. This paper introduces the overall hydrometric network in China, classifies the monitoring technology commonly applied in the country, and analyzes the comprehen-

Published by Copernicus Publications on behalf of the International Association of Hydrological Sciences.

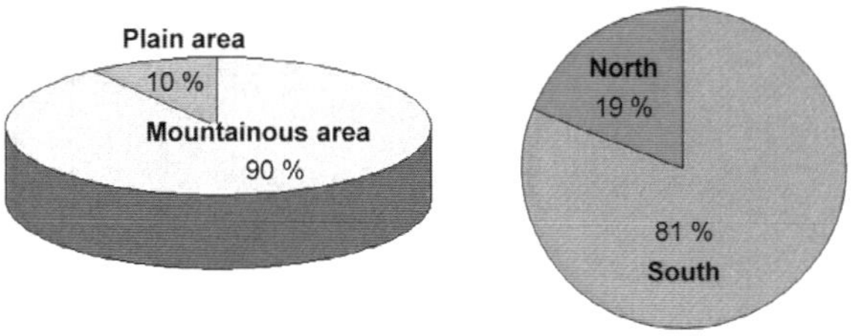


Figure 1. Distribution of water resources in China.

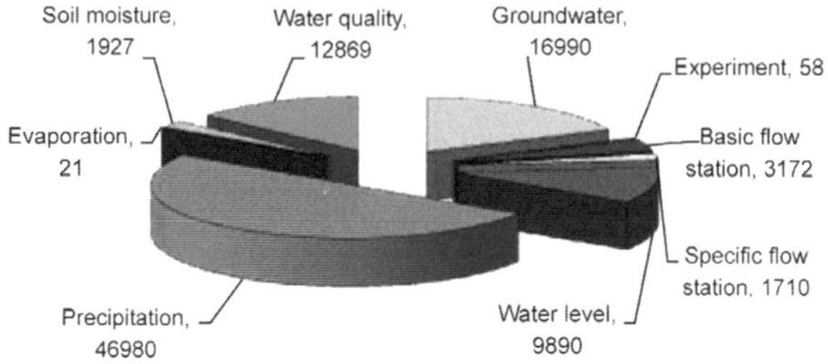


Figure 2. Structure of hydrological stations in China.

sive situation of water resources monitoring in China, taking the hydrometric network planning for interprovincial boundary rivers as an example. The distribution of monitoring stations at interprovincial boundary river sections is proposed to meet the needs of water resources management for provincial administrative divisions. The planning provides constructive guidance to the development of water resource monitoring of China.

## 2 Hydrometric network development

Water resource monitoring involves, for example, analyzing the water quantity, water quality, distribution of water status, and water exploitation, water utilization and conservation of water resources (Yao et al., 2013; Zhang et al., 2015). A hydrometric network is the basis for water resources monitoring, investigation, analysis and evaluation (Maidment, 1993; WMO Secretariat, 1988).

At present, water resources monitoring in China can be generally distinguished into two forms: dynamic monitoring for the natural water cycle and monitoring for water utilization. Where, dynamic monitoring for natural water cycle focuses on the monitoring of hydrological features of water bodies, including rivers, lakes, water channels, as well as regional groundwater. Monitoring for water utilization focuses on the monitoring of water supply, water use and water consumption processes through human activities, including water supply sources, water pipelines, sewage outlets and so on. For a long time, the dynamic monitoring of natural water cycle was given priority in the work of the hydrologic sector, while measurement of the water utilization status remains to be further improved. Following the implementation of stringent water management in recent years, monitoring for inter-administrative boundary water sections become a new task to meet the needs of water resources management for administrative divisions. In addition, for those areas and water bodies without hydrological stations, water survey is usually carried out at a regular base, including both regional water quantity and water quality investigations.

### 2.1 Dynamic water resources monitoring

After decades of development, China has established a comprehensive hydrological monitoring network all over the country, and accumulated a large amount of hydrological data, which provides a strong basis for the management of water resources. The hydrometric network has been long set up according to the needs of flood defence, drought relief and disaster reduction in river basins. Currently, there are in total 93 617 different kinds of hydrological monitoring stations (Deng, 2015). The structure of the country's hydrological stations is shown in Fig. 2.

The dynamic monitoring of water resources based on existing hydrological stations is carried out, as follows.

i. Monitoring of surface water flow. Traditional hydrological monitoring stations are long deployed in line with the distribution of watersheds. The average density of national hydrological stations is 1966 $km^2$ $site^{-1}$, where 3172 stations are the basic monitoring stations of main river control sections, 1710 hydrological stations are for intermediate and small rivers. Moreover, there are 46 980 precipitation monitoring stations, 21 evaporation monitoring stations, and 1927 soil moisture monitoring stations. These provide a full coverage of major river basins all over the country. With the data collected, analysis of hydrological regime of rivers and calculation of runoff changes in basins can thus be made. Hydrological characteristics and water resources conditions of river basins can be evaluated to manage river flow and water storages, which provide a solid foundation for evaluation of regional water resources.

ii. Monitoring of groundwater level. There are 16 990 groundwater monitoring stations in total, where 12 898 stations are basic stations, which are monitored at a weekly base, and 4092 are supplemental stations, which are monitored seasonally. These stations form the basic groundwater monitoring network of China. Through monitoring, the dynamics of the regional groundwater level can be thus outlined, and groundwater storage changes can be estimated, which provide the basic needs of groundwater evaluation (Lin et al., 2005; Wang, 2014). However, most of the exist-

**Table 1.** Summary of the planned monitoring stations distributed by main river basins.

| Basin | Basin area ($10^4$ km$^2$) | Planning stations | Of which existing stations | Of which planned new stations | | | |
|---|---|---|---|---|---|---|---|
| | | | | Total | Automatic monitoring sites | Tour gauging sties | Stationary monitoring sites |
| Yangtze River | 180 | 248 | 87 | 161 | 4 | 133 | 24 |
| Yellow River | 368 | 85 | 60 | 25 | | 17 | 8 |
| Huaihe River | 33 | 89 | 34 | 55 | 2 | 40 | 13 |
| Haihe River | 32 | 199 | 103 | 96 | | 79 | 17 |
| Songliao River | 80 | 100 | 49 | 51 | | 43 | 8 |
| Pear River | 122 | 62 | 14 | 48 | | 36 | 12 |
| Taihu Lake | 27 | 58 | 12 | 46 | 33 | 9 | 4 |
| total | – | 841 | 359 | 482 | 39 | 357 | 86 |

ing groundwater monitoring stations are distributed in northern China, and many are water extraction wells and irrigation wells, which are manually monitored. The present network density and regional layout cannot meet the increasing demand for information on water resources management. To strengthen and improve groundwater monitoring, the Ministry of Water Resources and the Ministry of Land Resources jointly proposed a project to measure the groundwater state through automatic monitoring systems (Qiao, 2009; Yang and Zhang, 2008). The project plans to construct a total number of 20 401 automatic groundwater monitoring stations. Monitoring for groundwater overdraft areas and different groundwater sources are particularly strengthened. Effective monitoring will be achieved for most of the country's plains, which cover an area of 3.5 million km$^2$.

iii. Monitoring of water quality elements. There are more than 12 000 surface water quality monitoring sections, where, 596 sections focus on the monitoring of interprovincial boundary waters (buffer zones), 1507 sites are monitoring water supply sources, and 5064 sections are for the 4493 state-administrated water functional zones (areas defined according to the natural water conditions and regional development targets) across the country. In addition, water quality monitoring is carried out at both 4640 waste water outlets to rivers and 4115 representative groundwater monitoring wells.

To strengthen and promote the work of water quality monitoring, the Ministry of Water Resources completed the national water resources conservation monitoring planning in 2013. A total number of 16 392 surface water quality monitoring sections are planned, providing a full coverage of the country's major water bodies. It can provide a strong support to the country's water conservation and protection goals by 2020.

### 2.2 Monitoring for administrative regions

Following the step-wise implementation of a stringent water resources management, "three red lines" were formulated, i.e. a controlling index of the total amount of water exploited, efficiency of water use, and water pollutions for each administrative region. Among them, the controlling index of total exploited water for the country is 670 billion m$^3$ by 2020, and 700 billion m$^3$ by 2030. Thus, monitoring of water resources for trans-administrative boundary water bodies appeared to become essential for the implementation of stringent water resources management.

To examine and evaluate the government performance on rational water consumption and effective water conservation in the administrative regions, the Ministry of Water Resources has put forward the intensive monitoring for the trans-administrative boundary water bodies based on the monitoring network planning (Zhang et al., 2012). The planning focuses at the provincial administrative sections or at interprovincial boundary rivers. New hydrological stations were further deployed in comply with the existing hydrological monitoring network. Summaries of the planning are as follows, as an example.

i. Principle of the planning. Monitoring stations at provincial administrative sections of interprovincial boundary rivers are deployed to comply with following principles: (a) major river stems and their first-order tributaries, (b) rivers whose catchment area greater than 1000 km$^2$, (c) key interprovincial water transfer project or interbasin water diversion project channels, (d) water vulnerable areas or seriously polluted areas of the interprovincial boundary water bodies.

ii. Outcome of the planning. Based on above principles, a total number of 841 monitoring stations are deployed at provincial administrative sections of interprovincial boundary rivers. Summary of monitoring stations distributed by main river basins are shown in Table 1. It is shown that, among the planned stations, 359 sites are existing monitoring stations, accounting for 43 % of the total number, while 482 sites are new stations, accounted for 57 % of the planning. To promote the automatic monitoring capacity, 82 % of the new sites are either tour gauging or automatic monitoring stations. To meet the requirement of monitoring regulations of water resources, monitoring of water quality will be synchronized with monitoring of water flow for all the 841 sites. Sketch map of the planned hydrological monitoring stations for interprovincial boundary rivers is shown in Fig. 3.

In the planning, there are 367 interprovincial boundary rivers, whose drainage areas are above 1000 $km^2$. Monitoring stations for 333 interprovincial boundary rivers at their provincial administrative sections are implemented, which account for 91 % of the total number of 367 rivers. Among which, 213 rivers are deployed with existing monitoring stations, accounting for 58 % of the total number of interprovincial boundary rivers. Analysis is further made by the method of area-runoff depth correlation algorithm. It shows that, with the planned monitoring capacity, an average 86 % of total inflow and 88 % of total outflow for provincial administrative regions can be effectively examined. Remaining rivers without planned stations are mostly located in mountainous areas, which are deserted areas with little water management demands. Taking into account the work done for ungauged areas, the planning can meet the increasing needs of regional water management for provincial administrative divisions.

### 2.3 Monitoring for water use

Monitoring for water use is usually carried out in real-time through monitoring stations. However, for those areas without monitoring stations, it can also be made through water investigation.

i. Monitoring for water consumptions. In comparison with the dynamic monitoring of the natural water cycle, monitoring for water consumptions is relatively weak in China. The monitoring ability is poor when compared with those developed countries (Martin, 2008; Office of the Auditor General of Canada, 2010; Wang et al., 2014). At present, a statistic method, such as water consumption quotas, is widely used for the measurement of water consumption in China. Nearly 30 % of industrial and 70 % of agricultural water use have not yet been directly metered. Based on the current monitoring situation, following principles are set up to strengthen the real-time monitoring of water consumptions by focusing on those important users: (a) centralized public and industrial users with annual surface water consumptions above 300 million $m^3$, (b) licensed public and industrial users with annual groundwater consumptions above 50 million $m^3$, and (c) users taken water from vulnerable and sensitive water-bodies for specific purpose. Based on the principle above, a pilot project on comprehensive water resources monitoring and management for 38 cities in 22 provinces has been established by the Ministry of Water Resources. The project provides online monitoring for 2891 drinking water sources, large water users, groundwater overdraft areas and water functional zones since 2006. Starting from 2012, a state project on water resources monitoring capacity building has been put forward. Online monitoring of 8558 key water users at national level, including 5576 surface water users, 2886 groundwater water users, and 96 locations with conjunctive use of surface and groundwater have been further put into practice. In combination with the monitoring carried by local counties, around 80 % of the total licensed water thus can be effectively monitored.

ii. Water investigation. Water investigation is carried out at regular basis for those areas without monitoring stations. Investigation for water flow is conducted once a year in general, and investigation for waste water into rivers and lakes is generally conducted every 1 to 3 years. These investigations are carried out in accordance with the Technical Guideline for Water Resources Monitoring. The main technical work is as follows: (a) investigation of natural water quantity, including river channel flow, inflow and outflow of interprovincial boundary rivers, water storage and groundwater, (b) investigation of water exploitation and utilization, including surface water consumptions, groundwater exploitation, groundwater recharge, mining drainage survey, and (c) specific investigations, including industrial and domestic wastewater into rivers and lakes, irrigation return flow.

Analysis shows that, a comprehensive hydrometric network has been developed in China. The spatial distribution of hydrological stations is by far rational and practical. The network distribution and monitoring elements are basically consistent with regional economic and social development patterns. However, due to the spatially and temporally unbalanced economic development of the country, the hydrological monitoring network has also been unevenly distributed. It is clear that, the monitoring network for water utilization is still insufficient. Particularly, those monitoring stations for inter-administrative boundary rivers are inadequate and monitoring for water exploitation and consumption are poor. Demands for timely and fully understanding of the regional water situation cannot be satisfied. There is an urgent need to accelerate the implementation of the designed monitoring network.

## 3 Application of monitoring technology

Traditional hydrological monitoring technology and methodology are fully applied in the monitoring of water resources.

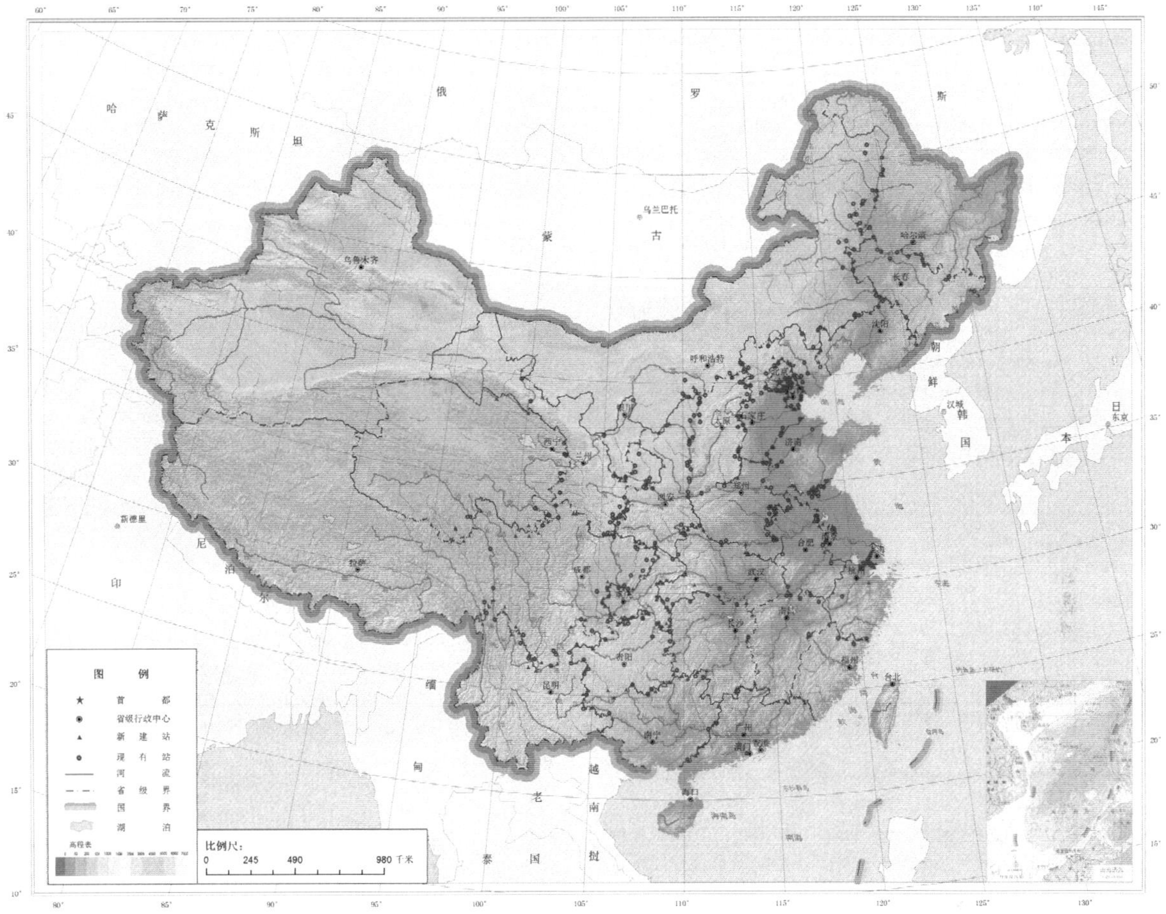


**Figure 3.** Sketch map of the planned hydrological monitoring stations for interprovincial boundary rivers.

With the rapid development of modern science and technology, automatic monitoring technology has also been widely introduced in water resource monitoring (Yao et al., 2013; Zhang et al., 2003). Some examples of technology of main hydrologic elements monitored are as follows:

i. Monitoring of water levels. Currently, automatic monitoring of water levels is extensively applied in the country. According to statistics, more than 90 % of the surface water levels in the country are automatically monitored. However, about 90 % of groundwater monitoring is still manually done. The capacity of automatic monitoring of groundwater will be significantly improved after the implementation of the state automatic groundwater monitoring system. For automatic water level monitoring, the float type, pressure type and bubble type water level meters are the main instruments applied for both surface water and groundwater. According to the technical regulations on water level monitoring, some manual checks of the water level is still necessary for the automatic monitoring sites. It is mainly used to cross check the automatic measurement instruments by manual readings.

ii. Monitoring of open channel flow. Open channel flow monitoring methods in China is done through the velocity-area method, hydraulics methods, tracer methods and volumetric methods (Yang et al., 1993). Velocity-area method is the most commonly used monitoring technology. Where, the point velocity-area method is the most widely applied one. Nearly 80 % of the current national basic hydrological stations utilize the point velocity-area method, whose most popular measurement instrument is the rotor type flow meters. Application of the Acoustic Doppler Current Profilers (ADCP) in China has constantly been increasing over the last decade (Liu et al., 2006). Up to 50 % of the hydrological stations in the recent monitoring planning for small and medium-size rivers uses the velocity-area method of current profiler type. In addition, when steady hydrological condition of the monitoring cross sections can be achieved, and reliable and stable water level-flow correlation can be established, then people are encouraged to make use of the rating-curve instead of the direct flow measurement.

iii. Monitoring of pipe flow. Monitoring for pipe flow can also be distinguished as velocity-area method and volumetric method in principle. For the monitoring of industrial and domestic pipe flow, both methods are extensively used. Types of popular instruments include the electromagnetic flow meter, acoustic flow meter, water meter and others. For the monitoring of irrigation water, above monitoring technology can generally be applied. While for the measurement of groundwater exploitations, indirect measurement such as electricity use, irrigation water meter and water consumption quotas are the typical methods utilized.

iv. Monitoring of water quality. Water quality monitoring can generally be divided into two kinds, one is the manual monitoring, and the other is automatic measurement (Chen et al., 2014). At present, no matter if it is surface water, groundwater, a natural water body or an artificial pipeline, manual sampling in combination with laboratory analysis is still the main method for water quality monitoring. Limited parameters are monitored through automatic measurement of the electrode method and online analysis technology.

v. Monitoring of precipitation, evaporation and water storage. For the monitoring of water storage, water levels are automatically monitored, and the storage capacity of reservoirs and other water bodies are usually obtained through the water level-volumetric correlations. For the monitoring of precipitation, automatic monitoring has been realized for all stations; while pan evaporation will remain to be manually monitored in general.

## 4 Discussions and suggestions

Water resource monitoring is important for better understanding of regional water conditions and for evaluation of water resources. In this paper, the overall hydrometric network in China is introduced, and the classification of the monitoring network and monitoring technology commonly applied in the country are presented. As an example, the monitoring network planning for interprovincial boundary rivers is demonstrated to meet the increasing demand of water management. It can be seen that in recent years, much attention has been paid to the monitoring of water resources in China. The water resources monitoring provides a reliable scientific basis for water exploitation, utilization, configuration, protection and conservation. Strengthening the monitoring of water resources is not only the requirement of implementing stringent water resources management, but it also is an important part of the development of hydrological work.

In recent years, China has been facing serious water problems, including water shortage, water pollution, and inefficient water use, while water resource monitoring still was fundamentally weak. Current monitoring cannot meet the actual needs of water management yet. In view of the outstanding problems in current water resources monitoring, the following suggestions are made on the establishment of a systematic monitoring network:

i. Traditional hydrological monitoring mainly focused on the monitoring of hydrological elements of the natural water cycle, including its spatial and temporal distributions, to better understand regional water quantity and quality conditions. Future monitoring of water resources should be put forward to comply better with the regional water status. Synchronized water quantity and water quality monitoring, regional water balance assessment in combination with measurements and investigations, should be implemented in water resource monitoring.

ii. Considering the relatively weak water resources monitoring, especially the monitoring of water exploitation and consumption in the country, there is an urgent need to accelerate the implementation of monitoring planning of the interprovincial boundary river sections, as well as the planning of the water resources monitoring capacity-building and automatic groundwater monitoring. Thus, the monitoring network can be optimized, network density can be improved, and water resources monitoring can be strengthened.

iii. Taking into account the cost of human resources for stationary monitoring stations, it is required to actively promote automatic monitoring to increase the water resources monitoring. Monitoring frequency and accuracy could also been improved, to satisfy the growing demand for water resources management.

## 5 Data availability

The "Directory of rivers to be carried out for water allocation (monitoring and assessment) during the period of 2013–2015", which is the basic data for the work of hydrometric monitoring planning on inter-provincial boundary river sections is avaible in the Supplement. Access to the hydrometric network data in part1 of the manuscript is available on the government web (http://www.mwr.gov.cn/zwzc/hygb/swqknb/, Bureau of Hydrology of Ministry of Water Resources, 2015).

**The Supplement related to this article is available online at doi:10.5194/piahs-93-85-2016-supplement.**

**Acknowledgements.** The work is supported by Zhang ShuAn from the Bureau of Hydrology of the Ministry of Water Resources. Special thanks to Henny A. J. van Lanen from the Wageningen University for his comments and revision to improve the paper.

## References

Bureau of Hydrology of Ministry of Water Resources: 2014 Annual Report of National Hydrological Statistics, Shangdong Hydrological Press, Shangdong, China, http://www.mwr.gov.cn/zwzc/hygb/swqknb/, 2015.

Chen, Z. Z, Yin, J., and Liang, S. S: Outlook for automatic water quality monitoring applied in Yangtze Estuary, Water Purification Technology, 33, 81–83, 95, 2014.

Deng, J.: 2014 Annual Report of Hydrological Development, Shandong Hydrological Publishing House, Shangdong, China, 37–40, 2015.

Lin, Z. D., Liu, H. S., Jiang, Y. X., Cheng, Y. L., and Zhang, H.: Technical standard for groundwater monitoring, Chin Waterpower Press, Beijing, China, 2005.

Liu, D. S., Jin, X. P., Zhu, X. Y., Zhou, F. Z., and Wei, J. C.: Code for discharge measurement of acoustic Doppler current, China Waterpower Press, Beijing, China, 2006.

Maidment, D. R.: Handbook of Hydrology, McGraw-Hill, 1993.

Martin, G. (Ed.): Water Framework Directive, translated by: International Economic and Technical Cooperation Center of the Ministry of Water Resources, China Waterpower Press, Beijing, China, 2008.

Office of the Auditor General of Canada: Network design for water quality monitoring of surface freshwaters: a review, 2010 Fall Report of the Commissioner of the Environment and Sustainable Development, 1–17, 2010.

Qiao, G. J.: Study on the functional area of groundwater in China, China Hydrology, 29, 90–93, 2009.

Wang, R.: Real time monitoring and analysis of shallow freshwater resources, Resources and Environment, 18, 115–116, 2014.

Wang, X. K., Qu, Y., Zhi, Y., and Ren, Y. F.: The monitoring system for long-term urban ecosystem study in USA and China, Advances in Earth Science, 29, 617–623, 2014.

Wang, X. J., Zhang, J. Y., Shamsuddin, S., He, R. M., Xia, X. H., and Mou, X. L.: Potential impact of climate change on future water demand in Yulin city northwest China, Mitigation and Adaptation Strategies for Global Change, 20, 1–19, 2015.

WMO Secretariat: Water resources assessment activities: handbook for national evaluation, UNESCO Paris, World Meteorological Organization, Geneva, 1988.

Yang, Y. C., Chen, H. F., Wang, B. C., Zhu, X. Y., and Wang, Y. C.: Code for liquid flow measurement in open channels. China Waterpower Press, Beijing, China, 1993.

Yang, J. Q. and Zhang, S. A.: General discussion on groundwater monitoring and prediction studies in China, in: Proceedings of the Symposium on Water Ecological Monitoring and Analysis, Shandong Atlas Publishing House, Shandong, China, 207–211, 2008.

Yao, Y. X., Zhang, S. A., Yang, J. Q., Fang, R., and Yang, G. L. (Eds.): Water resource information monitoring and transmission technology, Hohai University Press, Nanjing, China, 2013.

Zhang, J. Y., Zhu, C. N., Cui, J. J., Tang, Z. S., and Xu, Z. C.: Technical specification for hydrologic data acquisition system, China Waterpower Press, Beijing, China, 2003.

Zhang, S. A., Yang, J. Q., Lin, Z. D., Fang, R., Yang, G. L., Wei, Y. H., and Li, S. B.: Discussion on the outline of water resources monitoring network planning in China, in: Proceedings of the Conference on Academic Hydrology 2012, Hohai University Press, Nanjing, China, 337–343, 2012.

Zhang, S. A., Zhang, L. Z., Fang, R., Ding, F., Jing, Y., Yuan, Y. P., and Dai, N.: Technical guideline for the monitoring of water resources quantity, China Waterpower Press, Beijing, China, 2015.

Proc. IAHS, 374, 93–99, 2016
proc-iahs.net/374/93/2016/
doi:10.5194/piahs-374-93-2016


Open Access
PIAHS

# Conjunctive operation of river facilities for integrated water resources management in Korea

**Hwirin Kim**[1], **Cheolhee Jang**[2], **and Sung Kim**[2]

[1]Han River Flood Control Office, Ministry of Land, Infrastructure and Transport, Seoul, South Korea
[2]Hydro Science and Engineering Research Institute, Korea Institute of Civil Engineering and Building Technology, 283 Goyangdae-ro, Ilsanseo-gu, Goyang-si, Gyeonggi-do, 411-712, South Korea

*Correspondence to:* Hwirin Kim (hydro@korea.kr)

Published: 17 October 2016

**Abstract.** With the increasing trend of water-related disasters such as floods and droughts resulting from climate change, the integrated management of water resources is gaining importance recently. Korea has worked towards preventing disasters caused by floods and droughts, managing water resources efficiently through the coordinated operation of river facilities such as dams, weirs, and agricultural reservoirs. This has been pursued to enable everyone to enjoy the benefits inherent to the utilization of water resources, by preserving functional rivers, improving their utility and reducing the degradation of water quality caused by floods and droughts. At the same time, coordinated activities are being conducted in multi-purpose dams, hydro-power dams, weirs, agricultural reservoirs and water use facilities (featuring a daily water intake of over 100 000 $m^3 day^{-1}$) with the purpose of monitoring the management of such facilities. This is being done to ensure the protection of public interest without acting as an obstacle to sound water management practices. During Flood Season, each facilities contain flood control capacity by limited operating level which determined by the Regulation Council in advance. Dam flood discharge decisions are approved through the flood forecasting and management of Flood Control Office due to minimize flood damage for both upstream and downstream. The operational plan is implemented through the council's predetermination while dry season for adequate quantity and distribution of water.

## 1 Introduction

The construction of dams in Korea was motivated by the main purposes of either water supply or power generation. The construction of hydropower dams started in the 1960's. According to the 5-year economic development plan starting in 1962, the Seomjin River multi-purpose dam was the first to be built, and since 1972, multi-purpose dams such as the Soyang River and Andong dams have been built subsequently. Large-scale multi-purpose dams have been built continuously in order to meet the rapid increase of water demand, as well as to prevent flood damage, until the 1980's. In the 1990's, however, mid-scale multi-purpose dams like Buan, Miryang, and Hoengseong dams were built based on environmental considerations due to the objections from environmental organizations. In addition, there have been difficulties in the construction of multi-purpose dams, for instance, the nullification of the Youngweol dam construction project, following the start of the construction of Jangheung (Tamjin) multi-purpose dam in 1996. However, the construction of five new dams, including that of Buhang, is currently in progress.

Looking at the current status of the nation's dams Fig. 1, there are seventeen multi-purpose dams, fourteen water supply dams, eight hydropower dams, and two flood control dams (excepting of estuary barrages, agricultural reservoirs, pumped-storage hydroelectricity etc.). Dams are managed (operated) by different organizations depending on their original purpose such as water supply, flood control, or hydropower, Multi-purpose dams, domestic and industrial water dams are managed by the Ministry of Land, Infrastructure and Transport (K-water), hydropower dams by the Ministry of Knowledge Economy (Korea Hydro & Nuclear Power Co., Ltd.), and agricultural reservoirs by the Ministry

Published by Copernicus Publications on behalf of the International Association of Hydrological Sciences.

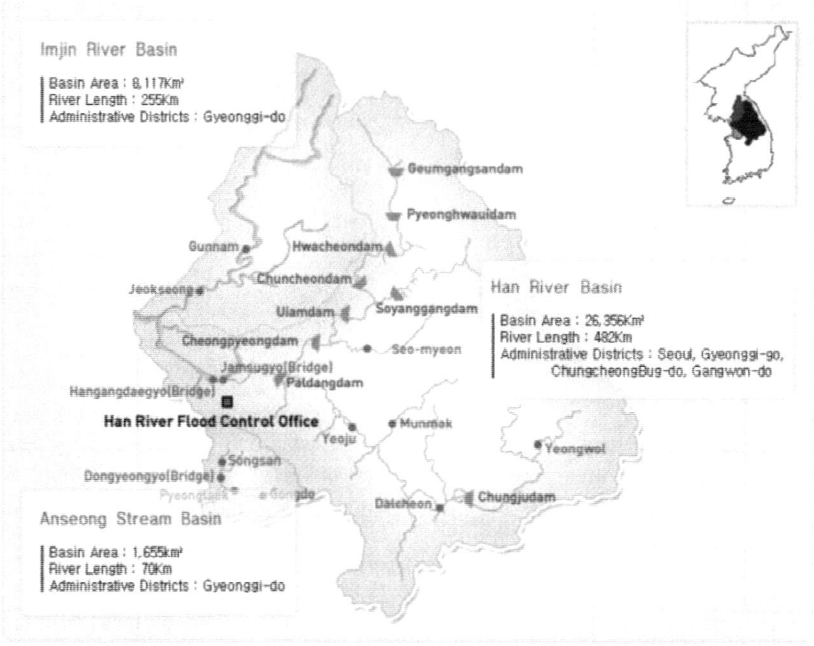


**Figure 1.** Dam locations in Han river (from the Han River Flood Control Office website: http://www.hrfco.go.kr/eng/service.do).

**Table 1.** Dams for conjunctive operation per watershed.

| Watershed | Multi-purpose dam | Hydropower dam | Water supply dam | Flood control dam | Weir |
|---|---|---|---|---|---|
| Amount | 17 | 8 | 14 | 2 | 16 |
| Han River (17) | Soyang River, Chungju, Hoengseong | Hwacheon, Chooncheon, Uiam, Cheongpyeong, Paldang, Gwesan, Doam | Gwangdong, Dalbang | Dam of Peace, GunNam | Ipo, Yeoju, Gangcheon |
| Nakdong River (24) | Andong, Imha, Hapcheon, Namkang Milyang, Guneui | | Yeongcheon, Angye, Gampo, Woonmoon, Daegok, Sayeon, DaeAm, Sunam, Yeoncho, Gucheon | | Changnyung Haman, Hapcheon, Changnyung, Dalsung, Gangjeong Goryeong, Chilgok, Gumi, Nakdan, Sangju |
| Geum River (5) | Daecheong, Yongdam | | | Sejong, Baekjae, Gongju | |
| Seomjin River (8) | Seomjin River, Jooam (main), Jooam (Flood Control Reservoir) | Boseong River | Soouh, Pyunglim | | Jooksan, Seungchon (Youngsan River) |
| Others (3) | Buan, Boryeong, Jangheung | | | | |

**Table 2.** The current status of dams in the Han River Watershed.

| Category | Name of the Dam | Watershed area ($km^2$) | Normal high water level (Elevation) (m) | Total storage capacity (million $m^3$) | Flood control (million $m^3$) | Water supply (million $m^3$ $yr^{-1}$) | Annual Energy Production (GWh $yr^{-1}$) |
|---|---|---|---|---|---|---|---|
| Total | | 64 198 | – | 8104.8 | 1984.5 | 4885.5 | 2687.5 |
| Multi-purpose dam | Subtotal | 9560 | – | 5736.9 | 1125.5 | 4712.5 | 1203.8 |
| | Soyang River | 2703 | 193.5 | 2900.0 | 500.0 | 1213.0 | 353.0 |
| | Chungju | 6648 | 141.0 | 2750.0 | 616.0 | 3380.0 | 844.1 |
| | Hoengseong | 209 | 180.0 | 86.9 | 9.5 | 119.5 | 6.7 |
| Hydropower dam | Subtotal | 50 883 | – | 1744.2 | 269.0 | – | 1507.0 |
| | Hwacheon | 3901 | 181.0 | 1018.0 | 213.0 | – | 326.0 |
| | Chooncheon | 4736 | 103.0 | 150.0 | 14.0 | – | 145.0 |
| | Uiam | 7709 | 71.5 | 80.0 | 16.0 | – | 161.0 |
| | Cheongpyeong | 9921 | 51.0 | 185.5 | 19.0 | – | 271.0 |
| | Paldang | 23 800 | 25.5 | 244.0 | 7.0 | – | 485.0 |
| | Gwesan | 671 | 135.6 | 15.3 | – | – | 11.0 |
| | DoAm | 145 | 707.0 | 51.4 | – | – | 108.0 |
| Flood control dam | Subtotal | 7418 | – | 2701.6 | 2701.6 | – | – |
| | Dam of Peace | 3227 | – | 2630.0 | 2630 | – | – |
| | GunNam | 4191 | – | 71.6 | 70.6 | – | – |
| Water supply dam | Subtotal | 154.4 | – | 18.7 | – | 40.6 | – |
| | Gwangdong | 125 | 672.0 | 11.0 | – | 26.0 | – |
| | Dalbang | 29.4 | 112.0 | 7.7 | – | 14.6 | – |
| Agricultural dam | Subtotal | 374 | – | 14.5 | – | 132.0 | – |
| | Gangneung | 109 | – | 12.7 | – | 26.0 | – |
| | Idong | 93 | – | 21.0 | – | 34.0 | – |
| | Gosam | 71 | – | 16.1 | – | 30.0 | – |
| | Geumkwang | 48 | – | 12.1 | – | 21.0 | – |
| | Giheung(Singal) | 53 | – | 11.7 | – | 21.0 | – |

for Food, Agriculture, Forestry and Fisheries (Korea rural community corporation). The conjunctive operation of each dam/weir in Korea is executed as per Table 1. Water supply dams are categorized into Taebaek region, Pohang region, Woonmoon, Ulsan region, Geoje region, Yeosu region, Pyunglim, etc.

## 2 Dams and weirs of the Han River watershed

The current status concerning the dams/weirs of the Han River watershed is shown in Table 2, for each dam and the watershed area, normal high water level, total storage capacity, dam type, dam height, etc.

There are 3 multi-purpose dams in the Han River Watershed (Soyang River, Chungju, and Hoengseong dams) with a total storage capacity of 5736.9 million $m^3$. There are 7 hydropower dams in the Han River Watershed, with a total power generation capacity of 494.8 MW $yr^{-1}$. Gwangdong and Dalbang dams are water supply dams, providing 41.0 million $m^3$ $yr^{-1}$. There are five agricultural reservoirs with over 10.0 million $m^3$ of effective storage capacity in the Han River Watershed, providing a water supply of 132.0 million $m^3$ $yr^{-1}$. The dam of Peace and GunNam flood control reservoir are flood control dams. The flood control capacity of the dam of Peace is 2630.0 million $m^3$. Agricultural reservoirs like Gosam, Idong, Giheung, and Geumkwang reservoirs are located in the Ansung-Cheon upstream watershed.

As shown in Table 3, 16 multi-purpose weirs are built nationwide. There are 3 weirs in operation in Han River Watershed. The Gangcheon, Yeoju, and Ipo weirs have been built and are currently in operation in the Han River watershed, thus allowing to secure water and to prevent the lowering of the groundwater level.

## 3 The conjunctive operation of dam and weirs

The conjunctive operation of river facilities is classified under the following frameworks: flood season, drought season, and river water quality. During the flood season, facility managers should comply with water level restrictions imposed by flood risk to their facilities. They must obtain prior approval from the flood control office (FCO) within the jurisdiction in case of the need of gate operation for flood control. If the FCO judges that issuing a notification for flood watch or a warning alert for a given river is required, or if flood-related damage is likely to happen, the FCO takes the necessary mea-

**Table 3.** Current status of weirs in the Han River Watershed.

| Category | Unit | Gangcheon Reservoir | Yeoju Reservoir | Epo Reservoir |
|---|---|---|---|---|
| Watershed area | $km^{-2}$ | 10 972 | 11 115 | 11 803 |
| Height | m | 8 | 8 | 6 |
| Design Flood | $m^3 s^{-1}$ | 16 070 | 16 070 | 16 070 |
| Maximum water level | elevation m | 38.5 | 33.5 | 28.5 |
| Management water level | elevation m | 38.0 | 33.0 | 28.0 |
| Minimum water level | elevation m | 35.2 | 31.7 | 25.3 |
| Total Storage capacity | million $m^3$ | 9.0 | 11.0 | 14.0 |
| Drought water level | elevation m | 33.7 | 31.5 | 25.0 |

sures considering the overall flood situation of a given basin, including weather and river conditions, as well as the storage capacity of the relevant dams and weirs. For example, the FCO directs the facility managers to use the storage volume of each facility down to the lowest operating water level, for flood control. During the drought season, facility managers operate according to the conjunctive plan of operation. Based on the site-specific Flowing Water Volume for River Maintenance (River Act Article 51), which can represent the state of these sites in terms of river discharge, the FCO instructs facility managers to take the necessary measures to prevent a potential situation of drought. In this sense, the FCO considers all factors contributing to the water balance of the entire basin such as climate and river conditions, storage capacities of dams and weirs, and the levels of usage of agricultural reservoirs. In addition, storage is also used to secure clean water through the improvement of the quality of river water.

In order to establish good cooperation between the relevant organizations related to the utilization of water resources, and to set the basic criteria necessary for their conjunctive operation, the Regulation Council for the conjunctive operation of dams and weirs, comprising the central council and a local council for each watershed (Han River, Nakdong River, Keum River, Yeongsan River) discuss and establish agreements on a regular basis. Central council is held in February every year, each watershed's council is held every December. The council consists of government organization, local governments, facility managers and specialists from private sector related with water management. The council works out conjunctive operation plan, decides matters concerning its change and assessment and other matters needed for effective conjunctive operation and mediates opinions of stakeholders related with conjunctive operation. For example, Han River Regulation Council are consisted of three representatives from Ministry of Land, Infrastructure and Transport, one representative from Ministry of Environment, five representatives from Local Governments, three representatives from each Dam operation Company and six specialists (Fig. 2).

The Minister of Land, Infrastructure and Transport can limit the usage by the users or control the permitted levels of water, in case the streamflow is under any of the conditions below, affecting the downstream management:

1. Difficulty to ensure in-stream flow according to the standard limit;

2. Possibility of negatively affecting the public benefits if the permitted water levels are not controlled due a prolonged drought;

3. If water users cannot use the water within the expiry date or if they have used less than the quota permitted by the Decree of the Ministry of Land, Infrastructure and Transport;

4. Difficulties using other allowed streamflow.

The adjustment of the permitted water, must be deliberated by the National River Management Commission. Before adjusting the permitted water, a Regulation Council for streamflow can be formed and act according to the residents' opinions, according to the presidential decree. The Regulation Council is under the relevant Flood Control Office, at the Ministry of Land, Infrastructure and Transport, and it is formed with less than 20 members including one chairman.

Once the annual cooperative plan is formulated, the following information is established either quarterly and monthly: matters concerning water levels and the plan of supply at certain facilities, the current and planned water usage at water intake facilities, as well as matters concerning water surface conditions and its management plan at dams and weirs. The data needed to establish cooperative plans are collected and utilized jointly by related organizations. The cooperative plans can be modified according to the resolutions emitted by the council for conjunctive operations in case of any changes in climate, rivers, electricity or intake conditions, or if there is a request of modification from the related organization. The FCO within the jurisdiction evaluates performance of conjunctive operation by season. Related organizations can request emergency drainage in case of emergency such as climate change, change of river discharge and water quality, and safety management.

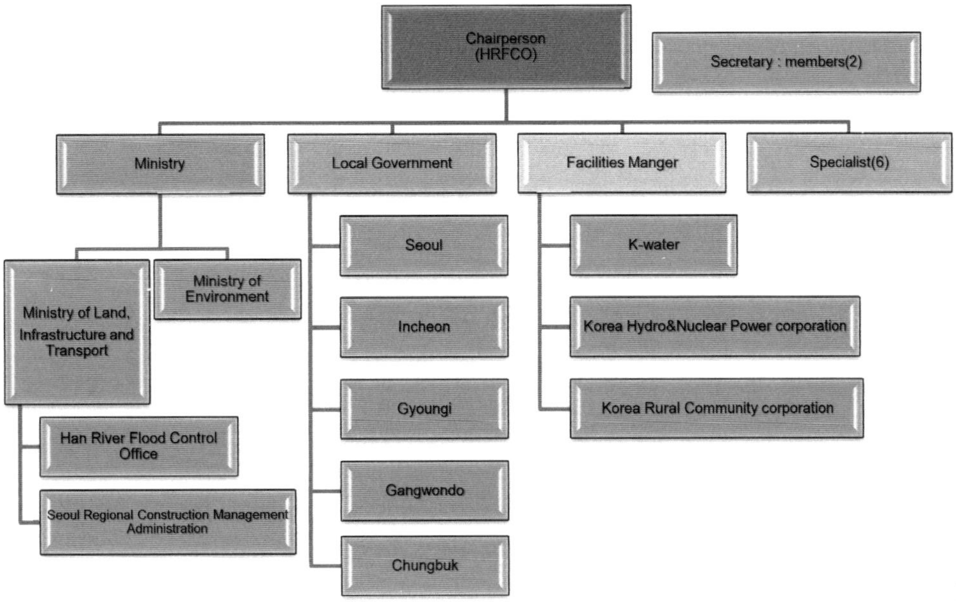


**Figure 2.** Regulation Council for the conjunctive operation of dams/weirs Council(18): MOLIT(3), MOE(1), Local Government(5), Dam operation(3), Specialist(6).

## 4 Operation of flood handling dam/weir and the case of 2006

The water level of dams is limited during the flood season in order to ensure flood control. A restricted water level is maintained for 3 months from 21 June to 20 September. The total flood control capacity of the Han River watershed is 1461.1 million $m^3$. According to multi-purpose dam control regulation, article 14, when the dam level needs to be lowered below the flood season restricted water level, and attending the flood situation up and downstream, it is possible to set a discharge whose amount and time are to be approved by the General Director of Flood Control Office, Ministry of Land, Infrastructure and Transport. The Design Flood Level defines a dam's maximum level that can be observed during flood control. Normal high water level is the maximum water level that can be observed in the segment that is utilized for water supply. Restricted Water Level for the flood season is the level that has been set in order to ensure flood control during the flood season, and it is the maximum level to maintain when there is no flood inflow. Regarding the dams for which there is no set Restricted Water Level, a normal high water level sets the reference. Minimum Storage Level is the lowest level at which the dam's water can be used. The flood control capacity is the storage capacity between the dam's Design Flood Level and Restricted Water Level for the flood season.

The procedure for discharge at multi-purpose and hydropower dams during the flood season is as follows: the dam manager requests approval for according to the discharge plan and the Flood Control Office, Ministry of Land, Infrastructure and Transport analyse the state of all related streams and decides to approve, or not, accordingly. Each dam manager is requested to revise the expected inflow and discharge plans, depending on the effects of downstream rainfall. The assessment of the downstream water level and dam's hydrologic characteristics such as dam's upstream rainfall, inflow, discharge, minimum storage level, and flood control capacity must be established. In case of controlling a movable weir to maintain the management level as per usual, this can be done by the facility manager (K-water) without any further approval or reporting procedure required, however, the facility manager is required to broadcast a security alert, when necessary. During the flood season, approval is needed in the

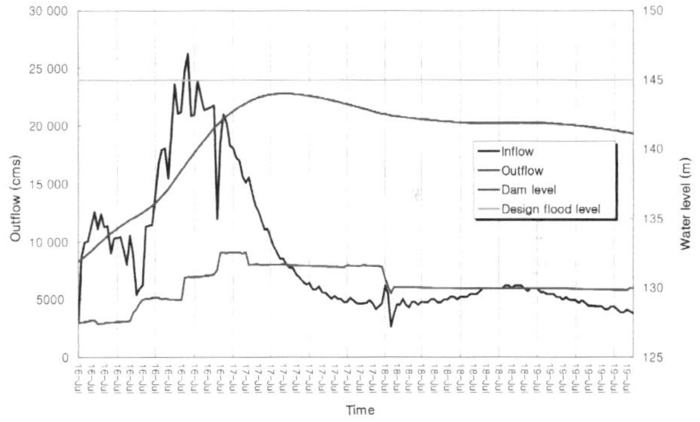


**Figure 3.** Flood control by Chungju dam in 2006 case.

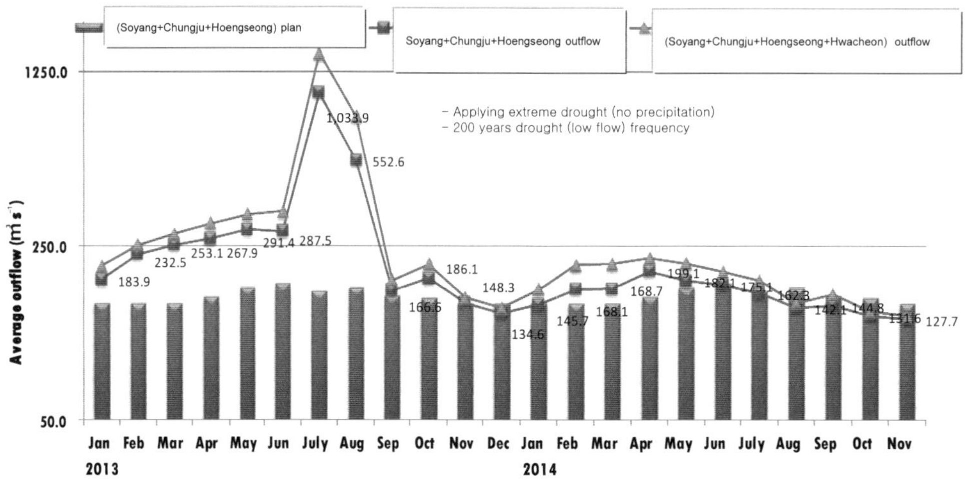


**Figure 4.** Providing real water demands by considering drought status and conjunctive operation.

beginning and the end stages of flood control, as well as during the maximum discharge (complete closure) at the movable weir's control floodgate. As aforementioned, the water level in the dam is controlled during the flood season in order to ensure flood control capacity.

Recently, a pre-release has not been executed because the water level of Soyang River, Chungju, Hwacheon, and Hoengseong dams has been maintained below the restricted water level. Thus, we would like to present the case of 2006, which is remembered as the most urgent and difficult dam operation performed during the flood season. The flood of July started with Ewiniar typhoon proceeding northward towards the Korean peninsula. Ewiniar dissipated when it reached the land, but the vapors within developed a seasonal rain front causing considerable rain to fall in the Han river watershed on 12 July. For this reason, the entrance to Jamsugyo Bridge, over the Han River was prohibited. The release began at the Chungju dam, on the Namhan River to ensure the security of the dam and the downstream area (Hwirin Kim et al., 2015). Due to continuous a rain front and localized heavy rain, at 02:15 LT (GMT +9) on 16 July, K-water requested the discharge of Chungju dam to be increased from 08:00 LT from 3000 to 5000 $m^3 s^{-1}$. Five hours later, concerned with the integrity of the Chungju dam, a request was made to increase the discharge to 10 000 $m^3 s^{-1}$. However, catastrophic effects in Yeoju and Seoul (Han River) areas were predicted in the case of the Chungju dam discharging at 10 000 $m^3 s^{-1}$ when it was below the restricted water level. Considering the status downstream, the release had to be delayed. At 13:00 LT, K-water received an approval request for a discharge at the Chungju dam. The decision was intensely discussed between upstream and downstream regions' counterparts. The dam upstream area requested increasing its discharge amount to avoid increasing to water level at the dam. An inundation of 500 households was anticipated, with 1500 people in the need to evacuate urgently. In contrast, in the downstream area of the dam, 51 households had already been inundated, and 163 people had been evacuated. However, 20 000 people were to be evacuated if the river level increased. For this reason, they requested decreasing the discharge amount. With the dam upstream area requesting an increased discharge, and the dam downstream area requesting a decreased discharge, Han River Flood Control Office, from the Ministry of Land, Infrastructure and Transport started revising the best possible discharge amount attending to the dam's integrity. The safety at the Chungju dam was top priority. Furthermore, they assessed the constantly changing state of flood in the dam upstream and downstream areas in order to avoid later inundations. Flood forecasting based on various scenarios was established and, accordingly, they reported to K-water on the best amount and period of discharge, following on discussions with decision makers (at 15:00 LT). Fortunately, the expected result was very accurate, and, at 144 m the maximum water level of Chungju dam did not lead to an emergency situation (08:00 LT), 1 m only from the Design Flood Level (145 m, see Fig. 3, Hwirin Kim et al., 2006).

The Jamsugyo Bridge was inundated during six days and emerged at around 04:10 LT on 21 July. The highest flood since the 1990's flood was successfully handled. If there was more rainfall, if the flood forecasting was not accurate or the dam operation improper, 2006's flood may have had catastrophic consequences such as those of 1990.

## 5 Preventing water shortage with dam/weir management and the case of 2015

Due to the severe drought of 2015, discussions concerning the conjunctive operation of the Han River Watershed emergency case took place between the Regulation Council for streamflow of the Han River watershed and the relevant offices took place. The first meeting of relevant offices was attended by the Water Resources Development Division, Ministry of Land, Infrastructure and Transport, the Han River Flood Control Office, Ministry of Land, Infrastructure and Transport, the Ministry of Environment, the Ministry of Agriculture and Forestry, the Chief of forecasting control division, the Han River Flood Control Office, the Ministry of Land, Infrastructure and Transport, K-water, Korea Rural Community Corporation, and Korea Hydro & Nuclear Power Co., Ltd in 4 March 2015. In the meeting, they agreed to execute conjunctive operations at the hydropower dams in order to lengthen water supply for a longer period. Concerns were expressed over the decrease of agricultural water from Soyang River and Chungju dams due to rainfall shortage. Keeping the minimum discharge ($124\,m^3\,s^{-1}$) of Paldang dam and considering Hwacheon dam's electricity release, they decided to decrease the discharge amount of Soyang River and Chungju dams, and established a cooperative structure to share data between Korea Hydro & Nuclear Power Co., Ltd. and K-water for conjunctive operation (Fig. 4).

Secondly, the Regulation Council for streamflow of the Han River Watershed took place. It approached regulations and handling the water intake according drought stage. Intake limitations were approved over a range of levels, and natural intake control through the promotion of water savings is being handled by K-water per agreed. Local government showed interest and requested cooperation for joint countermeasures. The majority voted in favor of intake regulations; and the conditional approval regarding the flexible intake limitation was to collect opinions from the Regulation Council. Intake limitation depending on the situation profile was determined and adopted.

## 6 Conclusions

Best efforts have been made in guaranteeing the facility manager's rights when it does not concern water management, not disturbing the public benefits, and giving all citizens an even usage benefit of water resources by decreasing damages caused by flood, water shortage, and the worsening of water quality, as well as promoting the downstream maintenance of functional usage through downstream facilities' conjunctive operation such as heightened agricultural reservoirs, dams, and weirs. Recently, the Regulation Council for Streamflow and Regulation Council for Conjunctive Operation for Dams and Weirs held meetings to assess various drought alleviation measures including adjusting dam discharge level, building response guildelines and action plans, and imposing a quota on water intake. Emergency duty is also in operation to stand ready for different emergency scenarios.

The current dam storage rate in Korea is below average; therefore, the country needs to handle drought and flood at the same time. Conjunctive operation of dams and weirs is urgently needed now more than ever. In order to protect lives and property, the government needs to be alert at all times and must handle the situation wisely with the newest forecasting systems and based on the know-how of several years.

## 7 Data availability

The data that support the findings of this study are available from the corresponding author upon reasonable request.

## References

Hwirin, K. and Jaehwa, N.: The July 2007 Flood Characteristics and Future Counterplan, Korean Society of Civil Engineering, 54, 128–136, 2006.

Hwirin, K., Junho, C., and Hyoseob, J.: Dam and Weir Operation for the preparing Flood, Korea Water Resources Association, 48, 8–23, 2015.

Proc. IAHS, 374, 101–104, 2016
proc-iahs.net/374/101/2016/
doi:10.5194/piahs-374-101-2016


Open Access
PIAHS

# Development and implementation of a monitoring and information system to increase water use efficiency in arid and semi-arid areas in Limarí, Central Chile (WEIN)

**Erich Berger, David Balmert, and Jürgen Richter**

ribeka GmbH, 53332 Bornheim, Germany

*Correspondence to:* Erich Berger (berger@ribeka.com)

Published: 17 October 2016

**Abstract.** The project WEIN was funded by the Federal Ministry of Education and Research (BMBF I Berlin, Germany) in the framework of the high-tech strategy of Germany's program "KMU-Innovativ". The project started in 2012 and was completed in 2014. In the scope of the project, an integrated system for analysis, monitoring and information at river basin level was developed, which provides relevant information for all stakeholders that are concerned with water resource issues. The main objective of the project was to improve water use efficiency and hence ensure the agricultural production in the region. The pilot region, in which this system was implemented, is the semi-arid Limarí basin in Northern Central Chile. One of the main parts of the project was the development and implementation of a web- and app-based irrigation water ordering and accounting system for local farmers.

## 1 Introduction

The growing world population, the rapidly increasing demand for high quality food and for water for agriculture use, the need to secure the provision of drinking water demands innovative and holistic solutions and products to increase the water use efficiency. These solutions are especially required in arid and semi-arid areas. Additional factors such as economic development, population growth and climate change deteriorate the situation. Therefore the long term, sustainable resource management at river basin level requires a complete, standardized and for all staff members operable management system. In the framework of the project WEIN a complex management system for water resources in the semi-arid region Limarí, Central Chile, was developed and implemented.

The aim of the project was to optimize usage of the scarce water resources subsequently to ensure the agricultural production.

The project was carried out in collaboration with SEBA Hydrometrie GmbH, ITT (TH Köln) and the local Chilean partners ULS (University La Serena, Chile) and CEAZA (Centro de Estudios Avanzados en Zonas Aridas, La Serena, Chile).

## 2 The project region: Limarí basin, Central Chile

The Limarí basin is located in the northern semi-arid region of Coquimbo (IV.) (Fig. 1). The Limarí catchment area covers about 1.3 million ha, where of 65 000 ha are used for agricultural production (INE, 2007). Despite the little amount of annual precipitation (average 146 mm p.a.) there is a lot of agricultural land use (Oyarzun, 2010). Because of this little average rainfall, the agricultural production is strongly dependent on the hydro-geological processes and the snow melt in the Andes. Therefore, this region is prone to the impact of the global warming, since the increased glacial melting and reduced snowfall will reduce the discharge from the mountains in the summer (Barnett et al., 2005). This could lead to an increasing number of droughts (Young et al., 2010). To catch up with all these factors, an efficient water management is indispensable. Figure 1 shows the Limarí river basin including main rivers, the reservoirs and the irrigation area alongside the Limarí.

Published by Copernicus Publications on behalf of the International Association of Hydrological Sciences.

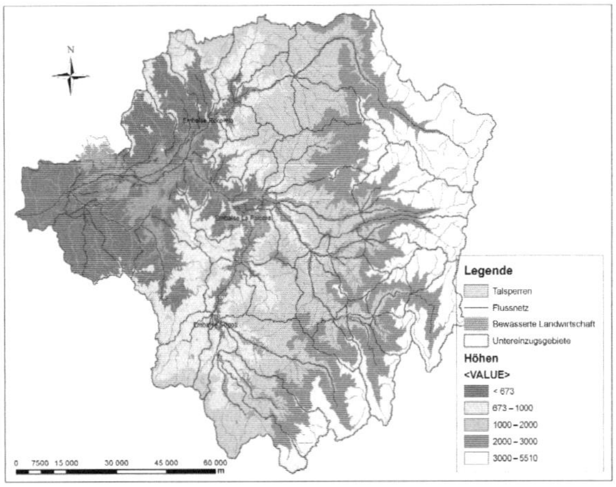


**Figure 1.** Limarí river basin (created by Kretschmer, 2011).

## 3 The monitoring and information system

The information and management system developed during the WEIN project consists on the one hand of the installed optimized SEBA sensor systems which measure relevant climate and water run off data, and on the other hand of the modular software system based on the ribeka water resource management system GW-Base.

All water used for irrigation purposes is distributed and managed in a complex reservoir and channel system. The water either derives directly from the local rivers (e.g. the Limarí) or from different reservoirs (e.g. Embalse la Paloma, Embalse Recoleta) which are dependent on the annual runoff due to the snowmelt. The Water distribution is managed by different water associations, each responsible for a channel system deriving from a reservoir or a certain river section. All local Farmers possess water rights which allow them to order corresponding water amounts from the responsible association. The farmers are entitled to certain water annual irrigation water quotas which are recalculated every year depending of the annual water availability. Hence, in addition to the water monitoring system a management system for water resources and the farmer's water rights (and corresponding irrigation water quotas) was developed and implemented. Orders for irrigation quotas are sent via smartphones by the farmers are automatically processed and carried out (Fig. 3). All relevant data (measurements, monitoring data, irrigation quotes) are automatically deployed and processed in the modular software system based on resource management system GW-BASE. Individual software modules were customized for these special usage scenarios and applications (e.g. water authorities, farmers, irrigation associations). For the first time, a monitoring and management system is available, which provides the possibility to use all relevant data for an integrated water management. The communication scheme and interaction of all results in the newly developed WEIN system are shown in Figs. 2 and 4.

### 3.1 The monitoring system

To account the available resources holistically, an extensive, continuous collection of the relevant data is needed. The innovative and customized sensor technology ensures this. The surface water and irrigation channels are monitored by newly developed, robust and cost-effective water flow and water level sensors. Rainfall and the lack of infrastructure – especially in the mountain region of the Andes – require a special adaptation of the installed sensors. Beside the problem of data transmission, rainfall and snowfall sensors have to work without an external power supply. The development of an energy-efficient and maintenance-free sensor combination which stands the rough climate conditions was needed.

All relevant pre-existing data concerning the Limarí basin was researched and imported to GW-Base and GW-Web. This data formed the basis for the whole monitoring system and is available for all project partners.

The water availability is quantified on the basis of climate and water runoff data. On irrigation system level, the channel outflow and reservoirs levels are measured continuously, in order to estimate water consumption and water losses due to leakage and evaporation. The estimation of future water availability and demand is carried out by an integrated review of the measurement results, historical hydro-meteorological and socio-economic data in combination with customized hydrological models.

The monitored data includes environmental parameters like precipitation and evaporation, water levels and quantities like discharge and runoff, as well as water quality parameters. All monitored data was imported to the GW-Base database.

### 3.2 The information and management system

To account the available resources holistically, an extensive, continuous collection of the relevant data is needed. The innovative and customized sensor technology ensures this. The surface water and irrigation channels are monitored by newly developed, robust and cost-effective water flow and water level sensors. Rainfall and the lack of infrastructure – especially in the mountain region of the Andes – require a special adaptation of the installed sensors. Beside the problem of data transmission, rainfall and snowfall sensors have to work without an external power supply. The development of an energy-efficient and maintenance-free sensor combination which stands the rough climate conditions was needed.

All relevant pre-existing data concerning the Limarí basin was researched and imported to GW-Base and GW-Web. This data formed the basis for the whole monitoring system and is available for all project partners.

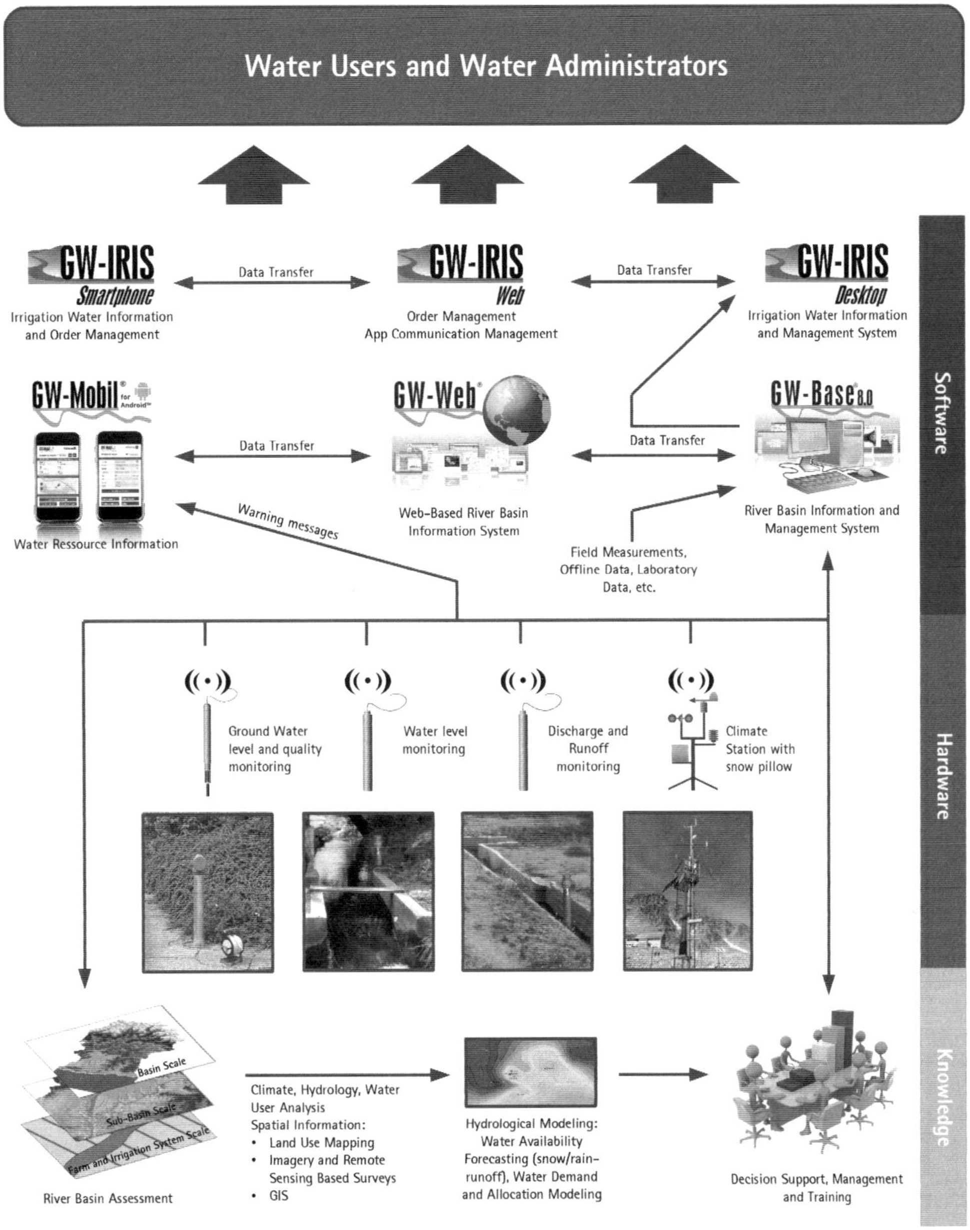


**Figure 2.** Communication scheme of the complete, newly developed WEIN system.

The water availability is quantified on the basis of climate and water runoff data. On irrigation system level, the channel outflow and reservoirs levels are measured continuously, in order to estimate water consumption and water losses due to leakage and evaporation. The estimation of future water availability and demand is carried out by an integrated review of the measurement results, historical hydro-meteorological and socio-economic data in combination with customized hydrological models.

The monitored data include environmental parameters like precipitation and evaporation, water levels and quantities like discharge and runoff, as well as water quality parameters. All monitored data was imported to the GW-Base database.

## 4 Conclusions

An integrated system for analysis, monitoring, information and resource management at river basin level, which allows efficient and demand-oriented irrigation water management,

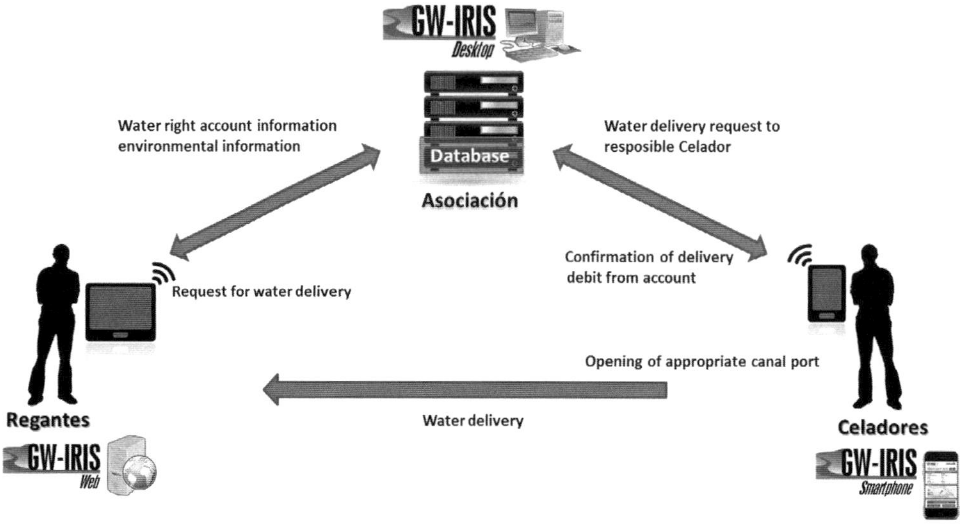


Figure 3. GW-IRIS communication scheme.

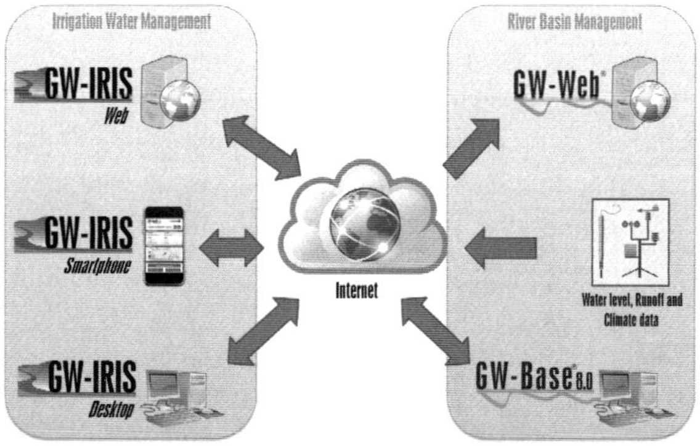


Figure 4. The complete software system. Communication scheme of the water monitoring and management system GW-Base and the newly developed irrigation administration system GW-IRIS.

is an essential tool for all concerned stakeholders. The system is a substantial contribution to increase water use efficiency in the Limarí region and therefore is able to ensure the agricultural production.

For the first time, a monitoring and management system is available, which provides the possibility to use all relevant data for an integrated water management. The worldwide demand for water will increase rapidly because of the growing world population and climatic changes. This collaborative product development ideally fits to this increasing demand.

## 5 Data availability

Additional research information can be found at:

- BMBF (Bundesministerium für Bildung und Forschung): www.bmbf.de, www.foerderportal.bund.de
- ribeka GmbH: www.ribeka.com
- Official Project Website: www.hidro-limari.info/

**Acknowledgements.** The WEIN project was funded by the Federal Ministry of Education and Research (BMBF, Berlin, Germany) in the framework of the high-tech strategy of Germany's program "KMU-Innovativ" (funding/project number: 02WQ1256A).

## References

Barnett, T. P., Adam, J. C., and Lettenmaier, D. P.: Potential impacts of a warming climate on water availability in snow-dominated regions, Nature, 438, 303–309, 2005.

INE: Censo Agropecuario: http://www.inecoquimbo.cl (last access: 18 April 2011), 2007.

Oyarzun, R.: Estudio de caso: Cuenca del Limarí, Región de Coquimbo, Chile, Compilación Resumida de Antecedentes, Centro de Estudios Avanzados en Zonas Aridas, La Serena, Universidad de la Serena, 2010.

Young, G., Zavala H., Wandel, J., Smit B., Salas, S., Jimenez, E., Fiebig, E., Espinoza, R., Diaz, H., and Cepeda, J.: Vulnerability and adaptation in a dryland community of the Elqui Valley, Chile, Climatic Change, 98, 245–276, 2010.

Proc. IAHS, 374, 105–112, 2016
proc-iahs.net/374/105/2016/
doi:10.5194/piahs-374-105-2016


Open Access
PIAHS

# Construction and evaluation of a Toolbox for the formulation of the Hydrologic component of the Basin Management Plans in Colombia

**Victor H. Garzón[1], Ricardo Garzón[1], Pedro M. Avellaneda[1], Erasmo A. Rodríguez[1], and Leonardo Alfonso[2]**

[1]Department of Civil and Agricultural Engineering, Universidad Nacional de Colombia, Bogotá, Colombia
[2]Hydroinformatics Chair, UNESCO – IHE, Delft, the Netherlands

*Correspondence to:* Victor H. Garzón (vhgarzonr@unal.edu.co)

Published: 17 October 2016

**Abstract.** The Colombian Ministry of Environment, Housing and Territorial Development issued the Policy for Integrated Water Resources Management in 2010, with the formulation of Basin Management Plans (POMCA) as a key component. The main purpose of these plans is to support land use planning and sustainable management of renewable natural resources. Since the announcement of the policy; however, advances in the formulation of POMCAs have been limited, due to the need of generating knowledge and the necessary information to formulate such plans. To contribute to the generation of such knowledge, we developed HidroCHEP: a Toolbox to support the formulation and hydrologic characterization of Colombian basins. In this paper, we report the design, architecture, implementation and use of the toolbox, to understand the climatic variability of the country and to improve predictions in ungauged basins. The use of the toolbox is illustrated and evaluated through its application in three pilot river basins (Pamplonita, Gualí and Ceibas) each with different level of available hydrologic and climatic data. This paper focuses on the use of the toolbox for the development of one of the products required by the technical guide for the formulation of POMCA, in the Pamplonita river basin. Main advantages and disadvantages of its application are evaluated. It is demonstrated that the toolbox has the potential to support the formulation of POMCAs in the country and to contribute to integrated national water resources management.

## 1 Introduction

Four of 214 large basins in the world (with more than 100 000 km$^2$) are located in Colombia, namely the Magdalena, Guaviare, Casanare and Meta rivers. The country has also three basins between 50 000 and 100 000 km$^2$ in the Cauca, Inírida and Putumayo rivers, and over 700 000 watersheds with areas smaller than 10 km$^2$. The country's water network is complemented by the richness of groundwater and surface water storage, among lakes, ponds, reservoirs, marshes and swamps which reach 17 000 water bodies (IDEAM, 2002). This water wealth represents for Colombia an asset that requires a good characterization for its proper management.

According to the National Water Study, ENA (IDEAM, 2010), over 80 % of the municipalities are supplied by small water sources (e.g. streams or rivers with low regulation conditions and high vulnerability regarding droughts and water quality). These sources do not guarantee yearly water availability, so it is essential to understand the status and dynamics of these systems in order to organize and to make a proper sustainable resource management. According to projections made by the ENA, about the situation of scarcity and vulnerability for the period 2015–2025, Colombia would face serious difficulties to supply water to the population and other users in some regions "if we do not take important measures of conservation, planning and management of natural resources as well as pollution abatement" (IDEAM, 2010, 38 pp.).

In March 2010, the Ministry of Environment, Housing and Territorial Development issued the Policy for Integrated Water Resources Management (PNGIRH), which presents recommendations for planning and management of water resources, including objectives, strategies, and indicators of implementation, monitoring and evaluation. A key component of the PNGIRH is the formulation of Basin Management Plans (POMCAs). The purpose of a POMCA is to evaluate and plan the management of natural resources at the basin level, maintaining a balance between socio economic use and the conservation of the physical and biotic components of the basin.

In this context, the environmental authorities are responsible to declare the planning state of a basin under their jurisdiction and to proceed with the formulation of POMCAs with partial or complete external technical support (National Decree 1729/2002, Article 19). According to the Colombian Institute of Hydrology, Meteorology and Environmental Studies (IDEAM), there are 39 basins and 343 sub-basins subject to management and eligible for POMCA formulation (IDEAM, 2004).

Colombia has the urgent need to generate knowledge and information to support the decision making, planning, management and sustainable use of water resources, which is also essential for water quality and hydro-ecological management (Sauquet et al., 2008), the productivity of stream and floodplain ecosystems (Harris et al., 2000). It is well known that numerous ecological processes in rivers are sensitive to their variations (e. g. Biggs and Close, 1989; Richter et al., 1996; Poff et al., 1997; Cattanéo, 2005; Beechie et al., 2006; Monk et al., 2006, 2007, 2008; Olden et al., 2006).

Data are significant to understand the complexities and the inter – relations of multiple physical and social aspects occurring in a river basin (Blöschl et al., 2013). Restrictions on the use and accessibility of basic information for hydrological purposes, added to the limited spatial and temporal coverage, are two challenges when modeling water resources in Colombian basins. In addition, the country needs more skilled professionals able to: apply and validate computational tools developed around the globe, understand the role of mathematical modeling to study physical processes, and evaluate different scenarios. The large climatic variability, in both temporal and spatial scales, has been a relevant challenge for the earth sciences in recent decades (Poveda, 2004), and this variability confuses the ability to understand the behaviour of water resources in the country.

It is evident the need to develop and consolidate technical supporting tools for professionals in the country, who are responsible for carrying out basin management. Proper understanding of these techniques, along with suitable computational tool, will allow manager to analyze different scenarios and support their decision-making process. To this end, the development of information systems that supports dynamically the decision making (IDEAM, 2010) is needed. This is the motivation for Universidad Nacional de Colombia, to construct a toolbox called HidroCHEP, which aims to covering different aspects from hydrological sciences, in order to support the diagnostic phase of the water component in any POMCA.

In this paper, we describe the design, architecture, implementation and use of HidroCHEP. Section 2 describes the design and construction of HidroCHEP, along with its use interface through a website. Section 3 presents the case studies implemented with emphasis on the Pamplonita pilot case study. Finally, in Sect. 4, a discussion and conclusions related to the use of HidroCHEP are presented.

## 2 HidroCHEP: a toolbox for basin management plans

The hydrological toolbox for the formulation of a POMCA, HidroCHEP, is an aggregate of computer applications, case studies and methodologies for formulating the hydrological component of POMCAs. As part of PNGIRH, the Ministry of Environment and Sustainable Development (MinAmbiente) published a technical guide for the formulation of POMCAs (MinAmbiente, 2014), which includes the description of six required stages: readiness, diagnosis, prospective and environmental zoning, formulation, implementation and monitoring, and evaluation. The characterization of the watershed is part of the diagnosis stage and has different topics and activities with defined products. The toolbox has a wide range of computer applications (most of them from free software), aimed to support the development of the products required by each topic presented in the technical guide.

HidroCHEP also includes a compilation of data sources for catchment characterization. It integrates tools and data sources with each POMCA required product, in order to complete the activity, facilitating the data collection, tools use and the development of the products.

The general architecture of HidroCHEP is indicated in Fig. 1, and it is based on adaptation and complementation of the architecture of similar existing hydrological toolboxes (e.g. eWater, 2012; CSDMS, 2008; Global Water Partnership, 2013). The toolbox is web-based, and three main sections are recognized, namely POMCA's guidelines, tools and case studies, described below.

### 2.1 Guidelines for the construction of POMCA's topic-based products

Each topic and activity related with the hydrological component of a POMCA has several required activities and associated products to each activity in order to understand the basin's water resources. The POMCA's topic section of the toolbox provides some guidelines to support the user in the construction of each product. The guide includes the description of the topic, its use for the construction of other products, the tools and necessary information, and a short explanation of the procedure to achieve the required products.

**Figure 1.** Main components of HidroCHEP's architecture.

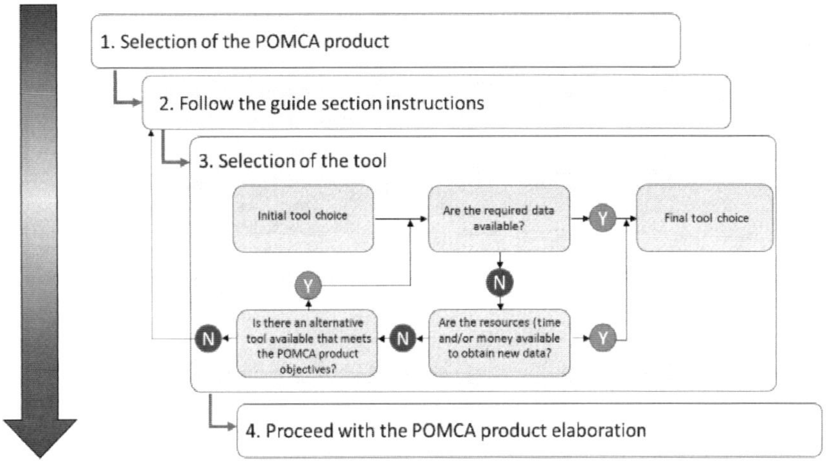


**Figure 2.** Flowchart of the use of HidroCHEP for the elaboration of a POMCA product.

Topics include hydrography, climate, hydrology, morphometry, and slopes. Each of these topics is divided into activities and each activity is briefly described. For instance, for the case of hydrography, two activities are recognized, namely (1) identification, description and river network drafting, and (2) characterization of drainage systems and patterns. Each activity is described following a structure that includes a description (where objectives, uses and products are mentioned), data needs, tools, and a step–based suggested procedure necessary to complete the activity.

## 2.2 Hydrological tools repository

One of the objectives of HidroCHEP is to collect different tools, essentially computer applications, which supports the construction of the products required by the POMCA. These are grouped into nine main categories: (1) data processing, tools to process and analyze the information typically used in hydrological applications according to the availability of the data in Colombia; (2) climate characterization, group of tools that allow the characterization of the meteorological variables required by the technical guide of the Ministry; (3) estimation of indexes, set of tools for the calculation of indices typically employed to characterize the hydrological behavior of a watershed; (4) water budget models, tools that allow to estimate water budgets in a watershed with different levels of aggregation and amount of information; (5) mathematical modeling: tools that supports the implementation of mathematical modeling for the simulation of the hydrological components of the watershed; (6) model's calibration and validation, tools used for calibration and validation of models within the framework of mathematical modeling in hydrology; (7) hydrography: set of tools that allow the characterization of the hydrographic network of a watershed; (8) water resources management, tools that include components of supply, demand, and water resources management, in order to evaluate different scenarios and hydrological responses; and (9) estimation of environmental flows, tools that allow to estimate environmental flows using different methods.

In addition, there is a category dedicated to the compilation of sources of information for obtaining the necessary meteorological and catchment characterization data, including that of global databases, as those produced by the national hydrometeorological network and national services.

## 2.3 Case studies

This section of HidroCHEP presents the case studies, showing the used tools for the development of the products re-

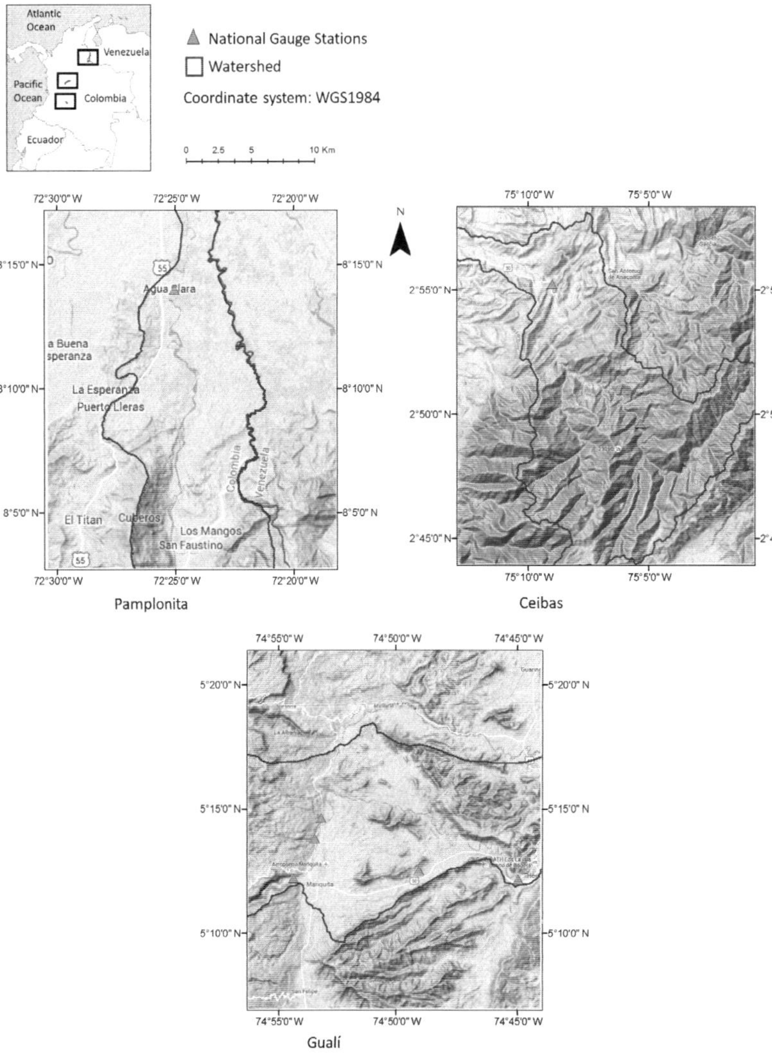


**Figure 3.** Location of the Pamplonita, Ceibas and Gualí catchments in Colombia.

quired by the technical guide, along with the results of the application.

Each case study follows a methodological framework as illustrated in Fig. 2. The user starts to consult the guide for the development of a specific product required by the technical guide, where several available tools are displayed. The user can make an initial selection of a tool, which is recommended for a minimal data information available; otherwise a different tool may be selected when there are more information.

## 3 Application of HidroCHEP for selected case studies

The actual version of HidroCHEP has three case studies of Colombian basins: Pamplonita, Gualí and Ceibas river basins. The locations of theses basins are shown in Fig. 3. Each case has different levels of available monitoring and hydrometeorological information, thus illustrating the toolbox capabilities. This section presents, for the Pamplonita river basin, the use of the toolbox for the development of one of the products required by the technical guide.

The Pamplonita river basin is a transboundary basin between Colombia and Venezuela, characterized by low hydroclimatological instrumentation and the need to quantify their surface water resources. It is the main source of water supply for the city of Cúcuta (sixth most populous city in Colombia) and for rice crop irrigation.

**Figure 4.** HidroCHEP interface showing a guide for a POMCA selected product.

## 3.1 Selection of the POMCA's topic

From the POMCA's topics included in HidroCHEP, it is possible to choose one out of five themes related to hydrological characterization: hydrography, climate, hydrology, morphometry and slope characterization. Each option has specific products. In the following section, the process to achieve product 06 "Making water budget at basin and sub-basin level", from the Hydrology topic (Fig. 4), is presented.

According to the guide shown in HidroCHEP, hydroclimatologic information of the study area is required. Hydrometeorological records are available for this basin (IDEAM, 2015).

## 3.2 Selection of the tool

One of the tools available to make a water budget in HydroCHEP is the software SAGA GIS (Conrad et al., 2015), a GIS with different interpolation options, and an administration and raster calculator. These features are adequate to meet the goals at this phase of the POMCA development (Fig. 5).

This software requires raster images of each variable considered for the water budget: a balance between rainfall and evapotranspiration.

## 3.3 Making the POMCA's product

As evapotranspiration data in the basin are not available, Turc's equation (Turc, 1961) is used to make an annual estimation, and Hargreaves's equation (Hargreaves and Samani, 1985) is used to make a monthly estimation. This is done using the raster calculator option of SAGA. The results are shown in Fig. 6.

Using the available information of precipitation and temperature, the monthly water budget from 1995 to 2014 is estimated. A raster calculator is programmed for several groups of images for each month, with the results shown in Fig. 6. Inter and intra annual variability of three of the main hydrological variables can be observed, which could help understanding the behavior of the Pamplonita river basin.

**Figure 5.** HidroCHEP interface showing the support material for the use of a tool.

## 4 Discussion

There are many tools and models globally available, which have been designed to achieve different hydrological purposes. Each model has different data requirements, and its implementation on real cases may not be straightforward for inexperienced novice user. Having large data sets does not necessarily mean that proper hydrological information is available (Gupta et al., 2008). As reported by Ackoff (1989) and Bellinger et al. (2004), it is necessary to understand that data conducts to information and do not represent the same. The latter embodies the understanding of a relationship of some sort, possibly cause and effect. In the same sense, information produces knowledge and, this one, in turn, understanding of the hydrological systems.

It is clear that, besides the need to have data available to understand the catchment behavior, techniques to exploit information from individual catchment studies are needed, which is required for the planning and management of water resources that is aimed by the PNGIRH in Colombia. In this context, HidroCHEP provides the linkage between data sources and tools, making its use efficient and helpful to understand the behavior of a basin.

Hence, one of the main contributions of HidroCHEP is to provide an easy access to different data sources including global and national hydrometeorological data, which allows the user to include not just traditional rainfall runoff data (and other hydrological data, e.g., soil moisture, evaporation, wind speed), but also other kinds of data related to the behavior of the catchment (e.g., vegetation, topography, drainage structure). The use of new data sources, such as satellite data, combined with field measurements and expert judgment of local hydrologist is also encouraged.

The interface of HidroCHEP enables the use and understanding of the tools that are provided, and at the same time it gives recommendations about their use in the formulation of POMCAs. The toolbox makes the information easy to be accessed by stakeholders, water professionals, and graduate and undergraduate students. Also, as the toolbox is composed mainly by free software, it facilitates the growth of a community of users and developers who can maintain and enhance its features.

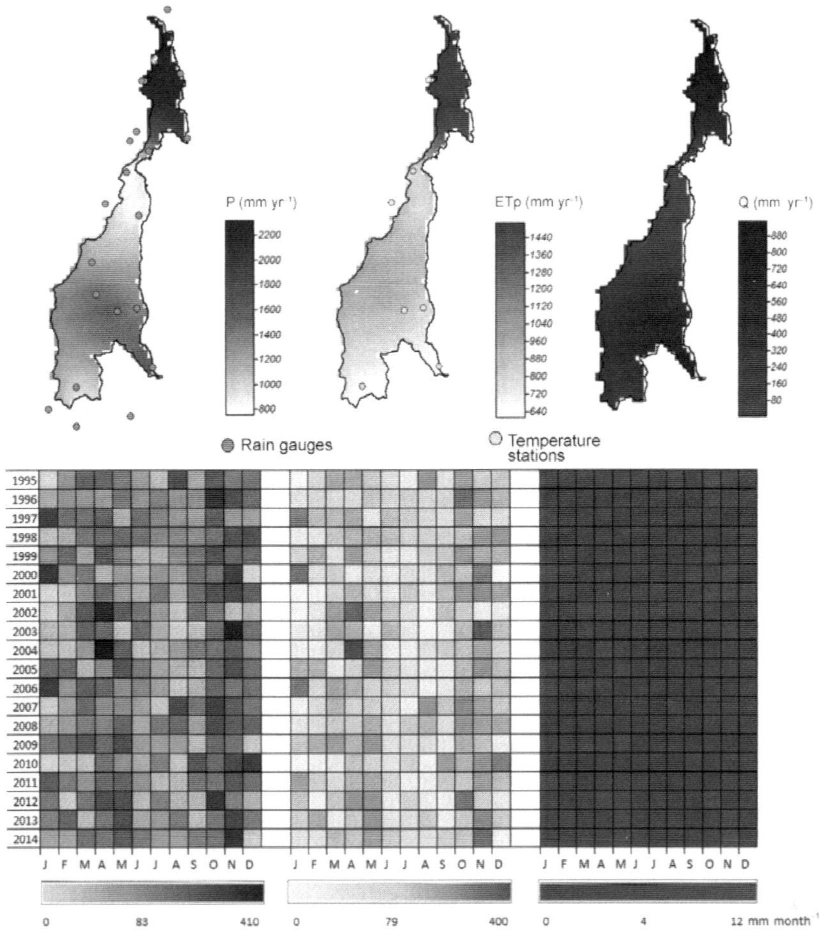


**Figure 6.** Annual and monthly water budget components for the Pamplonita's river basin using SAGA GIS.

With these contributions, it is expected that HidroCHEP provides a way to progress the formulation of POMCAS in the country and to advance the understanding of the national water resources. It is expected that stakeholders can gain confidence on their decisions, as they will be more informed for integral management.

HidroCHEP is under development. In this regard, a section dedicated to the users' feedback is foreseen, including a discussion forum, WIKIs, and the opportunity for scientist and professionals from the hydrologic sciences to share their own experiences.

## 5 Toolbox availability

All the elements of the Toolbox HidroCHEP can be accessed at the HidroCHEP website https://sites.google.com/site/hydrotoolbox/. The tools, documentation and case studies are available on this website (in Spanish).

**Acknowledgements.** The toolbox described in this paper was developed at the Universidad Nacional de Colombia, with the support of UNESCO IHE, and the Colcuencas project (Piloting Colombia's New IWRM Policy in Key Catchments) under the programmatic cooperation between the Directorate-General for International Co-operation (DGIS) of the Dutch Ministry of Foreign Affairs and UNESCO-IHE, also called the DGIS UNESCO-IHE Programmatic Cooperation (DUPC1). Any opinions, findings and conclusions or recommendations expressed in this material are those of the authors and do not necessarily reflect the views of the Universidad Nacional de Colombia, UNESCO IHE, or the Dutch Ministry of Foreign Affairs.

## References

Ackoff, R.: From data to wisdom, Journal of Applied System Analysis, 16, 3–9, 1989.

Beechie, T., Buhle, M., Ruckelshaus, M., Fullerton, A., and Holsinger, L.: Hydrologic regime and the conservation of salmon life history diversity, Biol. Conserv., 130, 560–572, doi:10.1016/j.biocon.2006.01.019, 2006.

Bellinger, G., Castro, D., and Mills, A.: Data, information, knowledge and wisdom, available at: http://www.systems-thinking.org/dikw/dikw.htm (last access: April 2015), 2004.

Biggs, B. J. F. and Close, M. E.: Periphyton biomass dynamics in gravel bed rivers: the relative effects of flows and nutrients, Fresh. Biol., 22, 209–231, doi:10.1111/j.1365-2427.1989.tb01096.x, 1989.

Blöschl, G., Sivalapan, M., Wagener, T., Viglione, A., and Savenije, H.: Runoff prediction in Ungauged Basins: Synthesis across Processes, Places and Scales, Cambridge University Press, Cambridge, ISBN 978-1-107-02818-0, 2013.

Cattanéo, F.: Does hydrology constrain the structure of fish assemplages in French streams? Local scale analysis, Archiv für Hydrobiologie, 164, 345–365, doi:10.1127/0003-9136/2005/0164-0345, 2005.

Colombian Institute of Hydrology, Meteorology and Environmental Studies (IDEAM): Colombia's State Resources and Environmental Profile 2001, Bogotá – Colombia, 2002.

Colombian Institute of Hydrology, Meteorology and Environmental Studies (IDEAM): Zoning and coding watershed in Colombia, Bogotá – Colombia, 2004.

Colombian Institute of Hydrology, Meteorology and Environmental Studies (IDEAM): National Water Study ENA 2010, Bogotá – Colombia, 2010.

Colombian Institute of Hydrology, Meteorology and Environmental Studies (IDEAM): Monthly hydrometeorological records: Selected years, 1995–2014, Retrieved from: National environmental information system, 2015.

Community Surface Dynamics Modeling System (CSDMS): CSDMS Model Repository, CSDMS Web Site, retrieved from: http://csdms.colorado.edu/wiki/Model_download_portal (last access: April 2015), 2008.

Conrad, O., Bechtel, B., Bock, M., Dietrich, H., Fischer, E., Gerlitz, L., Wehberg, J., Wichmann, V., and Böhner, J.: System for Automated Geoscientific Analyses (SAGA) v. 2.1.4, Geosci. Model Dev., 8, 1991–2007, doi:10.5194/gmd-8-1991-2015, 2015.

eWater Cooperative Research Centre: eWater Toolkit, eWater web site, retrieved from: http://toolkit.ewater.org.au/ (last access: April 2015), 2015.

Global Water Partnership: Integrated Water Resources Management Toolbox (IWRM Toolbox), retrieved from: http://www.gwp.org/en/ToolBox/ (last access: April 2015), 2013.

Gupta, H., Wagener, T., and Liu, Y.: Reconciling theory with observations: elements of a diagnostic approach to model evaluation, Hydrol. Process., 22, 3802–3813, doi:10.1002/hyp.6989, 2008.

Hargreaves, G. and Samani, Z.: Simplified coefficients for estimating monthly solar radiation in North America and Europe, Departmental paper, Dept. of Biol. And Irrig. Engrg, Utah State University, Logan, Utah, 1994.

Harris, N. M., Gurnell, D. M., Hannah, D. M., and Petts, G. E. : Classification of river regimes: a context for hydro-ecology, Hydrol. Process., 14, 2831–2848, 2000.

Ministry of Environment and Sustainable Development (MinAmbiente): Technical Guide for the formulation of Basin Management Plans (POMCAS), Bogotá – Colombia, 2014.

Monk, W., Wood, P., Hannah, D., Wilson, D., Extence, C., and Chadd, R.: Flow variability and macro invertebrate community response within riverine systems, River Res. Appl., 22, 595–615, doi:10.1002/rra.933, 2006.

Monk, W., Wood, P., Hannah, D., and Wilson, D.: Selection of river flow indices for the assessment of hydroecological change, River Res. Appl., 23, 113–122, doi:10.1002/rra.964, 2007.

Monk, W., Wood, P., Hannah, D., and Wilson, D.: Macro invertebrate community response to inter-annual and regional river flow regime dynamics, River Res. Appl., 24, 988–1001, doi:10.1002/rra.1120, 2008.

Olden, J., Poff, N., and Bledsoe, B.: Incorporating ecological knowledge into ecoinformatics: an example of modelling hierarchically structured aquatic communities with neutral networks, Ecological Informatics, 1, 33–42, doi:10.1016/j.ecoinf.2005.08.003, 2006.

Poff, N., Allan, J., Bain, M., Karr, J., Prestegaard, K., Richter, B., Sparks, R., and Stromberg, J.: The natural flow regime: a paradigm for river conservation and restoration, BioScience, 47, 769–784, doi:10.2307/1313099, 1997.

Poveda, G.: Hydroclimatology of Colombia: A synthesis from the Inter-Decadal scale up daytime scale, Revista Académica Colombiana, 28, 201–222, 2004.

Richter, B., Baumgarthner, J., Powell, J., and Braund, D.: A method for assessing hydrologic alteration within ecosystems, Conserv. Biol., 10, 1163–1174, doi:10.1046/j.1523-1739.1996.10041163.x, 1996.

Sauquet, E., Gottschalk, L., and Krasovskaia, I.: Estimating mean monthly runoff at ungauged locations: an application to France, Hydrol. Res., 39, 406, doi:10.2166/nh.2008.331, 2008.

Turc, L.: Evaluation de besoins en eau dírrigation, ET potentielle, Ann. Agron, 12, 13–49, 1961.

Proc. IAHS, 374, 113–116, 2016
proc-iahs.net/374/113/2016/
doi:10.5194/piahs-374-113-2016


Open Access
PIAHS

# Hydrology and water resources in Caspian Sea

**Kourosh Haddadi Moghaddam**

International Sturgeon Research Institute (ISRI), P.O. Box 41635-3464, Rasht, Iran

*Correspondence to:* Kourosh Haddadi Moghaddam (khkhmoghadam@gmail.com)

Published: 17 October 2016

**Abstract.** Precipitation is the main driver of the water balance variability of the water over space and time, and changes in precipitation have very important implications for hydrology and water resources. Variations in precipitation over daily, seasonal, annual, and decadal time scales influence hydrological variability over time in a catchment. Flood frequency is affected by changes in the year-to-year variability in precipitation and by changes in short-term rainfall properties. Desiccation of the Caspian Sea is one of the world's most serious ecosystem catastrophes. The Persian Sturgeon (*Acipenser persicus*) caught under 10 m depth using bottom trawl net by research vessel during winter 2012, summer and winter 2013 and spring 2014 in east, central and west of southern parts of Caspian Sea, then, their diets were investigated. During 136 trawling in the aimed seasons, Persian sturgeon with 1 to 2 years old and $179.67 \times 0.2$ g (body weight) and $29.97 \pm 0.4$ cm (Total length) captured. Examination of stomach contents in the sturgeon specimens revealed that the food spectrum was composed of bony fishes (*Neogobius* sp., *Atherina* sp. and *Clupeonella delicatula*), invertebrates belonging to the family Ampharetidae polychaeta worms including (*Hypanai* sp. and *Nereis diversicolor*), various crustaceans (*Gammarus* sp. and *Paramysis* sp.). Investigation on stomach contents of sturgeon *Acipenser persicus* caught under 10 m depth in 2012 to 2013 surveys showed that there is significant difference in the consumed food. The most food diversity have been observed in winter 2013, also Polychaeta is the primary consumed food and crustacean is the secondary one ($P > 0.05$), no new types of food (such as bony fishes or benthics) have been observed on food chain of *Acipenser persicus* and shows no significant difference ($P > 0.05$).

## 1 Introduction

Prey abundance, feeding and food habits are important issues for both effective Caspian Sea fisheries management and conservation biology in aquatic environment, so Sturgeons one the most important and valuable fishes in the world and Caspian Sea is one of their most important habitat's which 90 % of total world Caviar belongs to it (Ivanov et al., 1999). In recent years sturgeon stock and catch has been decreased in their most important habitats (Pourkazemi, 1996; Ivanov et al., 1999; Moghim et al., 2006). Biological and feeding factors have great influence an sturgeon catch to determine these factors, we should investigate on sturgeon feeding competition and other relations between different habits of fishes in various places and time is of high importance (Moghim et al., 2006) therefore should navigate the previous studies of researchers in this regards. Initial studies carried out on sturgeon feeding in north, middle and south Caspian Sea (Shorygin, 1937; Salnikov et al., 1975; Begenal, 1978; Moiseev and Filatova, 1988 and Polyaninova and Molodtseva, 2002). During sturgeon stock assessment at southern part of Caspian Sea in 2006 to 2008, diets of Persian sturgeon investigated in different depths and various seasons. Although some studies conducted in Iran up to now, they were cyclical (Kashentseva, 2005; Hashemyan et al., 2005) considering that too ecological changes happened in Caspian Sea (Haddadi Moghaddam et al., 2003; Roohi et al., 2007) the previous studies don't include all dietary information of sturgeon. So, the aims of the study were (i) determine seasonal change on Persian sturgeon (ii) determine more details on sturgeon food composition and identification of food index to better understand Persian sturgeon feeding condition.

Published by Copernicus Publications on behalf of the International Association of Hydrological Sciences.

**Table 1.** Number, seasons and means body weight (BW), total length (TL) and condition factor in Persian Sturgeon (Mean ± SD).

| Mean condition factor | Mean Total length (cm) | Mean body weight (g) | Seasons | Number |
|---|---|---|---|---|
| 0.31 ± 0.01 | 25.5 ± 0.43 | 48.9 ± 0.6 | Winter 2012 | 41 |
| 0.75 ± 0.25 | 35.07 ± 0.8 | 322.64 ± 0.3 | Summer 2013 | 20 |
| 0.48 ± 0.08 | 25.91 ± 0.1 | 294.81 ± 0.7 | Winter 2013 | 32 |
| 0.38 ± 0.07 | 33.4 ± 0.4 | 52.34 ± 0.3 | Spring 2014 | 15 |

## 2 Materials and methods

The study carried out during March, February 2012, March, February 2013, July, August 2013 and April, May 2014. According to the area of regions, 34 station selected through stratified random sampling design (including 21 stations in east, 5 stations in middle and 8 station in west). The sampled stations were located between (32°2′ N, 50°29′ E and 37°2′ N , 53°4′ E).Trawling conducted in each trawling, GPS and map with scale (1, 100 000) used to determine swept distance and geographical conditions of stations. Also bottom trawl (trawl with 9 m upper rope) and research vessel (250 hsp) were applied in this regard.

During days for 30 min (07.00–16.00) in each station and the speed was 25 to 3 Knots along the coast .After catching and measuring of mean total length and body weight, the digestive system of samples removed (Tavakoli et al., 2007) In order to remove the digestive tract, an anal incision created along the middle ventral line up to gill. So, the digestive system removed completely from ventral region (Chugunova, 1963).

Fish diets were determined from stomach contents and the ingested food evaluated by enumeration (Biswas, 1993). The manual for Caspian Sea invertebrate identification key applied for identifying primary and secondary foods in stomach content (Pryshtin et al., 1988). For sturgeon diet assessment, prey items were taxonomically classified and four indices – the prey occurrence index $F$ (%), the prey frequency index Cn (%), gastro somatic index GSI (%) and condition factor (CF), were calculated as follows:

- $F$ (%) $= 100 \times (N_p/N_i)$ where
- $N_p$ = the number of stomachs with a specific prey
- $N_i$ = the total number of non-empty stomachs. Values of ($F > 50$ %) indicate main prey;
- $10\,\% < F < 50\,\%$, secondary food items; and $F < 10\,\%$, occasional prey and
- Cn (%) $= 100 \times (N_i/N_p)$
- $N_i$ = is the number of the prey item in all nonempty stomachs
- $N_p$ = the total number of food items in all stomachs in a sample
- GSI (%) $= 100 \times (W_s/W_c)$
- $W_s$ = weight of stomach
- $W_c$ = weight of carcass (g)
- CF = W/TL3
- $W$ = weight (g)
- TL = total length (cm) (Euzen, 1987; Saborowski and Buchholz, 1996).

## 3 Results

The diet of megrim is based on benthic, teleosts and bottom, Crustacea and polychaeta. In this study different foods observed in Persian sturgeon stomach including various teleosts such as *Neogobious* sp., *Atherina* sp., *Clupeonella cultriventris* and benthic invertebrates like Ampharitidae (Hypania sp. and Hypaniolla sp.) *Nereis diversicolor*, Crustacean (Paramysis sp. and Gammarus sp.). In this case 115 immature Persian Sturgeons in 1 to 2 years old captured for feeding examination 90 % of caught sturgeon had food in stomach. comparative study of condition factor in sturgeon caught under 10 m depth showed the highest condition factor in eastern region (0.75 ± 0.25) in summer 2013 and the lowest one (0.26 ± 0.13) in western region of Southern part of Caspian Sea in spring. The results showed no significant differences in condition factor of Persian sturgeon ($P > 0.05$) (Table 1).

GSI comparison in Persian sturgeon during different seasons showed that the mean GSI of Persian sturgeon was highest in winter 2013 (109.6 ± 14.7) and lowest (51.83 ± 3.8) in summer 2013. The results showed no significant difference in GSI in various regions and seasons ($P > 0.05$). Investigation of food frequency percent showed that Persian sturgeon used crustacean as basic food in winter 2012, but in order seasons consumed teleostes as secondary food. Statistical results indicate that there is not significant difference in the foods consumed in different seasons and regions ($P > 0.05$) (Table 2).

## 4 Discussion

Coastal zone is important nursery habitats for many fishes where they can feed upon a concentrated food supply. Study

**Table 2.** Prey frequency percent in Persian Sturgeon.

| Seasons | Mean food (%) | Secondary food (%) |
|---|---|---|
| Winter 2012 | Ampharitidae (80 %) | Crustacea (20 %) |
| Summer 2013 | Ampharitidae (70 %) | Gobiidae (80 %) |
| Winter 2013 | Ampharitidae (85 %) | (Gobiidae) (15 %) |
| Spring 2014 | Ampharitidae (50 %) | (Gobiidae (25 %), Atherindiae (25 %)) |

on frequency and type of consumed prey in Persian sturgeon shows no regular pattern in their feeding behaviour under 10 m depth and Ampharitidae is their primary food. This could be due to unavailability of their food resources and factors like depth, temperature, type of bed and organic material rate. Teleosts consist the basic parts of food consumed by younger fishes (Hashemyan et al., 2005; Mirzajani et al., 2003; Holchick, 1989). But in lower ages considering the depth and environment geographical, conditions, polychaeta consumed by sturgeon. Examination of feeding intensity during recent surveys do not show any type of new food, but the rate and percent of ingested food has been changed and increased during warm seasons. Although factors such as type of bed, organic material rate and physico-chemical conditions can influence on feeding variety in aquatics (Holchick, 1989; Tavakoli et al., 2007; Gegadeesan and Ayyakkannu, 1992; Waldman, 1995). Basic changes occurred in ecological condition of Caspian Sea, which is due to increase in Caspian Sea water level up to 2.2 m from 1995 and elevation of oxygen level because entering urban and industrial pollution that leads to slugging of bed. These factors caused to specialization of food for some species and more large preys such as various species of teleosts observed in sturgeon stomach contents (Salnikov et al., 1975). Regarding these changes and new ecological condition of Caspian Sea, The results of study don't indicate any new feeding condition for Persian sturgeon. According to studies carried out under 10 m depth in east, middle and west of southern part in Caspian Sea (Gasemove, 1994) and considering the gradual warming of Caspian Sea water, it was determined that Sturgeon like to use larger preys which influence on the SI and condition factor of one's caught in warm seasons. Teleosts and benthic diversity become warmer seasons of year. Factors such as food frequency in region, food resources and population have the most influence on feeding rate and condition factor in these fishes (Nikolskii, 1963). Studies conducted on SI in 300 Persian sturgeon in eastern, northwest and south parts of Caspian Sea (Salnikov et al., 1975; Zolotarev et al., 1996) showed that SI was higher when Sturgeon consumed teleosts. The changes may be due to competition for using different types of food, sea waves and type of bed that fish inhabited (Graber, 1990; Kostyuchenko, 1994; Chechun, 1998). Study on variety in winter 2013 and spring 2014 is more than other seasons, which may be due to change in southern parts of Caspian Sea showed similar results by Cortest and Graber (1990). These modifications reflected changes in the availability of prey which influenced fish diet composition and were probably related to the loss of biodiversity in the Black Sea benthic communities which became dominated by some opportunistic species like polychaetes by Gomoiu et al. (2004).

## 5 Data availability

Analyses of Variance (ANOVA) were performed using the STATISTICA 6.1 program applied for data availability. The results identified by _SD and seasonal changes in stom ach content of Persian sturgeon determined by relative frequency. All statistical analyses were carried SAS version 9.0 software (SAS Institute, inc., USA).

**Acknowledgements.** This study is part of a project entitled "sturgeon stock assessment in southern parts of Caspian Sea" that was conducted with financial support of Iranian Fishes Research Organization (IFRO) at International Sturgeon Research Institute (ISRI). We would like to thank all the colleagues who helped and supported us in obtaining Persian sturgeon from Caspian Sea with the grant agreement number: 48/689.

## References

Begenal, G.: Methods for assessment for fish production in freshwater, Blackwell science public, UK, 190–194, 1978.

Biswas, S. P.: Manual of Methods in Fish Biology, South Asian Publication, New Delhi, India, 89, 23–25, 1993.

Cortest, E. and Graber, S. H.: Diet and feeding habits and estimates of daily ration of young, *Negaprionbrerirostris* (Poey), CopelaBul, UK, 4, 204–218, 1990.

Chechun, T. Ya.: Nutrition of the Acipenseridae of the Sea of Azov under Present Conditions, J. Appl. Ichthyol., 38, 155–158, 1998.

Chugunova, N. I.: Age and Growth Studies in Fish, Biol. Bul., 7, 201–210, 1963.

Euzen, O.: Food Habits and Diet Comparison of Some Fish of Kuwait, Biol. Bull., 58, 11–20, 1987.

Gegadeesan, P. and Ayyakkannu, K.: Seasonal variation of benthic fauna in marine zone of Coleroon estuary inshore water, southeast coast of India, Mar. Sci., 15, 223–229, 1992.

Gasemove, F.: Animals and biotic productions of Caspian Sea, Iranian fisheries publication, Iran, 450–455, 1994.

Gomoiu, M. T., Secrieru, D., Paraschiv, G., Opreanu, P., and Puschiaza, D.: Someremarks on the ecological state of benthic populations recorded during the IAEA, Mar. Geol., 9, 12–20, 2004.

Haddadi Moghaddam, K., Parandavar, H., Pajand, Z., Khoshghalb, M. B., Chubian, F., and Jooshideh, H.: Study on feeding habits of sturgeons in the Caspian Sea up to 10 m depth (Guilan-Province), Iranian fisheries Research Organization publication, Iran, 69–73, 2003.

Hashemyan, A., Khoshbavar, H. A., Rostami, R., and Taleshian, H.: A comparative analysis of feeding habits of sturgeon fish in shallow coastal waters of Mazandaran and Golestan provinces, south Caspian Sea, J. Fisheries Sci., 14, 157–166, 2005.

Holchick, M.: The fresh water fishes of European. Iranian fisheries Research publication, Iran, 401–422, 1989.

Ivanov, V. P., Vlasenko, A. D., Khodorevskaya, R. P., and Raspopov, V. M.: Contemporary status of Caspian sturgeon (Acipenseridae) stock and its conservation, J. Appl. Ichthyol., 15, 103–105, 1999.

Kostyuchenko, V. A.: Biology and state of fishing of the Sea of Azov Sturgeon's, J. Appl. Ichthyol., 21, 174–187, 1994.

Kashentseva, L. N.: The status of feeding of stellate Sturgeon population in the Caspian Sea, status of commercial species stock in the Caspian Sea and their management, J. Great Lake Res., 4, 23–35, 2005.

Mirzajani, A. R., Yosefzad, E., Sayadrahim, M., and Abdolmaleki, S. H.: Investigation onmeiofauna and substrate characteristic in the southern Caspian Sea, J. Fisheries Sci., 11, 98–107, 2003.

Moghim, M., Parafkandeh, F., Tavakoli, M., and Behruz Khoshghalb, M.: Investigation biological and statistical on sturgeon fish in 2001–2003, Iranian fisheries Research Organization publication, Iran, 61–65, 2006.

Moiseev, P. A. and Filatova, Z. A.: Animals and Bioproductivity of the Caspian Sea, Nauka publication Moscow, 105–114, 1988.

Nikolskii, G. V.: The Ecology of Fishes, Cold spring Harbor, New York, 340–345, 1963.

Polyaninova, A. A., and Molodtseva, A. I.: The benthos sturgeon feeding relationship for the Caspian Sea, J. Ichthyol., 38, 151–157, 2002.

Pourkazemi, M.: Caspian Sea sturgeon conversation and fisheries past, present and future, J. Appl. Ichthyol., 22, 1–4, 1996.

Pryshtin, Y. A., Wingraduva, V. N., Kundakuva, V., Kuna, M. C., and Vastakhuva, A. V.: Atlas of invertebrates Caspian Sea, Iranian fisheries Research Organization publication, Tehran, Iran, 345, 1988.

Roohi, A., Naderi, M. F., Vahedi, M., Qasemi, F., Roshantabari, F., Parafkande, B., Hasanzadekiabi, and Shiganova, T.: Strategy of Mnemoiopsis leidyi combating in the Caspian Sea, Iranian fisheries Research Organization publication, Iran, 30–78, 2007.

Saborowski, R. and Buchholz, F.: Annual changes in the nutritive state of North-Sea, J. Fish Biol., 49, 173–194, 1996.

Salnikov, N. E., Kukuradze, A. M., and Kirilyuk, M. M.: Prospects for Sturgeon Culture in the Northwestern Black Sea under Conditions of Complex Use of Water Resources, J. Appl. Ichthyol., 8, 87–94, 1975.

Shorygin, A. A.: Standard methods for water and wastewater, J. Fish. Biol., 5, 74–85, 1937.

Tavakoli, M., Jooshideh, H., Fadaeei, B., Keymaram, B., Parafkandeh, M., Moghim, M., Bahmani, H., Parandavar, Z., Pajand, K., HaddadiMoghaddam, F., Chubian, D., Kor, A., Halajian, F., Chakmeduze, S., Hasanzadeh, H., Taleshian, and Abassi, K.: Stock assessment of sturgeon fishes in the Southern part of Caspian Sea (Iranian water), Iranian fisheries Research Organization publication, Iran, 136–153, 2007.

Waldman, J. R.: Sturgeon and paddle fish Convergence of Biology, politics and greed, J. Fish. Biol., 20, 20–21, 1995.

Zolotarev, P. N., Shlyakhov, V. A., and Akselev, O. I.: Food Supply and Feeding of the Russian Sturgeon *Acipensergueldenstaedtii* and the Starred Sturgeon *Acipenserstellatus* in the North western Black Sea under Modern Ecological Conditions, J. Appl. Ichthyol., 36, 357–362, 1996.

Proc. IAHS, 374, 117–121, 2016
proc-iahs.net/374/117/2016/
doi:10.5194/piahs-374-117-2016


Open Access
PIAHS

# Development of seasonal flow outlook model for Ganges-Brahmaputra Basins in Bangladesh

**Sazzad Hossain[1], Raihanul Haque Khan[2], Dilip Kumar Gautum[2], Ripon Karmaker[1], and Amirul Hossain[1]**

[1]Flood Forecasting and Warning Center, Dhaka, Bangladesh
[2]Regional Integrated Multi-hazard Early Warning System, Thailand

*Correspondence to:* Sazzad Hossain (sazz176@yahoo.com)

Published: 17 October 2016

**Abstract.** Bangladesh is crisscrossed by the branches and tributaries of three main river systems, the Ganges, Bramaputra and Meghna (GBM). The temporal variation of water availability of those rivers has an impact on the different water usages such as irrigation, urban water supply, hydropower generation, navigation etc. Thus, seasonal flow outlook can play important role in various aspects of water management. The Flood Forecasting and Warning Center (FFWC) in Bangladesh provides short term and medium term flood forecast, and there is a wide demand from end-users about seasonal flow outlook for agricultural purposes. The objective of this study is to develop a seasonal flow outlook model in Bangladesh based on rainfall forecast. It uses European Centre for Medium-Range Weather Forecasts (ECMWF) seasonal precipitation, temperature forecast to simulate HYDROMAD hydrological model. Present study is limited for Ganges and Brahmaputra River Basins. ARIMA correction is applied to correct the model error. The performance of the model is evaluated using coefficient of determination ($R_2$) and Nash–Sutcliffe Efficiency (NSE). The model result shows good performance with $R^2$ value of 0.78 and NSE of 0.61 for the Brahmaputra River Basin, and $R^2$ value of 0.72 and NSE of 0.59 for the Ganges River Basin for the period of May to July 2015. The result of the study indicates strong potential to make seasonal outlook to be operationalized.

## 1 Introduction

Bangladesh is located at the downstream end of the three major river basins, collectively known as Ganges-Brahmaputra-Meghna (GBM) basin. The total catchment area is approximately 1.6 million km$^2$ of which only about 7.5 % lies within Bangladesh and the rest, 92.5 % lies in India, China, Nepal and Bhutan. It is estimated that an average flow of 1 009 000 million m$^3$ passes through these river systems during the monsoon season and monsoon precipitation is the major source flow (FFWC, 2014). The rivers in these three catchments mostly are originated from Himalayas mountain ranges and the pattern of flow, timing and intensity is governed by the quantity and distribution of precipitation, its form (rain or snow) and seasonality (ICIMOD, 2009). There are two distinct seasons in Bangladesh, a dry period from November to May and the wet from June to October. Over 80 % of the rainfall occurs during the monsoon or rainy season also known as flood season (WMO/GWP, 2003). The normal annual rainfall of the country varies approximately from 1200 mm in the west to over 5000 mm in the east (BWDB, 2011). Flood is a perennial problem in Bangladesh which causes huge economic losses each year. There are various factors, which contribute to the flooding in Bangladesh. One of the important factors is 80 % of the country is floodplain with 50 % of the country having elevation below 5 m of MSL. In a normal year, roughly 20 % of the country is inundated, while in years of severe flooding 60–70 % of the country gets affected by flood water. For instance, the duration of flood in 1998 was more than two months and 68 % of the country was under water (FFWC, 2004). High population density and extreme poverty aggravates the flood situation. Moreover, the country is also affected by hydrological drought from prevailing low flow in the river systems dur-

**Table 1.** Catchment Area of Brahmaputra and Ganges River.

| River Basin | Total catchment area (km$^2$) | Catchment Area (km$^2$) | | | | |
|---|---|---|---|---|---|---|
| | | India | Nepal | Bhutan | China | Bangladesh |
| Brahmaputra | 552 000 | 195 000 | – | 47 000 | 270 900 | 39 100 |
| Ganges | 1 087 300 | 860 000 | 147 480 | – | 33 520 | 46 300 |

ing the dry period. Thus, temporal variation in river flows has a significant impact on the different water usages such as irrigation, urban water supply, navigation, in stream flow requirement etc. Seasonal flow outlooks can therefore play big role in agricultural planning, selection of cropping pattern, irrigation scheduling and other areas like disaster management and longer-term water management planning.

There are some initiatives to predict medium to long range flow using climate data. Flood Forecasting and Warning Centre (FFWC) in Bangladesh issues 10 days probabilistic forecast at 38 locations along the major river systems using ECMWF medium range weather forecast (FFWC, 2013) with support from RIMES. Webster and Carlos (2004) predicted river discharge on 15–30-day time scale in the Brahmaputra and Ganges Basin in Bangladesh using ECMWF data. Seasonal outlooks for rainfall are available over the South Asian region under the South Asian Climate Outlook Forum (SASCOF) and the Bangladesh Meteorological Department also issues monthly outlook for the rainfall. Performance of longer range hydrological forecast (e.g. seasonal scale) largely depends on the accuracy of seasonal rainfall forecast. Hydrological modelling is used to generate flow using various meteorological inputs including rainfall, temperatures, soil moisture and evapotranspiration. ECMWF seasonal scale meteorological forecasts are an excellent source of forecast dataset to develop hydrological models. This study focuses on application of HYDROMAD model using ECMWF seasonal rainfall, temperature forecast data to generate flow for two large river basins- Brahmaputra and Ganges. The present study uses the mean of 41 set of ensembles available in the ECMWF seasonal forecast data set. This study also highlights potential dissemination aspects of seasonal flow outlook among the water users in Bangladesh.

## 2 Study area

The Ganges and Brahmaputra Basins are located approximately between latitude 21–31° N and longitude 73–96° E. The study area consists of two big river basins which are known as Brahmaputra and Ganges. Table 1 shows individual area in square km of these two river basins and corresponding area in respective co-riparian countries. The majority of the Ganges Basin is located inside India and major part of the Brahmaputra Basin is situated within China. The Ganges originates from the western side of the Himalayas,

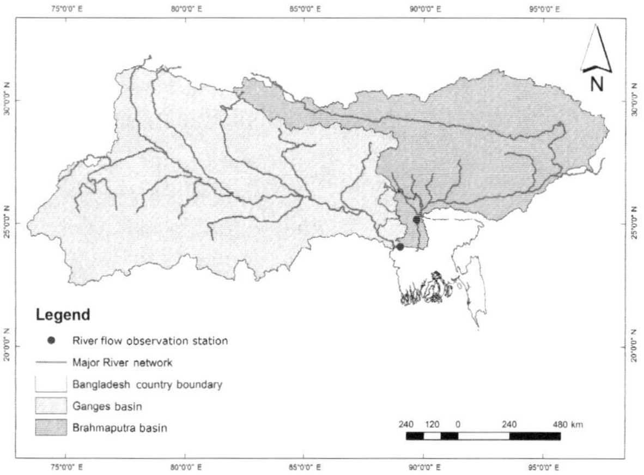


**Figure 1.** Ganges–Brahmaputra Basin map with respect to Bangladesh.

and it flows approximately 2000 km through India. It enters in Bangladesh from the western side of the country and flows towards southeast for about 250 km to meet confluence with the Brahmaputra. The Brahmaputra originates on the Chinese side of the Himalayas and flows about 1800 km through Tibet and India to enter Bangladesh from the northern part of the country. It flows for 275 km inside Bangladesh to join with the Ganges (Islam, 2002). Figure 1 shows the position of the Basins with respect to Bangladesh. The two river basins have distinct hydrological characteristics. Flood hydrograph of the Ganges shows that the flow starts to recede from at the end of September every year and minimum flow prevails in the river between March and April. Flow of Brahmaputra river starts rising in March due to melting of snow in the Himalayas while discharge of Ganges rises in July with the onset of monsoon. Usually, the flood flow of the Brahmaputra reaches peak in July and August, while the peaks of the Ganges occur in August and September. Though the two basins usually receive flood in two different time, synchronization of flow peaks are not uncommon and such cases cause major floorings in the country.

## 3 Data and methodology

The HYDROMAD model needs rainfall, temperature and potential evaporation to simulate model and observed flow

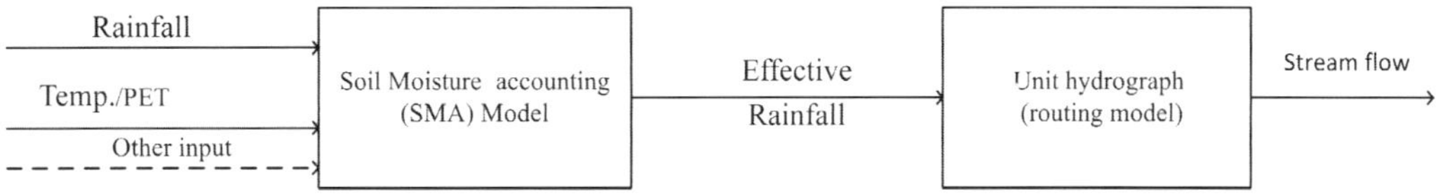


**Figure 2.** The modelling framework in the HYDROMAD package.

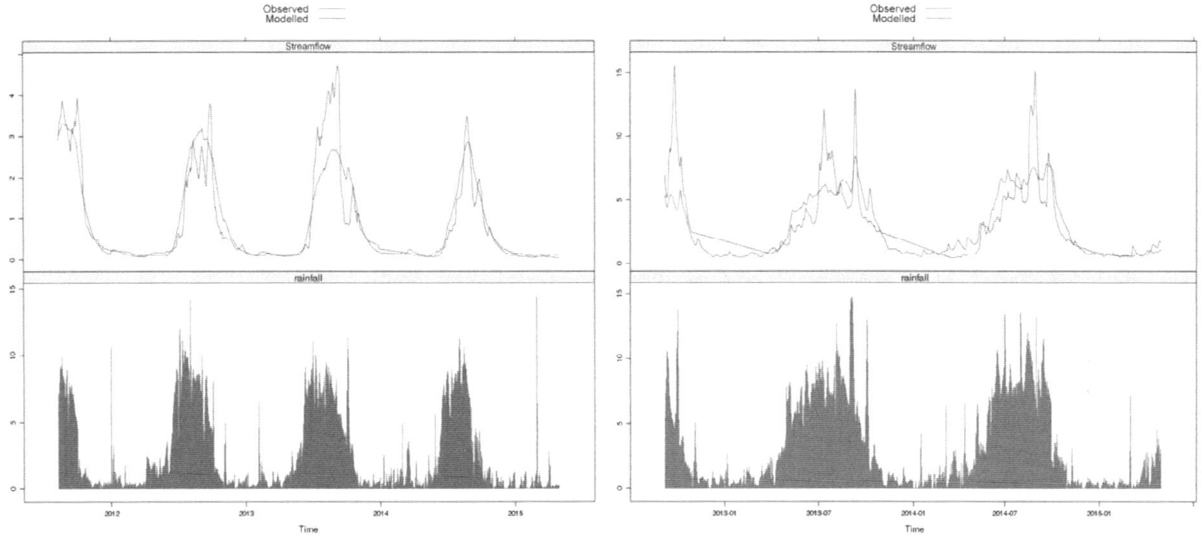


**Figure 3.** Observed and simulated discharge of without correction (Ganges-left and Brahmaputra-right).

data for calibration. The present study utilizes ECMWF seasonal (3 months and 6 months) rainfall and temperature forecast data for Ganges and Brahmaputra Basins. ECMWF seasonal forecast is ensemble forecast with 41 members, and it is the output of Atmospheric, Wave and Ocean models (Persson and Grazzini, 2011). The observed river discharge of the two basins has been taken from Bangladesh Water Development Board (BWBD) for calibration and validation. The HYDROMAD is a "R" language package for hydrological modelling and related data analysis. HYDROMAD is a spatially lumped model and adopts an empirical approach. It does not explicitly represent spatial variation over the catchment area (Andrews et al., 2011). The model can predict stream flow from time series of areal rainfall and temperature or potential evapotranspiration. The modelling framework in the HYDROMAD package is based on a two-component structure: (a) a soil moisture accounting (SMA) module; and (b) a routing or unit hydrograph module. Figure 2 explains the modelling frame framework of HYDRMAD Model. SMA model converts rainfall and temperature into effective rainfall – the amount of rainfall, which eventually reaches the catchment outlet as stream flow. The routing module then converts the effective rainfall into stream flow, which amounts to defining the peak response and shape of the recession curve. The major steps in developing hydrological model are to extract seasonal ensemble forecast of rainfall and temperature for the Ganges and Brahmaputra basins and compute ensemble mean for each grid, compute Mean Areal Precipitation (MAP) and Mean Areal Temperature (MAT) over the catchment. HYDROMAD model is simulated for calibration and validation period to generate seasonal flow outlook. Statistical criteria $R^2$, NSE have been calculated to test model performance. ARMA correction has been applied for the best fitting of the simulated data with the observed data set. Finally, a Web Based Decision Support System has been developed to disseminate the outputs to the water managers.

## 4 Result and discussion

The model was run from 2012 to 2014 as a calibration period, and the results are presented in Fig. 3. The results shows observed peaks both of the river basins are higher than the simulated value, which indicates that the model underestimated the peak flow. However, low flows simulated by the model have a better agreement with the observed data. The reasons for the difference in observed and simulated discharge could be due to uncertainly in rainfall forecast, the inconsistency in the observed flow data used for calibration etc. ARMA correction is applied for the purpose of improving the model performance. After applying correction, peak discharge as well

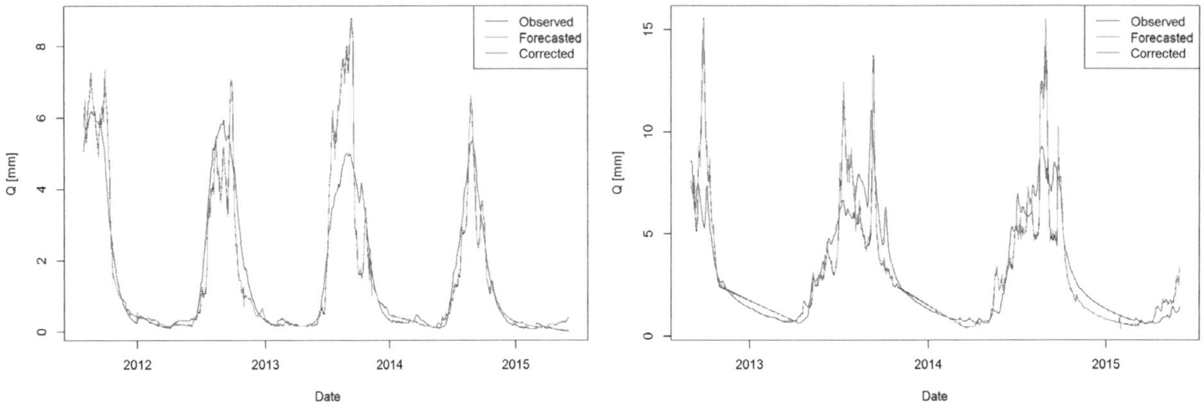


**Figure 4.** Observed and simulated discharge of with ARMA correction (Ganges-left and Brahmaputra-right).

as low flow, both of river basins show well matched with the observed data. Figure 4 shows hydrograph of observed and simulated discharge of Brahmaputra and Ganges after applying correction. The model performance is determined using coefficient of determination ($R^2$) and Nash–Sutcliffe efficiency (NSE) for both the basins. The model has been run since January 2015 for trail operation and $R^2$ and NSE has been calculated to check the model performance. The result shows good agreement between observed and simulated data, with $R^2$ value of 0.78 and NSE of 0.61 for the Brahmaputra River Basin, and $R^2$ value of 0.72 and NSE of 0.59 for the Ganges River Basin for the period of May to July 2015.

An initiative has been taken to disseminate seasonal flow outlook through a web based (http://sffs.rimes.int) system. Dissemination includes average seasonal flow, normal flow and forecasted flow with three distinct colours. The dissemination system automatically generates an outlook like "The flow of Ganges River is going to be below normal for the months of September, October and November". An emailing system has been embedded into the portal to disseminate the outlook to desired stakeholders.

## 5 Conclusions

The model has been simulated for a period of 2012 to 2014 for calibration and validation purposes and based on the results, is being simulated regularly with the outputs published to a website in every month. The first phase of the study was concluded in 2015, and the overall result is very promising in terms of the application of ECMWF precipitation forecast in developing seasonal river flow outlook across the GBM basin in Bangladesh. As rainfall forecast is affected by various climate drivers so, it is important to consider those factors for forecasting with hydrological models. At present, the model forecast output is experimentally disseminated through a web-based system to get feedback from different users specially water resources planners and disaster managers. The Ganges–Brahmaputra is a transboundary river basin and details upstream information is not available. HYDROMAD is a lump model and takes only few data to simulate the model. So, it might be a better approach to operate for seasonal flow model to incorporate more information of upper basin as well. Assessment and research on the application of seasonal flow outlook in different field of water resources management like irrigation, agriculture, disaster management etc. should be carried out to make this forecast meaningful.

## 6 Data availability

Hydrological data of the Ganges and Brahmaputra basin inside Bangladesh is accessible from http://www.hydrology.bwdb.gov.bd/. The observed rainfall data also available from Bangladesh Metrological Department (BMD) with special request, and ECMWF seasonal rainfall forecast available from Regional Integrated Multi-Hazard Early Warning System (RIMES) taken Result of the present study also available from http://sffs.rimes.int.

**Acknowledgements.** The present research work has been conducted for the increase of flood forecast lead-time of Bangladesh by the technical support of Regional Integrated Multi-hazard Early Warning System.

## References

Andrews, F. T., Croke, B. F. W., and Jakeman, A. J.: An open software environment for hydrological model assessment and development, Environ. Modell. Softw., 26, 1171–1185, doi:10.1016/j.envsoft.2011.04.006, 2011.

BWDB: Inception Report on Research and Prediction Modelling Through Upgrading of Flood Forecasting System by Increasing Lead Time, BWBD, Dhaka, 30, 2011.

FFWC: Annual Flood Report 2004, Flood Forecasting and Warning Center, BWBD, Dhaka, 6, 2004.

FFWC: Annual Flood Report 2013, Flood Forecasting and Warning Center, BWBD, Dhaka, 70, 2013.

FFWC: Annual Flood Report 2014, Flood Forecasting and Warning Center, BWBD, Dhaka, 1, 2014.

ICIMOD: Project Proposal on The Hindu Kush-Himalayan Hydrological Cycle Observing System, ICIMOD, Kathamundu, Nepal, 1, 2009.

Islam, S.: Hydrology of Floods in South Asia, SaciWATERs Workshop on Floods in South Asia, Dhaka, Bangladesh, 28–30 November, 2002.

Persson, A. and Grazzini, F.: User guide to ECMWF forecast products, Version 4.0, available at: http://www.ecmwf.int (last access: 15 April 2015), 2011.

Regional Integrated Multi-Hazard Early Warning System (RIMES): Bangladesh Seasonal Flow Outlook, available at: http://sffs.rimes.int, 2016.

Webster, P. J. and Carlos, H.: Prediction of monsoon rainfall and river discharge on 15–30-day Time scales, American meteorological society, 85, 1745–1765, 2004.

WMO/GWP: Integrated Flood Management Case Study Bangladesh: Flood Management, The Associated Programme on Flood Management, 1, 2003.

Proc. IAHS, 374, 123–128, 2016
proc-iahs.net/374/123/2016/
doi:10.5194/piahs-374-123-2016


Open Access PIAHS

# East African wetland-catchment data base for sustainable wetland management

**Constanze Leemhuis, Esther Amler, Bernd Diekkrüger, Geofrey Gabiri, and Kristian Näschen**
Department of Geography, Bonn University, Bonn, 53115, Germany
*Correspondence to:* Constanze Leemhuis (leemhuis@uni-bonn.de)

Published: 17 October 2016

**Abstract.** Wetlands cover an area of approx. 18 Mio ha in the East African countries of Kenya, Rwanda, Uganda and Tanzania, with still a relative small share being used for food production. Current upland agricultural use intensification in these countries due to demographic growth, climate change and globalization effects are leading to an over-exploitation of the resource base, followed by an intensification of agricultural wetland use. We aim on translating, transferring and upscaling knowledge on experimental test-site wetland properties, small-scale hydrological processes, and water related ecosystem services under different types of management from local to national scale. This information gained at the experimental wetland/catchment scale will be embedded as reference data within an East African wetland-catchment data base including catchment physical properties and a regional wetland inventory serving as a base for policy advice and the development of sustainable wetland management strategies.

## 1 Introduction

As one response to emerging fundamental questions on food security in Africa, the UNEP stresses the role of wetlands for agriculture in Africa (Frenken and Mharapara, 2002) leading to an encouragement of many African governments and NGOs for wetland farming practices to improve food security, reduce poverty and facilitate the diversification of rural livelihoods (McCartney et al., 2010). Wetlands, characterized by nutrient rich soils and year-round water and/or soil moisture availability, sustain especially smallholder farmers with opportunities to produce crops all year-round (Sakané et al., 2011). Nevertheless besides provisioning services like food, water and building material, wetlands provide a range of other regulating, cultural and supporting services (Rebelo et al., 2009). An increasing agricultural use of wetlands in Sub Saharan Africa could be observed during the recent decades, mainly driven by demographic growth, climate change and globalization effects (Bond, 2014; Ulrich, 2014). Degradation of overexploited upland fields, market opportunities and the need to earn cash income (Wood and van Halsema, 2008) are key influencing factors for a shift from upland agriculture to wetland agriculture. However, such an intensification of wetland use and the resulting increase in regional food production will only be achieved sustainably if the benefits derived from diverse wetland services can be reconciled with the requirements of increased food production. Wetlands are complex ecosystems in which groundwater and surface water interact. They are commonly defined by the presence of standing water for some period during the growing season, either at the surface or within the root zone (Lehner and Döll, 2004). According to a review paper of diverse global wetland studies (Amler et al., 2015) the global wetland extent is varying between 5.6 and 12 million km$^2$ with a range of uncertainty depending on the used wetland definition and the pragmatic difficulty of quantifying wetlands in aerial and satellite images. The most detailed global study on wetland extent is provided by Lehner and Döll (2004) estimating a global wetland extent of 9.17 million km$^2$ with about 1.3 million km$^2$ located in Africa. Wetlands in Sub Saharan Africa consist mainly of alluvial floodplains and inland valleys and cover an area of around 0.17 million km$^2$ in the East African countries of Kenya, Rwanda, Uganda, and Tanzania (Stevenson and Frazier, 1999). These two dominating inland wetland types cover more than 80 % of East African total wetland area. Alluvial floodplains are described by a well-defined river in a smoothly sloped valley characterized by oxbow depression,

Published by Copernicus Publications on behalf of the International Association of Hydrological Sciences.

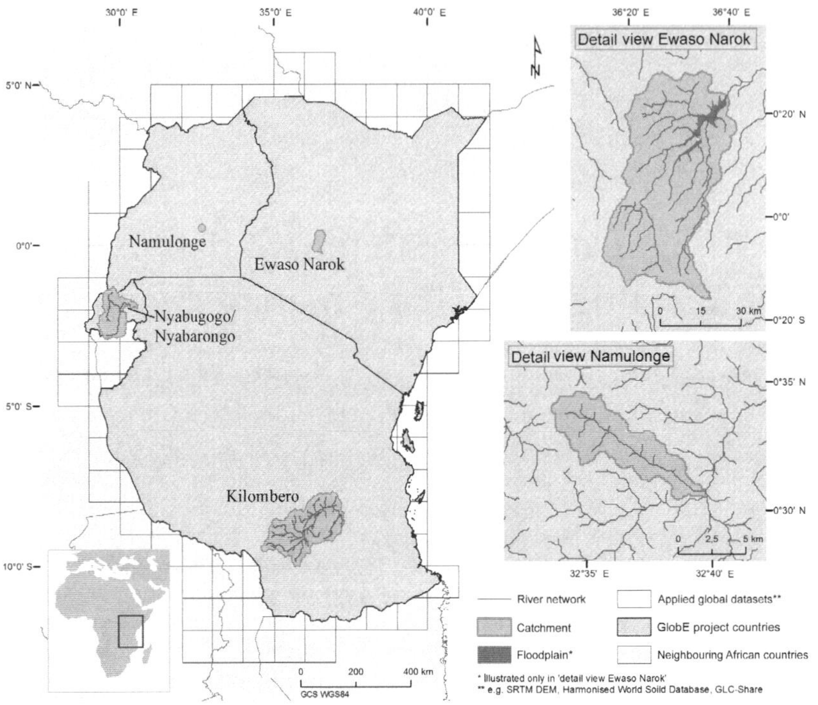


**Figure 1.** Nested experimental test site/regional wetland – catchment data base approach for East Africa comprising the countries Kenya, Uganda, Rwanda, and Tanzania. (http://gadm.org/, 2015 and Jet Proulsion Laboratory, 2015, modified)

natural levees and seasonal overbank flooding (Kotze et al., 2012). Inland valleys are seasonally water logged and often grass-covered linear depressions in head water zones of rivers comprising valley bottoms and hydromorphic fringes but excluding river floodplains (Mitsch and Gosselink, 2007; Rodenburg et al., 2014).

## 2 Methodology

### 2.1 Study sites

For our study (http://www.wetlands-africa.de/) we focus on the main East African freshwater geomorphic wetland types floodplain and inland valley in the countries Kenya, Rwanda, Uganda, and Tanzania. Besides the regional East African approach the scales of research are sub-divided into catchment, wetland and super test site scale (Fig. 1). Two floodplains and two inland valleys under semi-arid and humid climate conditions and their respective catchments are selected at the catchment-wetland scale (Table 1). For in-depth studies on the trade-offs and linkages of Ecosystem Services with water and matter fluxes under different land use systems, two smaller areas are selected as super test-sites for intensive in-situ monitoring campaigns. With an inland valley close to the Kampala-Jinja transect in Uganda and a smaller headwater catchment of the lowland Kilombero floodplain in Tanzania the two selected super test-sites cover our main priority wetland types of the overall research area in East Africa.

### 2.2 Wetland-catchment approach

The overall wetland water balance strongly determines the entire wetland ecosystem including all its functions and services (Dixon and Wood, 2003). Depending on the geomorphic setting of the wetland, besides precipitation, surface- and subsurface inflow has a strong impact on the wetland water balance. Therefore off-site factors like land use change, infrastructure measures including dams or river fragmentation and overall climate variability mainly determine the seasonal surface- and subsurface inflow and hence water availability of the wetland. In a nutshell, the off-site factors and their interactions with the physical state of the wetland determine the in- and outflow of the wetland and hence the water availability, as well as direct pressure for downstream ecosystem services. The physical state of the wetland comprises the onsite factors of physical wetland characteristics and its associated anthropogenic use. According to Kotze et al. (2012) these on-site and off-site factors influence wetland health on a scale of anthropogenic wetland degradation compared to its pristine state. Therefore wetlands cannot be described as isolated ecosystems (von der Heyden and New, 2003) and there is a strong need for a nested catchment-wetland approach to

**Table 1.** Wetland study sites and physical wetland and catchment characteristics, *super test-site of the project.

| Catchment/study site | Ewaso Narok | Nybarongo | Kilombero (Ifakara)* | Kyoga (Namulonge)* |
|---|---|---|---|---|
| Geomorphic wetland type | Alluvial floodplain | Alluvial floodplain and inland valleys | Alluvial floodplain | Inland valleys |
| Country | Kenya | Rwanda | Tanzania | Uganda |
| Wetland area [$km^2$] | 80.5 | 128.6 | 537.4 | 116.7 |
| Catchment area [$km^2$] | 2610 | 8900 num. headwater catchments | 40 240 | num. headwater catchments |
| Elevation [m a.s.l.] | 1600–2400 | 1200–3000 | 200–2500 | 1100 |
| Climate | semi-arid | humid | sub-humid | sub-humid |
| Annual precipitation [mm] | 700 | 1000 | 1400 | 1350 |

assess its overall state of degradation within the catchment context.

### 2.3 Regional East African wetland-catchment data base

The core idea of a regional East African wetland-catchment data base is to capture all direct and indirect drivers forcing wetland degradation on a wetland-catchment scale. In a causal network that is based on a combination of the DPSIR analysis and the Ecosystem Services framework (van Dam et al., 2013), the strong relationship between drivers and pressures allows an assessment of the degradation state of the wetland and its associated ecosystem services (Fig. 2). First of all this requires a harmonized, area-covering East African wetland map, which is based on standardised information about wetland location, type and extent applying remote sensing technology and global geo datasets (Fig. 1).

The large East African region with remote and hardly accessible wetlands is frequently cloud-covered due to its climatic conditions. Many wetlands in East Africa underlie a strong seasonality which needs to be considered and incorporated in a wetland map. These challenges in mapping of East African wetlands can be overcome by the combination of different sensor systems with a multitemporal data analysis approach. By now the publicly available, regularly recorded time series of medium to coarse resolution satellite data e.g. MODIS (NASA MODIS Data, 2015) and SPOT (VITO, 2015) are expected to be able to capture the distinct seasonality of wetlands, especially in comparison to their surrounding uplands. A map product which is compiled based on this rationale can be enhanced by data acquired from sensors with higher spatial resolution, e.g. Sentinel 2 (European Space Agency, 2015) which started its operational phase in 2015. In addition a regional East African catchment data base is needed to characterize the direct and indirect drivers for wetland degradation including demography, infrastructure measures and physical catchment properties in space. Physical catchment properties like e.g. topography, geology, soil, land use, and climate strongly determine water availability for each specific wetland and can be spatially represented by a regional hydrological model like e.g. the spatially distributed mesoscale hydrologic model mHM (Samaniego et al., 2010). Due to limited data availability the entire regional catchment data inventory has to be based on freely available global geo- and climate datasets like e.g. the Harmonised World Soil Database (Nachtergaele et al., 2012) and GLC-Share (Latham et al., 2014) for regional land use data. These freely available geo datasets are constantly evolving through improved methods in geostatistics in combination with increasing databases which lead to more precise and higher resolution datasets (e.g. Hengl et al., 2015). In order to select an optimum global geo- and climate dataset for the application of the mesoscale hydrological model simulating water availability on the regional East African scale the most suitable datasets are applied and assessed on the experimental catchment scale with the distributed process-based hydrological catchment model SWAT (Arnold et al., 1998). This ensemble modelling further serves as a model performance measure of the regional mesoscale hydrologic model that is using an ungauged modelling approach. On the experimental wetland scale small-scale hydrological processes and water related ecosystem services under different types of management are analyzed and wetland boundary conditions under varying climate and land use/infrastructure scenarios are simulated with the distributed hydrological catchment model SWAT. The multi-scale approach allows evaluating the agreement between regional and experimental wetland-catchment approach especially with respect to the significance of off-site factors on ecosystem services.

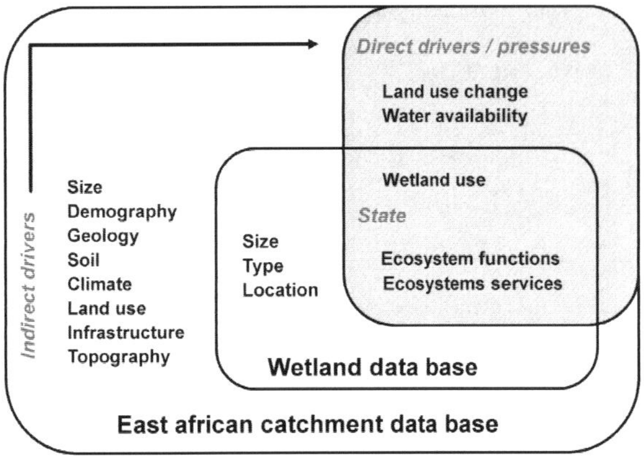


**Figure 2.** Conceptual model and data structure of the nested East African wetland-catchment data base.

## 3 Discussion

A combined DPSIR analysis/ Ecosystem Services framework can assist wetland policy and management by incorporating formal and informal knowledge about the functioning of the wetland system and analysis of trade-offs between human and ecosystem health (van Dam et al., 2013). Moreover the DPSIR approach allows a comprehensive mapping of the agriculture – wetland interactions identifying complex causal interrelations, multiple options and levels of response (Wood and van Halsema, 2008). There are a number of Sub Saharan African studies looking at wetland-agricultural interactions and their associated ecosystem services (Beuel et al., 2016; McCartney et al., 2010; Wood and van Halsema, 2008; McCartney and Houghton-Carr, 2009; Kotze et al., 2012). These studies are mainly conducted at the wetland scale and require a detailed on-site survey, always telling a unique local wetland story. However for regional environmental planning of sustainable wetland use a multiple scale analysis embedding the regional, catchment and wetland scale is needed. Rodenburg et al. (2014) have described a couple of studies that outline the regional or national potential of inland valleys with GIS analysis, spatial modelling and remote sensing techniques for identifying agricultural inland valley development potential. These studies map the inland valley agricultural potential but do not analyze the direct and indirect drivers and pressures leading to an exploitation of the wetland and hence a loss of ecosystem services and are furthermore lacking a study of hydrological wetland-catchment interactions. The experimental wetland-catchment framework serves as an overall reference data set to assess the analytic potential of the regional wetland catchment data base for sustainable regional and national wetland management. This includes in-depth hydrological process studies at the wetland scale looking at the impact of wetland management options on water related ecosystem services. Beyond the experimental wetland scale embedded catchment studies analyze the impact of indirect/- and direct drivers like e.g. climate variability and land use change on water availability, and hence the main direct drivers determining wetland water availability as a key pressure of water related ecosystem services of the wetland and down-stream users. Here water availability is one of the key direct drivers mapping regional hotspots of likely wetland degradation. For example land use change to subsistence and large scale agricultural farming leads to an overexploitation of water resources by river abstractions in one of our experimental test sites located in the Upper Ewaso Ng'iro North Basin, Central Kenya (Mutiga, 2010; Aeschbacher et al., 2005). This upstream catchment based direct driver has an immense impact on the water availability of the associated downstream wetlands like e.g. the Ewaso Narok wetland and hence on diverse water-related ecosystem functions and services. Here regional agricultural water management programs like promoting water efficient irrigation or crops and enforcing water abstraction permits for particular large scale agriculture farming by the local water authorities would help to regulate the uncontrolled water abstractions for crop irrigation and mitigate the direct impact on the water balance of the wetland. Due to the limited hydrological and meteorological data availability in East Africa, a spatially distributed mesoscale hydrological model that is based on global-data sets and embedded in a regional wetland-catchment data base is a promising tool to assess the off-site factors of wetland degradation at the regional scale. An assessment of model performance based on experimental catchment studies is needed to further evaluate the potential of the regional wetland-catchment data base for national and regional sustainable wetland management.

## 4 Conclusions

Wetlands perform a range of environmental functions and provide manifold socio-economic benefits to local communities and a wider population in developing countries (Dixon and Wood, 2003). For regional environmental planning a harmonized transnational comprehensive wetland-catchment data base is needed to assess wetland ecosystem services in a catchment context and identify hotspots of wetland degradation to foster further local sustainable wetland programs.

## 5 Data availability

Country borders in Fig. 1 were downloaded from http://gadm.org/download. The applied SRTM DEM in Fig. 1 was downloaded from USGS earth explorer (USGS, 2016) (http://earthexplorer.usgs.gov/) and modified with Arc SWAT where the authors of the article know that the SRTM DEM has errors which affect the modeling of river network and catchments. The delineation of the estimated floodplain in Fig. 1 is described in Beuel et al. (2016). The modified SRTM data as well as the approximate floodplain extent can be directly requested from the first author Constanze Leemhuis by e-mail.

**Acknowledgements.** This research is conducted under the GlobE: Wetlands in East Africa project, which is funded by the German Federal Ministry of Education and Research (FKZ: 031A250 A-H), with additional funding provided by the German Federal Ministry of Economic Cooperation and Development. Many thanks to our East African partners and all colleagues of the GlobE: Wetlands in East Africa project, who provided data and assistance.

## References

Aeschbacher, J., Liniger, H., and Weingartner, R.: River Water Shortage in a Highland – Lowland System, Mt. Res. Dev., 25, 155–162, doi:10.1659/0276-4741(2005)025[0155:RWSIAH]2.0.CO;2, 2005.

Amler, E., Schmidt, M., and Menz, G.: Definitions and Mapping of East African Wetlands: A Review, Remote Sensing, 7, 5256–5282, doi:10.3390/rs70505256, 2015.

Arnold, J. G., Srinivasan, R., Muttiah, R. S., and Williams, J. R.: Large area hydrologic modelling and assessment part I: model development, J. Am. Water Resour. Assoc., 34, 73–89, 1998.

Beuel, S., Alvarez, M., Amler, E., Behn, K., Kotze, D., Kreye, C., Leemhuis, C., Wagner, K., Willy, D. K., Ziegler, S., and Becker, M.: A rapid assessment of anthropogenic disturbances in East African Wetlands, Ecol. Indic., 67, 684–692, 2016.

Bond, J.: A holistic approach to natural resource conflict: The case of Laikipia County, Kenya, J. Rural Stud., 34, 117–127, 2014.

Dixon, A. and Wood, A.: Wetland Cultivation and Hydrological Management in Eastern Africa: Matching Community and Hydrological Needs Through Sustainable Wetland Use, Nat. Resour. Forum, 27, 117–129, 2003.

European Space Agency: Sentinel-2 User Handbook, available at: https://sentinel.esa.int/documents/247904/685211/Sentinel-2_User_Handbook last access: 28 September 2015.

Frenken, K. and Mharapara, I.: Wetland development and management in SADC countries, Proceedings of a sub-regional workshop held by FAO sub-regional office for East and Southern Africa (SAFR), Harare, Zimbabwe, 19–23 November, 2002.

GADM.org: Database of Global Administrative Areas, available at: http://www.gadm.org/ (last access: 24 August 2016), 2015.

Hengl, T., Heuvelink, G. B. M., Kempen, B., Leenaars, J. G. B., Walsh, M. G., Sila, A., Mc Millan, R. A., Mendes de Jesus, J., Tamene, L., and Tondoh, J. E.: Mapping Soil Properties of Africa at 250 m Resolution: Random Forests Significantly Improve Current Predictions, PLoS ONE, 10, e0125814, doi:10.1371/journal.pone.0125814, 2015.

Jet Propulsion Laboratory: U.S. releases enhanced Shuttle Land Elevation Data, available at: http://www.jpl.nasa.gov/news/news.php?release=2014-321 (last access: 24 August 2016), 2015.

Kotze, D. C., Ellery, W. N., Macfarlane, D. M., and Jewitt, G. P. W.: A rapid assessment method for coupling anthropogenic stressors and wetland ecological condition, Ecol. Indic., 13, 284–293, doi:10.1016/j.ecolind.2011.06.023, 2012.

Latham, J., Cumani, R., Rosati, I., and Bloise, M.: GLC-SHARE. Global Land Cover SHARE, available at: http://www.glcn.org/databases/lc_glcshare_downloads_en.jsp (last access: 25 August 2015), 2014.

Lehner, B. and Döll, P.: Development and validation of a global database of lakes, reservoirs and wetlands, J. Hydrol., 296, 1–22, doi:10.1016/j.jhydrol.2004.03.028, 2004.

McCartney, M. P. and Houghton-Carr, H. A.: Working wetland potential: an index guide for the sustainable development of African wetlands, Nat. Resour. Forum, 33, 99–110, 2009.

McCartney, M., Rebelo, L. M., Senaratna Sellamuttu, S., and de Silva, S.: Wetlands, agriculture and poverty reduction, IMWI Research Report 137, Colombo, Sri Lanka, International Water Management Institute (IWMI), 2010.

Mitsch, W. J. and Gosselink, J. G.: Wetlands, 4th edn., Wiley, New York, p. 600, 2007.

Mutiga, J. K., Mavengano, S. T., Zhongbo, S., Woldai, T., and Becht, R.: Water Allocation as a Planning Tool to Minimise Water Use Conflicts in the Upper Ewaso Ng'iro North Basin, Kenya, Water Resour. Manage., 24, 3939–3959, doi:10.1007/s11269-010-9641-9, 2010.

Nachtergaele, F., van Velthuizen, H., Verelst, L., and Wiberg, D.: Harmonised World Soil Database. Version 1.2, available at: http://www.fao.org/soils-portal/soil-survey/soil-maps-and-databases/harmonized-world-soil-database-v12/en/ (last access: 25 August 2015), 2012.

NASA MODIS Data: http://modis.gsfc.nasa.gov/data/, last access: 5 August 2015.

Rebelo, L. M., Finlayson, C. M., and Nagabhatla, N.: Remote sensing and GIS for wetland inventory, mapping and change analysis, J. Environ. Manage., 90, 2144–2153, doi:10.1016/j.jenvman.2007.06.027, 2009.

Rodenburg, J., Zwart, S. J., Kiepe, P., Narteh, L. T., Dogbe, W., and Wopereis, M. C. S.: Sustainable rice production in African inland valleys: Seizing regional potentials through local approaches, Agr. Syst., 123, 1–11, doi:10.1016/j.agsy.2013.09.004, 2014.

Sakané, N., Alvarez, M., Becker, M., Böhme, B., Handa, C., Kamiri, H., Langensiepen, M., Menz, G., Misana, S., Mogha, N., Möseler, B., Mwita, E., Oyieke, H., and van Wijk, M. T.: Classification, Characterisation, and Use of Small Wetlands in East Africa, Wetlands, 31, 1103–1116, doi:10.1007/s13157-011-0221-4, 2011.

Samaniego, L., Kumar, R., and Attinger, S.: Multiscale parameter regionalization of a grid-based hydrologic model at the mesoscale, Water Resour. Res., 46, 1–25, doi:10.1029/2008WR007327, 2010.

Stevensons, S. and Frazier, S.: Review on wetland inventory information in Africa, in: Global review on wetland resources and priorities for wetland inventory. Supervising Scientist Report 144, Wetlands International Publication 53, edited by: Finlayson, C. M. and Spiers, A. G., Supervising Scientist, Canberra, 1999.

Ulrich, A.: Export-oriented horticultural production in laikipia, kenya: Assessing the implications for rural livelihoods, Sustainability (Switzerland), 6, 336–347, 2014.

USGS: Earth Explorer, available at http://earthexplorer.usgs.gov/, last access: 24 August 2016.

van Dam, A., Kipkemboi, J., Rahman, M., and Gettel, G.: Linking Hydrology, Ecosystem Function, and Livelihood Outcomes in African Papyrus Wetlands Using a Bayesian Network Model, Wetlands, 33, 381–397, doi:10.1007/s13157-013-0395-z, 2013.

VITO: SPOT-Vegetation Programme, available at: http://www.vgt.vito.be/, last access: 5 August 2015.

von der Heyden, C. J. and New, M. G.: The role of a dambo in the hydrology of a catchment and the river network downstream, Hydrol. Earth Syst. Sci., 7, 339–357, doi:10.5194/hess-7-339-2003, 2003.

Wood, A. and van Halsema, G. E.: Scoping agriculture wetland interactions. Towards a sustainable multiple-response strategy, FAO Water Report 33, Food and Agriculture Organization of the United Nations, Rome, Italy, 155, p. 17, 2008.

Proc. IAHS, 374, 129–136, 2016
proc-iahs.net/374/129/2016/
doi:10.5194/piahs-374-129-2016


Open Access
PIAHS

# The relationship between irrigation water demand and drought in the Yellow River basin

**Yu Wang[1], Weihao Wang[2], Shaoming Peng[1], Guiqin Jiang[1], and Jian Wu[1]**

[1]Yellow River Engineering Consulting Co., Ltd., Zhengzhou, Henan, 450000, China
[2]Institute of Hydrology and Water Resources Hohai University, Nanjing, 210000, China

*Correspondence to:* Yu Wang (happytimes@126.com)

Published: 17 October 2016

**Abstract.** In order to organize water for drought resistance reasonably, we need to study the relationship between irrigation water demand and meteorological drought in quantitative way. We chose five typical irrigation districts including the Qingtongxia irrigation district, Yellow River irrigation districts of Inner Mongolia in the upper reaches of the Yellow River, the Fen river irrigation district and the Wei river irrigation district in the middle reaches of the Yellow River and the irrigation districts in the lower reaches of the Yellow River as research area. Based on the hydrology, meteorology, groundwater and crop parameters materials from 1956 to 2010 in the Yellow River basin, we selected reconnaissance drought index (RDI) to analyze occurrence and evolution regularity of drought in the five typical irrigation districts, and calculated the corresponding irrigation water demand by using crop water balance equation. The relationship of drought and irrigation water demand in each typical irrigation district was studied by using grey correlation analysis and relevant analysis method, and the quantitative relationship between irrigation water demand and RDI was established in each typical irrigation district. The results showed that the RDI can be applied to evaluate the meteorological drought in the typical irrigation districts of the Yellow River basin. There is significant correlation between the irrigation water demand and RDI, and the grey correlation degree and correlation coefficient increased with increasing crops available effective rainfall. The irrigation water demand of irrigation districts in the upstream, middle and downstream of the Yellow River basin presented different response degrees to drought. The irrigation water demand increased 105 million $m^3$ with the drought increasing one grade (RDI decreasing 0.5) in the Qingtongxia irrigation district and Yellow River irrigation districts of Inner Mongolia. The irrigation water demand increased 219 million $m^3$ with the drought increasing one grade in the Fen river irrigation district and Wei river irrigation district. The irrigation water demand increased 622 million $m^3$ with the drought increasing one grade in the downstream of Yellow River irrigation districts.

## 1 Introduction

The Yellow River Basin is one of the areas hardest hit by drought in history due to its low rainfall and high evaporation. Over the past 3710 years from 1766 to 1944 BC, the Yellow River basin experienced 107 droughts recorded in history. During the 25 years from 1950 and 1974, 17 droughts occurred, averagely once every one year and half, including 9 severe droughts. In the past 20 years, severe droughts occurred many times in the upper and middle reaches of the Yellow River Basin, resulted in a significant yield reduction and very difficult to get drinking water for local people. The drought in 1997 caused record days of Yellow River cutoff and a record cutoff river length. Since 2000, the Yellow River almost saw successive years of drought, severe drought occurred in 2008 and 2010, resulted in agricultural huge losses. Yellow River Basin is a resource-based water shortage area, contradiction in water supply and demand in arid years becomes more prominent, irrigation water is limited, in this sense, the quantitative study of the relationship between meteorological drought and irrigated water demand is of great significance for the rational organization of irrigation water

Published by Copernicus Publications on behalf of the International Association of Hydrological Sciences.

**Table 1.** Main irrigation districts and there characteristics index in the Yellow River basin.

| Name of irrigation district | Location | Effective irrigation area ($hm^2$) | Precipitation station number (set) | Evaporator station number (set) | Annual mean precipitation (mm) | Annual mean evaporation capacity (mm) | Groundwater buried depth (m) |
|---|---|---|---|---|---|---|---|
| Qingtongxia irrigation district | Xiaheyan to Shizhuishan reaches | 44.1 | 12 | 8 | 196.0 | 900.7 | 0.5–3.0 |
| Hetao irrigation district | Shuizuishan to Hekouzhen reaches | 55.9 | 15 | 10 | 200.2 | 1257.5 | 0.6–2.7 |
| Fen river irrigation district | Taiyuan basin of middle Shanxi at Fen River Basin | 44.2 | 18 | 9 | 491.5 | 899.4 | 1.0–30.0 |
| Wei river irrigation district | Guanzhong Plain of Shannxi at Wei River Basin | 83.5 | 19 | 12 | 543.8 | 1144.5 | 5.0–20.0 |
| Irrigation district in lower reaches of the Yellow River | Henan, Shandong below Huanyuankou reaches | 220.0 | 24 | 15 | 672.7 | 886.9 | 0.5–5.0 |

sources. Extensive research of drought assessment both at home and abroad has been carried out, fruitful results have been achieved, however, quantitative research of the relationship between drought and irrigation requirement is less. Take the main irrigation districts of Yellow River basin as the study areas; this paper presents the quantitative study of the relationship between irrigation requirement and meteorological drought, aims to providing a scientific basis for the Yellow River drought mitigation.

## 2 Overview of irrigation districts

Climatic conditions and Water resources in the Yellow River basin determine the development of agriculture relies heavily on irrigation, large and medium irrigation districts play a pivotal role in agricultural production. Agricultural bases in the Yellow River basin are concentrated in plains and channel basins. For example, Ningxia-Inner Mongolia Hetao Plain in the upper reaches of Yellow River is the successful model in building "oasis agriculture" in arid area, Fen-Wei basin in the middle reaches is one of key agricultural production bases, the irrigation district in the lower Yellow River, located at Huang-Huai-Hai plain, is the major grain producing area in China. This study takes the Qingtongxia irrigation district and Hetao irrigation district in the upper reaches, Fen river irrigation district and Wei river irrigation district in the middle reaches, and Yellow River irrigation district in the lower reaches as study areas. The main irrigation districts and their characteristics index are given in Table 1.

The data referred to in this study include: hydrology, meteorology, groundwater and crop parameters. Hydrological, meteorological data refer to such data cluster as precipitation, evaporation, temperature, pressure, sunshine duration, wind speed and relative humidity from 1956 to 2010 of main irrigation districts and neighbored 142 weather stations (including 88 precipitation stations and 54 evaporator stations) as released by National Climate Information Center of China Weather Bureau, the groundwater level data refers to groundwater monitoring and correlation study fruits of Yellow River (Chen and Liu, 2013; Sun et al., 2014; Wu, 2009; Zhao et al., 2009), crop parameter refers to the survey and statistic of comprehensive planning of water resources of Yellow River and CROP database of FAO (MOA, 2009).

## 3 Drought assessment of irrigation district

### 3.1 Indices and methods for drought assessment of irrigation district

Meteorological drought means the abnormal water shortages as caused by the imbalance of precipitation and evaporation, which can be characterized by precipitation, evaporation ratio or statistics. The agricultural drought means that output reduction or output failure may occur when normal growth of crop is affected due to crop water imbalance, water deficit as caused by environmental factors or human activities. If not considering the intervention of human activities, meteorological drought is the only driving factor of agricultural drought, so selection of rational meteorological drought index is the key for quantitative analysis of the relationship between meteorological drought and irrigation requirement.

Common meteorology drought index can be divided into single factor indicator, dual-factor indicator and multivariate indicators (Wang et al., 2011, 2012; Zhang et al., 2007). Single factor indicator includes precipitation or accumulated precipitation anomaly percentage, precipitation or accumulated precipitation standardized statistic standardized precipitation index (SPI), etc., the calculation of which is simple, only regarding statistic of precipitation as the characterization indicator of meteorology drought, incapable of reflecting the whole characteristic of meteorology drought; dual-factor indicator, such as $K$ index of aridity, reconnaissance drought index (RDI), and so on, such index uses the precipitation and evaporation ratio or the statistic thereof to character the meteorology drought, thus reflecting drought from the view of water balance. Multivariate index involves a typical Palmer drought index (PDSI) established based on soil water equilibrium principle, so the calculation method is more complicated.

In combination with the meteorological drought characteristics of Yellow River irrigation district, we can select RDI as the drought evaluation index of typical irrigation district. Compared with other indicators, RDI has the following advantages (Tsakiris and Vangelis, 2005; Eleanor and Simon, 2010; Ashok and Vijay, 2011): (1) capable of applying to drought assessment at river basin, region or other different levels; (2) suitable for drought assessment at any time or scale; (3) using the cumulative deficit between rainfall and evaporation to characterize the drought, having decisive physical significance; (4) compared with single factor index, RDI takes into account precipitation, evaporation factor, while the evaporation is affected by many climatic factors such as temperature, humidity, sunshine, so more sensitive to environmental changes, suitable for regional/basin drought evaluation under conditions of climate; (5) capable of establishing effective links with the soil water deficit; (6) since the data is relatively easy to be obtained, the calculation method is not complicated.

RDI calculation method: As for an evaluated area, assuming that there are $n$ evaluation periods, the ratio $x_i$ of precipitation $P_i$ to potential evaporation $\mathrm{PET}_i$ can be calculated following Eq. (1):

$$x_i = \frac{P_i}{\mathrm{PET}_i}, \quad i = 1, 2, l..., n. \tag{1}$$

Assuming $x_i$ is stochastic variable $x$ and subjected to $\Gamma$ distribution, the probability density distribution function $f(x)$, based on sample data, can be following Eq. (2):

$$f(x) = \frac{1}{\beta^\gamma \Gamma(x)} x^{\gamma-1} e^{-x/\beta}, \quad x > 0 \tag{2}$$

where $\beta > 0$, $\gamma > 0$ are respectively for scale parameter and shape parameter of Probability Density Function, which can be came at by maximum likelihood estimate method. As for the ration $x_i$ of precipitation to potential evaporation at some period, the probability stochastic variable $x$ is less than $x_i$ can be acquired based on Eq. (3); if $x_i = 0$, the probability which stochastic variable $x$ is less than $x_i$ can be estimated based on Eq. (4).

$$F(x < x_i) = \int_0^\infty f(x)\mathrm{d}x \tag{3}$$

$$F(x_i = 0) = k/n \tag{4}$$

where $k$ is for sample number when $x_i = 0$; $n$ is for total sample number, i.e. the number at evaluated period.

If normally standardized treatment of Probability Distribution Function $n$ $(x < x_i)$, we can get Eq. (5), the functional morphology parameter $z$ value is the purposed RDI value. The evaluation criteria of RDI see Table 2.

$$F(x < x_i) = \frac{1}{\sqrt{2\pi}} \int_0^\infty e^{-z^2/2}\mathrm{d}x \tag{5}$$

**Table 2.** Grading classification of meteorology drought based on RDI.

| Status | RDI | Grade |
|---|---|---|
| 0 | $-0.5 < \mathrm{RDI}$ | No drought |
| 1 | $-1.0 < \mathrm{RDI} \leq -0.5$ | Light drought |
| 2 | $-1.5 < \mathrm{RDI} \leq -1.0$ | Medium drought |
| 3 | $-2.0 < \mathrm{RDI} \leq -1.5$ | Heavy drought |
| 4 | $\mathrm{RDI} \leq -2.0$ | Special drought |

### 3.2 Drought evaluation of irrigation district based on RDI

RDI is used to have an evaluation of drought in main irrigation districts of upper, middle and lower reaches of Yellow River in 1956 to 2010, the results are given in Fig. 1. The irrigation districts experienced 86 droughts, including 11 severe droughts, which is in reasonable agreement with the typical drought conditions as recorded historically for Yellow River, for example, the oversized drought occurring in 1965, 1966, 2005, 2009 and 2010. From the view of drought conditions at various irrigation districts, Qingtongxia irrigation district experienced 18 droughts, most severe in 1980s; Hetao irrigation district experienced 21 droughts, most severe in 1980s; Wei River irrigation district experienced 20 droughts, most severe since 2000; Fen river irrigation district experienced 17 droughts, most severe in 1990s; Yellow River irrigation district in the lower reaches experienced 10 droughts, relatively severe in 1950 to 1960, and 1980s. From the view of drought changing trend, the RDI value of irrigation districts in the upper and middle reaches trends toward decrease, which shows that drought trends toward exacerbated; the RDI value of irrigation districts in the lower reaches trends toward increase, the drought may be somewhat alleviated.

## 4 Irrigation requirement of irrigation districts

### 4.1 Water demand for crop irrigation

Crop irrigation requirement means the irrigation quantity required for high yield during overall growing season under feasible soil moisture and fertility level, which can be inquired into through field water layer (or soil moisture) dynamic equilibrium in a simulation way. The water source available for crop includes effective precipitation, soil water and groundwater-recharged water quantity, the crop irrigation is intended to supplementing plant transpiration soil evaporation of water. Since dry crop is mainly planted along the Yellow River, the net irrigation requirement during growing season is subjected to water balance theory, and the water balance formulae is given following Eq. (6):

$$Q_i = \mathrm{ET}_{\mathrm{c}_i} - \mathrm{Pe} - \mathrm{Ge}_i + \Delta W \tag{6}$$

where $Q_i$ is for net irrigation requirement(mm) of $i$th crop; $\mathrm{ET}_{\mathrm{c}_i}$ is for evaporative capacity of $i$th crop transpiration (mm), $ET_{c_i} = K_{c_i} \times ET_0$, $ET_0$ is for evaporative capacity of reference crop transpiration, which can be calculated by Penman–Monteith formulae as recommended by FAO; $K_{c_i}$ is for crop coefficient, adopting different value depending on different growing phases; Pe is for monthly effective precipitation (mm) during growing season, which can be calculated based on monthly rainfall during crop growing season and available factor at irrigation district; $Ge_i$ is for the groundwater utilization quantity (mm) during $i$th crop growing season, the groundwater utilization is the function for the Setting Depth and evapotranspiration conditions under some soil texture and crop conditions: $Ge = f(H) \times ET_{c_i}$, of which $f(H)$ is for the coefficient of groundwater utilization, relating to the groundwater Setting Depth, for example, when groundwater level ranges from 1 to 3 m, Ge may take 30 to 10 % of $ET_{c_i}$, when groundwater level is more than 3 m, the crop groundwater utilization quantity can be ignored; $\Delta W$ is for the variation of soil moisture storage capacity of planned wet-layer soil during growing season, in case of study of crop water demand at long period, the variation of moisture storage capacity can be ignored.

## 4.2 Irrigation requirement in irrigation districts

The water demand in irrigation district relates to planted crops and sown area in irrigation district, the calculation of it needs to take into account the planting structure, sown area of leading crops, net ration of irrigation requirement for crop, irrigation water use factor and so on. In order to analyze annual change of water demand order in irrigation district and its response to drought index fluctuation, we can analyze irrigation requirement in irrigation district under identical crop planting structure, soil condition and irrigation water management model at 2010 level. The planting structure of crops in irrigation districts of upper, middle and lower reaches of Yellow River is given in Table 3.

From the view of decadal variations of irrigation water demand, the change of water demand of irrigation districts of Yellow River in 1956 to 2010 is opposite in direction to the drought index fluctuation, i.e. the irrigation requirement is large when drought index is small, but small when large; the maximum water demand in Qingtongxia irrigation district appeared in 1980s, minimum water demand in 1956 to 1969, the maximum water demand in Hetao and Fen river irrigation districts appeared after 2000, the maximum water demand in Wei river irrigation district appeared in 1990s, while that in Yellow River irrigation district in the lower reaches appeared in 1956 to 1969. The water demand of main irrigation districts of Yellow River in different periods is given in Table 4.

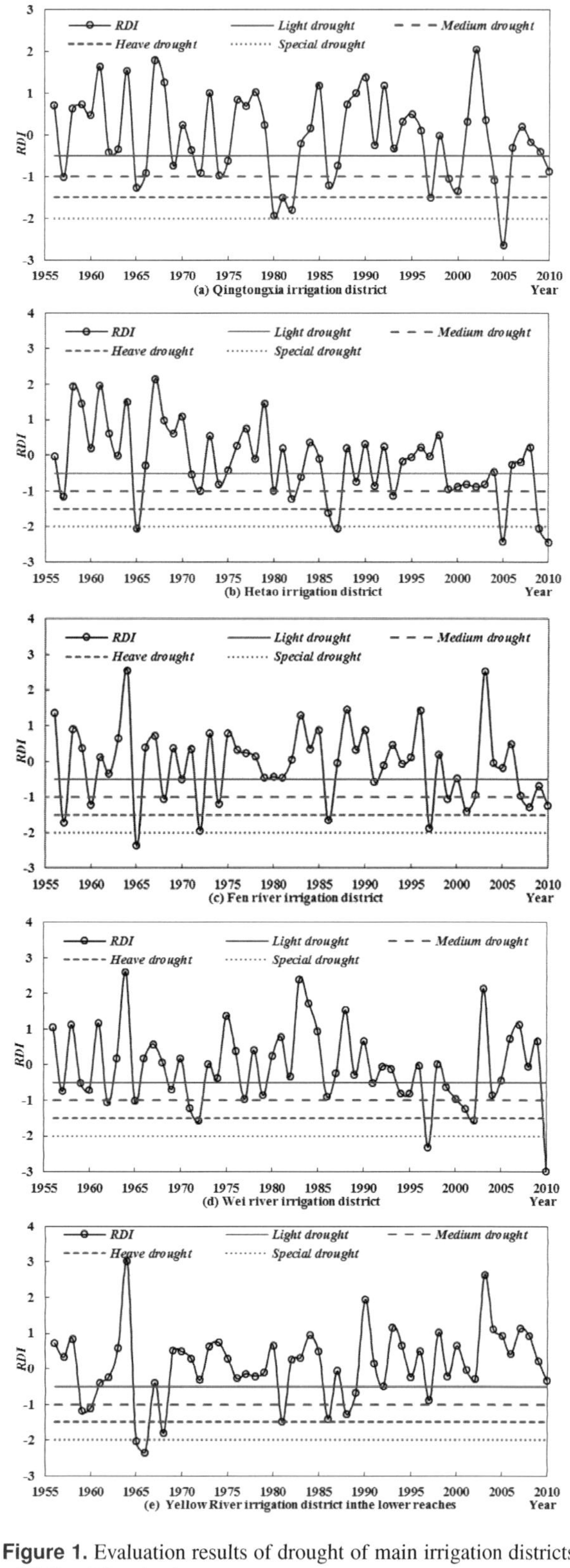


**Figure 1.** Evaluation results of drought of main irrigation districts of Yellow River basin.

**Table 3.** Agricultural planting structure of main irrigation districts of Yellow River in 2010.

| | Grain crop (%) | | | | Economic crop (%) | | | | | Forest and grassland farming (%) | |
|---|---|---|---|---|---|---|---|---|---|---|---|
| Name of irrigation district | Rice | wheat | corn | Coarse cereals | Cotton | Oil plants | Melon and fruit | Vegetables | Others | Forest fruit | Pasture |
| Qingtongxia irrigation district | 22.1 | 13.4 | 41.5 | 0.0 | 0.0 | 1.7 | 0.0 | 11.2 | 0.0 | 10.1 | 0.0 |
| Hetao irrigation district | 0.0 | 17.9 | 24.0 | 3.6 | 0.0 | 27.9 | 2.5 | 5.3 | 8.5 | 4.8 | 5.7 |
| Fen river irrigation district | 0.0 | 33.7 | 40.5 | 6.1 | 0.0 | 3.4 | 0.5 | 1.8 | 13.9 | 0.0 | 0.0 |
| Wei river irrigation district | 0.0 | 23.7 | 16.5 | 1.7 | 0.0 | 39.1 | 6.7 | 6.1 | 0.0 | 0.0 | 6.1 |
| Yellow River irrigation district | 6.8 | 29.5 | 28.3 | 5.2 | 0.3 | 4.8 | 5.4 | 7.9 | 2.2 | 9.6 | 0.0 |

**Table 4.** Irrigation requirement of main irrigation districts of Yellow River in different periods (unit: 0.1 billion $m^3$.

| Periods/year | Qingtongxia irrigation district | Hetao irrigation district | Fen River irrigation district | Wei River irrigation district | Yellow River irrigation district in lower reaches |
|---|---|---|---|---|---|
| 1956–1969 | 49.59 | 47.75 | 22.56 | 33.37 | 102.52 |
| 1970–1979 | 53.03 | 51.00 | 25.18 | 38.36 | 89.03 |
| 1980–1989 | 54.57 | 53.94 | 24.45 | 35.45 | 100.47 |
| 1990–1999 | 52.77 | 52.81 | 25.68 | 43.05 | 88.85 |
| 2000–2010 | 54.39 | 54.77 | 30.27 | 41.46 | 91.04 |

## 5 Relationship between irrigation requirement and drought in irrigation districts

### 5.1 Link mechanism between drought and irrigation requirement

From the view of overall growth of crop, the moisture source includes precipitation, groundwater utilization and overhead damping. At initial period of drought, the reduced available precipitation results in reduced soil moisture, the crop begins to adjust the guidance degree of its pore, thus reducing somewhat transpiration evaporative capacity and maintaining the body water equilibrium; when the drought goes forward, the soil moisture content drops continuously, when exceeding over its conditions and abilities, the moisture deficit occurs, simultaneously, drought development causes such climatic factors as air temperature, humidity, illumination, wind speed and so on developing toward the direction in favor of evaporation, in this way the transpiration evaporative capacity of crop increases, irrigation requirement is also large.

### 5.2 Analysis of correlation between drought and irrigation requirement

In order to make clear the relation between drought in irrigation district and irrigation requirement, we can use grey correlation analysis method to judge the relevance size according to the resemblance between discrete data of gray system, and the correlation degree is the measurement to reflect the size of the nearness level. We can use grey correlation analysis method to calculate the grey correlation degree between them, regarding drought index RDI of irrigation district as the reference sequence, regarding irrigation requirement in irrigation district as the comparison sequence. Prior to the correlation analysis, we may have a dimensionless treatment of reference sequence and comparison sequence, use standardized treatment.

$$S_i(t) = \frac{x_i(t) - \overline{x_i}}{\sigma} \tag{7}$$

where $i = 0, 1$ is for reference sequence and comparison sequence, $t = 1, 2, \ldots, N$.

The absolute error between comparison sequence and reference sequence at $t$ time is:

$$\Delta(t) = S_1(t) - S_0(t). \tag{8}$$

The maximum difference is:

$$\Delta_{\max} = \max|S_1(t) - S_0(t)|. \tag{9}$$

The minimum difference is:

$$\Delta_{\min} = \min|S_1(t) - S_0(t)|. \tag{10}$$

The correlation coefficient of two sequences at $t$ time is:

$$\xi(t) = \frac{\Delta_{\min} + 0.5\Delta_{\max}}{\Delta(t) + 0.5\Delta_{\max}}. \tag{11}$$

**Table 5.** Correlation relationship between irrigation requirement and RDI drought index of main irrigation districts of Yellow River.

| Name of irrigation district | Qingtongxia irrigation district | Hetao irrigation district | Wei River irrigation district | Fen River irrigation district | Yellow River irrigation district in lower reaches |
|---|---|---|---|---|---|
| Grey correlation degree $r$ | 0.596 | 0.614 | 0.694 | 0.671 | 0.719 |
| Related coefficient $R$ | 0.368 | 0.571 | 0.600 | 0.596 | 0.753 |

Correlation degree of two sequences is:

$$r = \frac{1}{N}\sum_{t=1}^{N}\xi(t). \quad (12)$$

In order to prove the relationship between irrigation requirement and drought at irrigation district, we can use related coefficient for demonstration; the correlation analysis between irrigation requirement and drought index of main irrigation districts of Yellow River from 1956 to 2010 and analytic results of related coefficient are given in Table 5.

The analytic result concerning grey correlation degree between irrigation requirement and RDI drought index of key irrigation districts of Yellow River shows that their grey degrees of association are both more than 0.5, which reveals that they two have high relevance; from the view of value change of grey correlation degree, the correlation degree from the upper to the lower reaches increases gradually, while that of Qingtongxia irrigation district is the least at 0.596, that of Yellow River irrigation district in the lower reaches is the maximum at 0.719, which shows that the relevance between irrigation requirement of irrigation district of upper reaches and RDI is weaker than the lower reaches. In order to judge the correlation nearness level between irrigation requirement and drought in irrigation districts, we can use correlation coefficient for demonstration, the correlation coefficient spans $[-1, 1]$, if $|R|$ value is more approaching to 1, which shows the linearly dependent level between irrigated requirement and RDI is higher, if $|R|$ value is more approaching to 0, which shows that their linearly dependent level is lower, when $|R|$ value is greater than 0.5, it can be believed that they have significant linearly dependent relationship. Through the analysis of correlation coefficient between irrigation requirement and RDI in main irrigation districts, the correlation coefficient of Qingtongxia is 0.368, less than 0.5, so is low correlation, the correlation coefficient of other irrigation districts is greater than 0.5, so is significant correlation, which shows that linearly correlativity exists between irrigation requirement and RDI, and the spatial variation law is basically consistent with grey correlation degree, correlation coefficient from upper to lower reaches trends toward increase.

From the view of main factors affecting irrigation requirement and drought, the annual mean precipitation of Qingtongxia irrigation district is only 196 mm, annual mean evaporative capacity is 900.7 mm, groundwater level maintains at 0.5 to 3 m, the total evapotranspiration of crop is large, efficient utilization quantity of rainfall is small, the moisture supplementation during growing season gives priority to irrigation and groundwater utilization, the irrigation requirement is less affected by meteorological factor fluctuation, and the correlation degree with RDI in the irrigation district is the least; annual mean precipitation of Hetao irrigation district is 200.2 mm, annual mean evaporative capacity is 1257.5 mm, permanent ground-water level ranges from 0.63 to 2.68 m, the available effective precipitation of crop is less while groundwater utilization quantity is large, the irrigation requirement is less affected by meteorological factors, and the correlation degree with RDI in the irrigation district is relatively small; the mean annual precipitation at Fen River and Wei irrigation districts in the middle reaches is 500mm or above, the groundwater level of most areas is more than 5m, the available groundwater utilization is limited, and its specific gravity increases by crop available rainfall, the impact of irrigation requirement by Irrigation District factors is gradually increased, the correlation degree and correlation coefficient between irrigation requirement and RDI are both larger; the precipitation of Yellow River irrigation district in the lower reaches is 600 to 800 mm, the effective rainfall serves as the main water source for growing season, therefore, irrigation requirement is susceptible to the influence by meteorological factors, the correlation degree and correlation coefficient between irrigation requirement and RDI is the maximal. Upon the comprehensive analysis of relationship between irrigation requirement and RDI, it shows that with the increase of annual mean precipitation (the increase of crop available effective precipitation), the incidence relation between irrigation requirement and RDI is enhanced.

### 5.3 Quantitative relation between irrigation requirement and drought index

Regression analysis is a sort of mathematical statistics method to treat the correlativity between variables (Chen et al., 1995). Regarding RDI drought index as the argument, and regarding irrigation requirement of irrigation district as the dependent variable, we can use least square method to establish regression equation for irrigation requirement and

**Table 6.** Drought index regression between irrigation requirement and RDI of main irrigation districts of Yellow River.

| Irrigation district | Irrigation requirement-drought index regression equation | $t$ test value |
|---|---|---|
| Qingtongxia irrigation district | $Q = -1.3593\,\text{RDI} + 53.279$ | 26.576 |
| Hetao irrigation district | $Q = -2.8136\,\text{RDI} + 51.774$ | 48.488 |
| Fen River irrigation district | $Q = -3.8019\,\text{RDI} + 25.648$ | 56.664 |
| Wei River irrigation district | $Q = -4.9569\,\text{RDI} + 38.413$ | 61.463 |
| Yellow River irrigation district in lower reaches | $Q = -12.431\,\text{RDI} + 92.292$ | 98.731 |

RDI of main irrigation districts of Yellow River to study the correlativity between irrigation requirement and RDI and find out the inherent law (Li, 2011; Tong, 2007; Chen, 2010; Wu and Zhao, 2010; Zhang et al., 2008). To prove whether or not there is significance in the equation of linear regression for irrigation requirement and RDI, we have a $t$ test of reliability. Under the $t$ distribution in which the degree of freedom is 53 ($n-2$, $n$ study order years' number 55) and confidence level $\alpha = 0.05$, the critical value of $t$ distribution, and the proven result is given as follows: the $t$ test value of regression equation is far greater than proven critical value 2.0091, which shows that there exists significant linear relationship between irrigation requirement and RDI. The equation for irrigation requirement and RDI of main irrigation districts to use regression analysis is given in Table 6.

From the view of regression relation between irrigation requirement and drought index of Yellow River, we discovered that with the decrease of RDI index (drought level increases), the irrigation requirement increases. The irrigation requirements of all irrigation districts are different in response to RDI change, for example, the response of irrigation requirement of irrigation district in the upper reaches to RDI change is week relatively, while that in the middle and lower reaches is strong relatively. When RDI index of Qingtongxia irrigation district in the upper reaches decreases by 0.1, the irrigation requirement increases 0.0136 billion $m^3$; when drought increases by one grade (RDI index decreases by 0.5), irrigation requirement on average increases 0.068 billion $m^3$; when RDI index of Hetao irrigation district decreases by 0.1, the irrigation requirement increases 0.0281 billion $m^3$; when drought increases by one grade, irrigation requirement on average increases 0.141 billion $m^3$; when RDI index of Fen River irrigation district in the middle reaches decreases by 0.1, the irrigation requirement increases 0.0380 billion $m^3$; when drought increases by one grade, irrigation requirement on average increases 0.190 billion $m^3$; when RDI index of Wei River irrigation district in the middle reaches decreases by 0.1, the irrigation requirement increases 0.0496 billion $m^3$; when drought increases by one grade, irrigation requirement on average increases 0.248 billion $m^3$; when RDI index of Yellow River irrigation districts in the lower reaches decreases by 0.1, the irrigation requirement increases 0.1243 billion $m^3$; when drought increases by one grade, irrigation requirement on average increases 0.622 billion $m^3$.

## 6 Conclusions

1. Meteorology drought is the major driving factor of agricultural drought. Based on the analysis of existing drought evaluation method and characteristic, we use RDI drought index to evaluate the drought condition in main irrigation districts of Yellow River in 1956 to 2010, the evaluation results are basically consistent with recorded drought of irrigation district of Yellow River, which shows that it is suitable for the drought evaluation of irrigation district of Yellow River. From the view of development trend of drought of irrigation district as evaluated by RDI, the drought degree of irrigation districts in the upper, middle reaches of Yellow River increases to some extent, the drought degree of irrigation districts in the lower reaches somewhat decreases.

2. In consideration of planting structure in the irrigation districts, sown area of leading crop, net irrigation requirement of crop and utilization coefficient of irrigation requirement and other factors, we use crop water balance equation to analyze and calculate the sequence irrigation requirement of main irrigation districts in 1956 to 2010, which shows that the irrigation requirement fluctuation is synchronized with and opposites in direction to the change of drought index (i.e. the irrigation requirement is large for the period when drought index is small).

3. The grey correlation degree and related data are applied to analyze the relationship between irrigation requirement of irrigated areas in Yellow River and drought index, which shows the significant relationship exists between them, increases with the increasing rainfall in irrigation district (the crop's efficient rainfall utilization also increases), the response of irrigation requirement to meteorological factors is enhanced, the correlation between irrigation requirement of middle and lower reaches and drought index is apparently higher than that of upper reaches. The regression analysis of irrigation requirement and drought index shows that with the reduction of RDI (drought grade increases), irrigation requirement increases, the response of irrigation requirement in different irrigation districts to drought index change is different, when drought is advanced with one grade (RDI index reduced by 0.5), irrigation requirement of Qingtongxia and Hetao irrigation districts in the upper reaches on average increases 0.105 billion $m^3$, that of Fen river and Wei river Districts in the middle reaches increases 0.219 billion $m^3$, that of Yellow

River irrigation district in the lower reaches increases 0.622 billion $m^3$.

## 7 Data availability

All the data sets we used in the article were publicly accessible, the meteorological data was released by National Climate Information Center of China Weather Bureau, accessed at http://www.data.cma.cn, and other data sets like characteristic of irrigation districts of Yellow River basin can be obtained on Water Resources Bulletin for Yellow River accessed at http://www.yellowriver.gov.cn/other/hhgb/ published by Yellow River Conservancy Commission.

**Acknowledgements.** This work was jointly supported by the project of the National Key Technologies R & D Program (Grant No. 2013BAC10B02) and the National International Scientific and Technological Cooperation of China (Grant No. 2013DFG70990).

## References

Ashok, K. M. and Vijay, P. S.: Drought modeling – A review, J. Hydrol., 403, 157–175, doi:10.1016/j.jhydrol.2011.03.049, 2011.

Chen, J. W.: Impact of climate change on irrigation requirement of typical crop of Heihe river basin, J. Irrig. Drain., 3, 69–73, 2010.

Chen, Y. C. and Liu, H. L.: Research on the relationship between groundwater of Qingtongxia irrigation district and soil salinization, Value Eng., 2, 311–313, 2013.

Chen, Y. M., Guo, G. S., Wang, G. X., Kang, S. Z., Luo, H. B., and Zhang, D. Z.: Water demand and irrigation of Chinese major crop, China Water Power Press, Beijing, 1995.

Eleanor, J. B. and Simon, J. B.: Regional drought over the UK and changes in the future, J. Hydrol., 394, 471–485, doi:10.1016/j.jhydrol.2010.10.003, 2010.

Li, Y. X.: Application of regression analysis in hydrological data processing, Natural Science, Chongqing University of Science and Technology, Chongqing, 5, 177–179, 2011.

New China Agricultural Statistics in 60 Years, China Agriculture Press, Beijing, Ministry of Agriculture of The People's Republic of China (MOA), 2009.

Sun, L. Y., Yang, S. Q., Li, F. J., Liu, D. P., and Chang, C. L.: Analysis of interaction of groundwater setting depth of Hetao irrigation district and concentration of Total Nitrogen, Water-Sav. Irrig., 7, 311–313, 2014.

Tong, L.: Study on the Response of agricultural water consumption in Shiyang river in northwest arid region to environmental changes, Xi'an, Doctoral Dissertation of Northwest Agriculture and Forestry University, Xi'an, 2007.

Tsakiris, G. and Vangelis, H.: Establishing a drought index incorporating evapotranspiration, Europe Water, 9/10, 3–11, 2005.

Wang, B., Wang, G. Z., Huang, J. B., Zhang, Z. Y., and Fu, Q.: Agricultural drought evaluation model and application, China Water Power Press, Beijing, 2011.

Wang, S. Y., Zheng, G. F., Yang, J., and Li, X.: Contrastive analysis of the application of several drought evaluation indexes in Ningxia, Chinese Deserts, 2, 517–524, 2012.

Wu, C. B.: Analysis of change law of groundwater setting depth of Fen river irrigation district, Groundwater, 2, 35–37, 2009.

Wu, P. T. and Zhao, X. N.: Impact of climate change on Chinese agriculture water and grain production, Agricult. Eng., 2, 1–6, 2010.

Zhang, Q., Deng, Z. Y., Zhao, Y. D., and Qiao, J.: Impact of global climate change on the agriculture in China northwest, J. Ecol., 3, 1210–1218, 2008.

Zhang, T. F., Wang, J. S., and Guo, J. Y.: Change characteristic of autumn drought index in northwest, Arid Reg. Res., 1, 87–92, 2007.

Zhao, J., Cai, M. K., Wei, X. M., Ning, Y. F., and Shen, B.: Analysis and prediction of Baojixia groundwater dynamic law, Groundwater, 6, 5–7, 2009.

Proc. IAHS, 374, 137–142, 2016
proc-iahs.net/374/137/2016/
doi:10.5194/piahs-374-137-2016


Open Access
PIAHS

# Seasonal forecast of Kharif flows from Upper Jhelum catchment

**Wolfgang Bogacki**[1] **and M. Fraz Ismail**[2]

[1]Department of Architectural and Civil Engineering, University of Applied Sciences, Koblenz, 56075, Germany
[2]National Engineering Services (Pvt) Limited, Lahore, 54000, Pakistan

*Correspondence to:* W. Bogacki (bogacki@hs-koblenz.de)

Published: 17 October 2016

**Abstract.** An operational hydrological forecast model was set-up based on the Snowmelt-Runoff Model (SRM) in order to forecast Kharif flows from Upper Jhelum catchment. Zone-wise degree-day factor functions were derived by diagnostic calibration and are applied according to a defined temperature rule when melting starts. While predicting the depletion of snow-covered area by SRM's modified depletion curve approach, scenario runs with temperature and precipitation of past years are carried out which are evaluated statistically to forecast the seasonal flow volume.

## 1 Introduction

Irrigated agriculture provides 90 % of Pakistan's food requirements, 22 % of its Gross Domestic Product and provides employment to 60 % of the population of the country (Euroconsult, 2011). Pakistan ranks 5th in rice export and 8th in wheat production in the world (FAO, 2011). Consequently, agriculture's share in water usage is about 97 %, which is well above the global average of about 70 % (Akram, 2009). The Indus Basin Irrigation System, comprising a canal command area of more than 14 million ha, is one of the largest contiguous irrigation systems of the world. On average, annual canal water diversion from River Indus and its major tributaries amounts to about 130 $km^3 a^{-1}$.

Flows in the Indus Basin mainly originate from the high mountain ranges of the Western Himalaya – Karakoram – Hindu Kush region. Though the flow regime also comprises glacier melt and runoff from rainfall during monsoon season, the predominant contribution comes from the melting of seasonal snow accumulated during the preceding winter and spring (Archer et al., 2010). Depending on the altitude of the catchment, flows originating from snowmelt start to rise in March to April. Average monthly flow volumes reach their peak in May to July.

Expected flows during the forthcoming Kharif cropping season (April–September) are forecasted by end of March. Based on these forecasts, the Indus River System Authority (IRSA) decides the provincial shares and the provincial irrigation departments subsequently determine the seasonal water allocation to the different canal command areas. A reliable seasonal forecast of the water resources to be expected from snowmelt is therefore of utmost importance for the agricultural production.

## 2 Study area

One of only two major reservoirs in Pakistan, the Mangla reservoir, is located at the foothills of the Western Himalayas on Jhelum river (Fig. 1). After a dam raising project completed in 2009, the storage capacity is 9.1 $km^3$, equivalent to about 35 % of mean annual flow in Jhelum river which was approx. 26 $km^3 a^{-1}$ during the period 2003–2011. The snowmelt contribution to annual flows is about 50 % on average, with variations between 45–60 % from year to year. While average monthly flow in Jhelum river starts rising already in February reaching its maximum in May (Fig. 2), peak flood events usually occur during monsoon season in July–September.

The catchment of Jhelum River upstream of Mangla dam has an area of about 33 500 $km^2$ and a mean hypsometric elevation of nearly 2400 m a.s.l. The lowest elevation at Mangla is about 300 m a.s.l. whilst the highest peak has an altitude of 6285 m. Due to its comparatively low altitude, according to the GLIMS glacier database (Paul and Frey, 2010) only

Published by Copernicus Publications on behalf of the International Association of Hydrological Sciences.

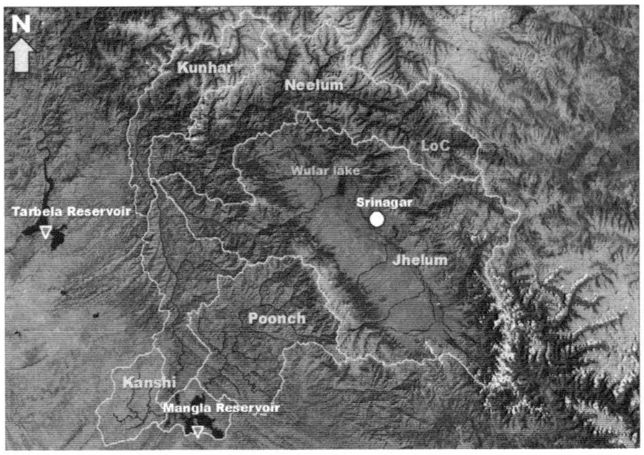


**Figure 1.** Landsat TM scene of study area.

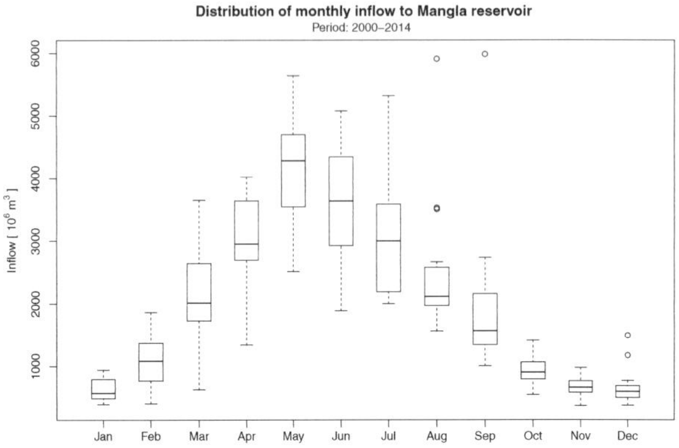


**Figure 2.** Distribution of monthly inflow to Mangla reservoir 2000–2014.

0.7 % of the catchment is covered by glaciers or perennial snow thus having no major impact to the flow regime.

## 3 Methodology

### 3.1 Snowmelt modelling approach

Snowmelt is a complex physical process that is primarily driven by the energy flux entering the snowpack. Two categories of snowmelt models exists: energy balance models attempt to quantify the components of the heat balance equation while temperature index models use air temperature as a principal predictor of melt rates. Temperature index models have been widely used as they offer generally a good model performance while data requirements are comparatively low which is of particular importance in remote and rugged mountainous catchments like in the Himalaya – Karakoram – Hindu Kush region. Among the class of temperature index models SRM (Martinec, 1975) has become very popular in conjunction with the increasing availability of remote sensing snow cover data, as it circumvents the error prone snow accumulation approach used by other models.

An Excel® version (Bogacki and Hashmi, 2013) of WinSRM (Martinec et al., 2011) was developed in order to allow for changes in the underlying code like the handling of heavy rainfalls that had to be adapted to the catchment characteristics. Furthermore, Excel® is well known to most engineers who can easily adapt the layout or data structure to their specific needs.

### 3.2 Data sources

SRM uses air temperature, snow covered area, and precipitation as daily input variables. The watershed is divided into elevation zones using their mean hypsometric elevation as reference for zone-wise degree-day calculation.

There are three high elevation climate stations in the Pakistani part of the Upper Jhelum catchment operated by WAPDA's[1] Snow and Ice Hydrology Project, namely Pir Chenasi at 2650, Shogran at 2930 and Saif-ul-Maluk at 3200 m a.s.l. However none of them has a continuous series of daily data over a longer period and permanent data accessibility that is needed for an operational forecast model. Thus, the WMO climate station at Srinagar airport located at an altitude of 1587 m a.s.l. was chosen as temperature base station, of which a full set of climatic data can be obtained online from the GSOD[2] data-base with a time-lag of about 2 days only. Degree-days in each elevation zone are calculated using a constant temperature lapse-rate of $-6\,^{\circ}\mathrm{C\,km^{-1}}$.

The MODIS (Terra) Snow Cover Daily product[3] with a spatial resolution of approx. 500 m is used to determine the snow cover of the catchment. As the sensor cannot detect snow below clouds, a cloud elimination algorithm is applied using temporal interpolation between two cloud-free days for each pixel. Afterwards the daily percentage of snow cover area in each elevation zone is calculated and smoothed by moving average.

As spatial interpolation of daily precipitation station data in mountainous regions is particularly difficult, the remote sensing based RFE 2.0 South Asia[4] daily rainfall product (Xie et al., 2002) is used. According to SRM's elevation band approach, the gridded data having a spatial resolution of approx. 10 km is mapped to the respective elevation zones.

[1] Pakistan Water and Power Development Authority

[2] Global Summary Of the Day. Download at: ftp://ftp.ncdc.noaa.gov/pub/data/gsod/ (NCDC, 2010)

[3] Moderate-resolution imaging spectro-radiometer ftp://n5eil01u.ecs.nsidc.org/SAN/MOST/MOD10A1.005/ (Hall et all., 2006)

[4] RainFall Estimates version 2.0 created by the NOAA Climate Prediction Center's FEWS-NET group (2011–2015) sponsored by USAID. Download at: ftp://ftp.cpc.ncep.noaa.gov/fews/S.Asia/data/

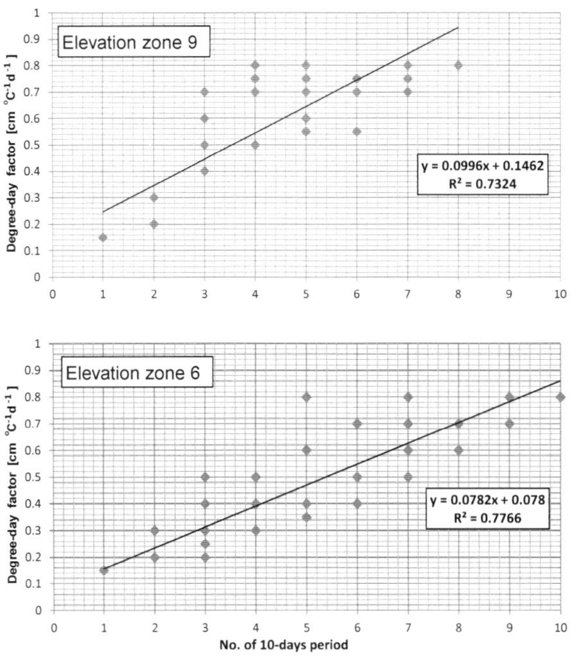


**Figure 3.** Increase of degree-day factors with time (10-days periods) after melting start for elevation zones 6 and 9. Degree-day factors are obtained by diagnostic calibration.

## 3.3 Model parameters

The most important model parameter controlling daily snowmelt is the degree-day factor [cm $°C^{-1}$ $d^{-1}$], which transforms the index variable degree-day [°C d] into actual melt [cm $d^{-1}$]. In a first step, best-fitting degree-day factors were obtained in 10-days intervals for each elevation zone by diagnostic calibration matching observed Mangla inflows with the simulated ones.

The degree-day factor generally increases during melt season as the snowpack becomes "ripe" due to accumulation of energy by the time (e.g. Martinec et al., 2011; Hock, 2003). This process happens later at higher elevation zones as temperatures are lower but the increase is quicker than in the lower zones, as energy input by solar radiation is longer in effect until actual melting starts. In order to develop a generalised rule as needed in the forecasting procedure, zone-wise degree-day factor functions (Ismail et al., 2015) where developed. The relation of degree-day factors versus time, i.e. the 10-days periods, was plotted for each elevation zone for all calibrated years, using the first period when the increase commences as starting point. Figure 3 shows the evolution of degree-day factors by time for elevation zones 6 (2500–3000 m a.s.l.) and 9 (4000–4500 m a.s.l.) respectively. Finally a degree-day factor function was developed by linear regression for each elevation zone of Upper Jhelum catchment (Table 1).

**Table 1.** Zone-wise degree-day factors depending on 10-days periods after melting start.

| | Elevation zone | | | | | | |
|---|---|---|---|---|---|---|---|
| Period | 1 – 5 | 6 | 7 | 8 | 9 | 10 | 11 |
| 1 | 0.25 | 0.31 | 0.27 | 0.32 | 0.35 | 0.34 | 0.34 |
| 2 | 0.32 | 0.38 | 0.36 | 0.40 | 0.44 | 0.49 | 0.49 |
| 3 | 0.39 | 0.45 | 0.45 | 0.48 | 0.53 | 0.65 | 0.65 |
| 4 | 0.46 | 0.52 | 0.54 | 0.56 | 0.62 | 0.80 | 0.80 |
| 5 | 0.53 | 0.59 | 0.62 | 0.64 | 0.71 | | |
| 6 | 0.59 | 0.66 | 0.71 | 0.72 | 0.80 | | |
| 7 | 0.66 | 0.73 | 0.80 | 0.80 | | | |
| 8 | 0.73 | 0.80 | | | | | |
| 9 | 0.80 | | | | | | |

The start of melting in each zone differs from year to year, depending on the actual temperature development during spring. Based on average 10-days period temperatures, a forecast rule was developed controlling when to start the respective degree-day factor function in a particular zone. These threshold temperature values decrease from 6 °C in the lower to 1 °C in the upper elevation zones.

The other model parameters required by SRM like temperature lapse-rate, recession coefficient, runoff coefficient for snow, lag-time, etc., were applied basin-wide and kept constant for all years. The values of these parameters were determined according to the methods described by Martinec et al. (2011) and slightly adjusted to achieve a good fit over the whole calibration period. Only the procedure of adjusting the recession coefficient for heavy rainfalls, which is hard coded in WinSRM had to be adapted to the catchment characteristics, as otherwise peak runoffs were heavily overestimated.

## 3.4 Forecasting approach

The main objective of the Upper Jhelum snowmelt runoff model is the operational forecast of water availability in the Kharif cropping season (April–September) but the model is also utilised for 10-daily forecasts in order to fine-tune irrigation water distribution during the season. Because medium-term weather forecasts were deemed more promising than long-term seasonal, in a first analysis 10-daily flow hindcasts were carried out using 15-days temperature and precipitation forecasts from NCEP[5] and ECMWF[6] weather forecast models obtained from the TIGGE[7] archive. However flow

[5]National Centers for Environmental Prediction, USA (http://www.ncep.noaa.gov/)

[6]European Centre for Medium-Range Weather Forecasts (http://www.ecmwf.int/)

[7]TIGGE (http://tigge.ecmwf.int/), the THORPEX Interactive Grand Global Ensemble, is a key component of the World Weather Research Programme to accelerate the improvements in the accuracy of 1 day to 2 week high-impact weather forecasts.

**Table 2.** Comparison of annual and Kharif (April–September) mean absolute volumetric error [%] of 10-daily flow forecasts using (a) scenario approach, (b) NCEP, and (c) ECMWF 15-days weather forecasts.

| | Annual error [%] | | | Kharif error [%] | | |
|---|---|---|---|---|---|---|
| Year | Scenario | NCEP | ECMWF | Scenario | NCEP | ECMWF |
| 2007 | – | – | – | 6.0 | 12.3 | 11.7 |
| 2008 | 10.1 | 9.0 | 26.7 | 7.4 | 8.6 | 22.4 |
| 2009 | 10.4 | 11.6 | 22.8 | 6.9 | 10.8 | 23.9 |
| 2010 | 13.9 | 21.3 | 32.1 | 9.1 | 17.2 | 27.2 |
| 2011 | 7.7 | 8.9 | 12.6 | 6.7 | 7.0 | 13.8 |
| Average | 10.5 | 12.7 | 23.6 | 7.2 | 11.2 | 19.8 |

predictions based on these weather forecasts were found inferior compared to a scenario approach as described below (Table 2). Thus for the seasonal forecasts only the scenario approach was taken into consideration.

When forecasting the three model variables snow-covered area, temperature, and precipitation have to be predicted. In order to estimate the future depletion of the snow-covered area SRM uses so called "modified depletion curves" which are derived from the conventional depletion curves of each elevation zone by replacing the time scale with the cumulative daily snow-melt depth (Martinec et al., 2011). The decline of the modified depletion curves depends on the initial accumulation of snow and represents the actual snow-water equivalent. When initial snow depth is low the modified depletion curve declines faster than in years when a lot of snow has accumulated.

End of March, when the seasonal forecast is carried out, an elevation zone showing already some decline in snow-covered area, and hence having also some cumulated degree-days, is chosen as "key zone". Comparing the relation of decline in snow-covered area versus cumulated degree-days with a statistical analysis of the modified depletion curves of previous years the actual amount of snow is estimated and the future depletion anticipated accordingly, while assuming similar snow conditions for all elevation zones.

While the snow-covered area and its depletion is calculated only once for the actual season to be forecasted, scenario runs are carried out with historic temperature and precipitation data-sets of various years. This results in an ensemble of predicted seasonal flows representing historic weather conditions from which by statistical analysis a forecast of "most likely" (median) as well as expected flows under "dry" or "wet" conditions can be derived.

## 4 Results and discussion

The final flow forecast model, i.e. using fixed model parameters as well as the developed degree-day factor functions and the start rule as described in Sect. 3.4, was validated for the series of years 2002–2011. The relative volume difference $D_\mathrm{v}$ and the coefficient of determination $R^2$ according to the Eqs. (1) and (2):

$$D_\mathrm{v} = \frac{V - V^*}{V} \times 100\,[\%] \tag{1}$$

$$R^2 = 1 - \frac{\sum_{i-1}^{n}(Q_i - Q_i^*)^2}{\sum_{i-1}^{n}(Q_i - \bar{Q})^2}\,, \tag{2}$$

These equations are used as model accuracy criteria, where $V$ and $V^*$ are the observed and the simulated annual flow volumes, $Q_i$ and $Q_i^*$ are the observed and the simulated daily discharge values, and $\bar{Q}$ is the average observed daily discharge.

The years 2002 and 2011 that had not been used for parameter calibration and determination of the degree-day factor functions resulted in an $R^2$ of 0.89 for each year and a $D_\mathrm{v}$ of 2.3 and −2.8 % respectively (NESPAK and AHT, 2012). The total series of years has an average $R^2$ of 0.83 and an average absolute volume error of 4.3 %. Figure 4 gives an example of simulated versus observed inflow hydrograph.

The forecast capability of the developed model was evaluated by hindcasts for the years 2000–2011 using the same procedures as developed for the operational seasonal forecasts. Meanwhile, real forecasts exist for the Kharif seasons of the years 2012–2015. Results are compared with IRSA's forecasts that are based on a statistical approach and with forecasts from the UBC[8] watershed model (Quick and Pipes, 1977) that is used by WAPDA's Snow and Ice Hydrology Project. Figure 5 shows a comparison of the relative error of the forecasted Kharif flow volumes by the different approaches. SRM in conjunction with the scenario approach generally predicts Kharif flows with an error of less than 10 %, in many years of less than 5 %. The year 2010 was exceptional, as there was an approx. 200 years flood event in the monsoon season.

[8]University of British Columbia Watershed Model

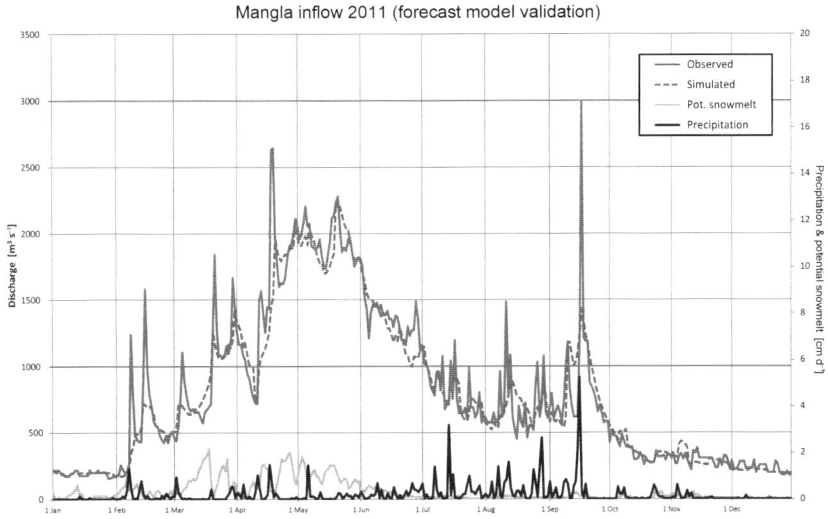


**Figure 4.** Results of validation of final flow forecast model (dashed line) compared to observed inflows to Mangla reservoir (solid line) for the year 2011. The graph also shows at the bottom rainfall and snowmelt depth [cm $d^{-1}$].

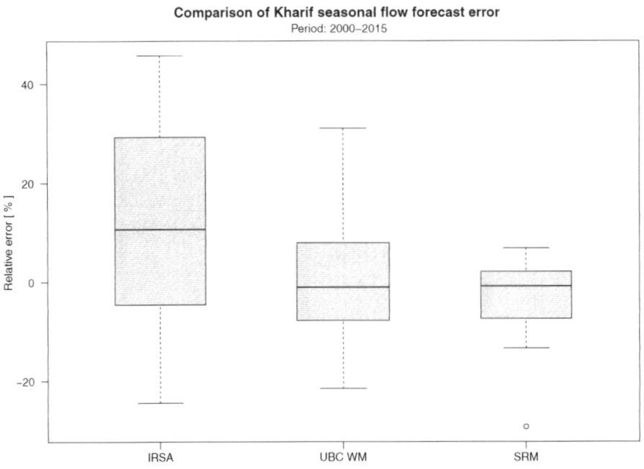


**Figure 5.** Statistics of relative error of Kharif flow volume forecasts by IRSA's statistical approach, UBC watershed model, and SRM. SRM 2000–2011 are hindcasts, all other forecasts. The series include the approx. 200 years flood in 2010.

The quite good accuracy of seasonal flow prediction in the Upper Jhelum catchment might be due to a combination of favourable factors. One is definitely SRM's good approximation of the actual snow situation at the start of the melting season by utilising the MODIS snow-cover remote-sensing data in conjunction with the modified depletion curve approach. Once the amount of available snow is known, it is most likely, that all snow is melted during the Kharif season.

In addition, the variability of the second important flow component, i.e. rainfall from westerly disturbances in spring and monsoon events in summer, is sufficiently considered by the scenario approach, which transforms the inter-annual variation of precipitation into a series of seasonal flows that can be statistically evaluated. The use of the gridded precipitation product RFE, although it sometimes over- or underestimates a single rainfall event considerably, might also be favourable in that context as the spatial distribution seems to be more fitting as any interpolation of station data in such mountainous regions.

In order to further improve the seasonal forecast capability of the scenario approach, research is under way on the early identification of cold/warm respectively dry/wet years e.g. correlated to the ENSO[9] status, which might allow for a more specific selection of a subset of corresponding historic years.

Considering that the error in predicting the Kharif flow volume is less than 10 % in 14 out of 16 years with a mean absolute error in this period of 6.4 % it can be stated that the combination of SRM with remote sensing data and the scenario approach has proven to be a reliable procedure for operational seasonal flow forecasting in the Upper Jhelum catchment. Furthermore this approach could be easily improved if the meteorological characteristics of the forthcoming Kharif season can be teleconnected to global climatic conditions like the ENSO status during the winter.

**Acknowledgements.** The study was carried out as part of the project "Upgrading of Tools, Water Resources Database, Management Systems and Models" under sub component "B1" of World Bank financed "PK Water Sector Capacity Building & Advisory Services Project" (WCAP): P110099. The authors thank National Engineering Services Pakistan (Pvt.) Ltd. (NESPAK), Lahore and AHT Group AG, Essen, Germany that they could be part of the

[9]El Niño Southern Oscillation

project team. They are highly grateful to civil engineering department of University of Engineering and Technology Lahore and civil engineering department of Hochschule Koblenz, University of Applied Sciences, for supporting subsequent research and wish also to express their high gratitude to Indus River System Authority (IRSA) and WAPDA's Snow and Ice Hydrology Project for sharing their forecast results.

## References

Akram, A. A.: Indus Basin water resources, Tiempo, Issue 70, http://www.environmentportal.in/files/IndusBasin.pdf (last access: 8 August 2016), 2009.

Archer, D. R., Forsythe, N., Fowler, H. J., and Shah, S. M.: Sustainability of water resources management in the Indus Basin under changing climatic and socio economic conditions, Hydrol. Earth Syst. Sci., 14, 1669–1680, doi:10.5194/hess-14-1669-2010, 2010.

Bogacki, W. and Hashmi, D.: Impact of Climate Change on the Flow Regime of the Mangla Basin. GWSP Conference "Water in the Anthropocene: Challenges for Science and Governance. Indicators, Thresholds and Uncertainties of the Global Water System", Bonn, Germany, May 2013, doi:10.13140/RG.2.1.1934.6167 (last access: 8 August 2016), 2013.

Euroconsult Pakistan (Pvt.) Ltd.: Handbook on Water Statistics of Pakistan, Water Sector Capacity Building and Advisory Services Project (WCAP), March 2011.

FAO: Pakistan and FAO achievements and success stories, http://www.fao.org/3/a-at014e.pdf (last access: 8 August 2016), 2011.

Hall, D. K., Salomonson, V. V., and Riggs, G. A.: MODIS/Terra Snow Cover Daily L3 Global 500 m Grid V005 [February 2000–September 2015], NSIDC Boulder, Colorado USA, updated daily, ftp://n5eil01u.ecs.nsidc.org/SAN/MOST/MOD10A1.005/, 2006.

Hock, R.: Temperature index melt modelling in mountain areas, J. Hydro., 282, 104–115, 2003.

Ismail, M. F., Rehman, H., Bogacki, W., and Noor, M.: Degree Day Factor Models for Forecasting the Snowmelt Runoff for Naran Watershed, Sci. Int. Lahore, 27, 1961–1969, 2015.

Martinec, J.: Snowmelt-Runoff Model for Stream Flow Forecasts, Nord. Hydrol., 6, 145–154, 1975.

Martinec, J., Rango, A., and Roberts, R.: Snowmelt Runoff Model User's Manual, WinSRM Version 1.14. Agricultural Experiment Station Special Report 100, New Mexico State University, Las Cruces, NM 88003, USA, 2011.

NCDC: Global Surface Summary of the Day – GSOD, Version 7 [January 2000–September 2015], National Climatic Data Center, NESDIS, NOAA, U.S. Department of Commerce, Asheville, North Carolina USA, ftp://ftp.ncdc.noaa.gov/pub/data/gsod/, updated daily, 2010.

NESPAK and AHT: Hydrological Flow Forecast Model for Mangla Catchment, Upgrading of Tools, Water Resources Database, Management Systems and Models Under Sub Component "B1" of WCAP, Final Report, 2012.

NOAA Climate Prediction Center's FEWS-NET group: RainFall Estimates version 2.0, ftp://ftp.cpc.ncep.noaa.gov/fews/S.Asia/data/, May 2001–September 2015.

Paul, F. (submitter), Frey, H., and Paul, F. (analysts): GLIMS Glacier Database, Boulder, CO, National Snow and Ice Data Center, 2010.

Quick, M. C. and Pipes, A.: UBC watershed model, Hydrol. Sci. Bull., 221, 153–161, 1977.

Xie, P., Yarosh, Y., Love, T., Jonowiak, J. E., and Arkin, P. A.: A real-time daily precipitation analysis over south asia. Preprint of the 16th Conference on Hydrology, Orlando, Florida, American Meteorological Society, Washington DC, USA, 2002.

Proc. IAHS, 374, 143–150, 2016
proc-iahs.net/374/143/2016/
doi:10.5194/piahs-374-143-2016


# A snow and ice melt seasonal prediction modelling system for Alpine reservoirs

**Kristian Förster[1,2], Felix Oesterle[1], Florian Hanzer[1,2], Johannes Schöber[3], Matthias Huttenlau[1], and Ulrich Strasser[1,2]**

[1]alpS – Centre for Climate Change Adaptation Innsbruck, Innsbruck, Austria
[2]Institute of Geography, University of Innsbruck, Innsbruck, Austria
[3]TIWAG, Tiroler Wasserkraft AG, Innsbruck, Austria

*Correspondence to:* Kristian Förster (kristian.foerster@uibk.ac.at)

Published: 17 October 2016

**Abstract.** The timing and the volume of snow and ice melt in Alpine catchments are crucial for management operations of reservoirs and hydropower generation. Moreover, a sustainable reservoir operation through reservoir storage and flow control as part of flood risk management is important for downstream communities. Forecast systems typically provide predictions for a few days in advance. Reservoir operators would benefit if lead times could be extended in order to optimise the reservoir management. Current seasonal prediction products such as the NCEP (National Centers for Environmental Prediction) Climate Forecast System version 2 (CFSv2) enable seasonal forecasts up to nine months in advance, with of course decreasing accuracy as lead-time increases.

We present a coupled seasonal prediction modelling system that runs at monthly time steps for a small catchment in the Austrian Alps (Gepatschalm). Meteorological forecasts are obtained from the CFSv2 model. Subsequently, these data are downscaled to the Alpine Water balance And Runoff Estimation model AWARE running at monthly time step. Initial conditions are obtained using the physically based, hydro-climatological snow model AMUNDSEN that predicts hourly fields of snow water equivalent and snowmelt at a regular grid with 50 m spacing. Reservoir inflow is calculated taking into account various runs of the CFSv2 model. These simulations are compared with observed inflow volumes for the melting and accumulation period 2015.

## 1 Introduction

Hydropower is a major contributor to carbon-free energy production in the European Alps where ideal conditions for this type of energy production prevail (Schaefli, 2015). In general, mountainous regions are subjected to strong altitudinal gradients, which is important for efficient hydropower production. Moreover, mountain ridges induce orographic lifting of air masses causing higher precipitation depths than in surrounding lowland regions (Koboltschnig and Schöner, 2011). Snow and ice are accumulated and seasonally released as melt water in spring and summer making mountainous headwaters very important for hydrological regimes of various European rivers and thus for hydropower as well (Viviroli et al., 2007).

Reservoirs are not only relevant for hydropower generation but also for flood control and water supply. Thus, hydropower management depends on various factors that need to be addressed in day-by-day operation. Typically, long-term observations and short-term meteorological and hydrological predictions serve as a basis for this task. However, predicting several components of the water balance components at various spatial and temporal scales is challenging in this environment as meteorological fields such as temperature or precipitation greatly vary in space and time, even at short time scales (Klemeš, 1990).

In general, two types of seasonal prediction strategies are available for hydrologic applications (Yuan et al., 2015): On one hand, common methods in hydrology are based on the Ensemble Seasonal Prediction, building upon procedures which resample historical meteorological data. This method

Published by Copernicus Publications on behalf of the International Association of Hydrological Sciences.

is well suited where initial conditions can be seen as a "land surface memory" that last over a certain time period (Yuan et al., 2015). On the other hand, climate models based on (dynamical) seasonal methodology rely on physical modelling of the coupled atmosphere-ocean system. This is important as atmosphere-ocean systems interact at a broad range of scales, influenced by sea surface temperature anomalies as well as associated surface air pressure fields such as El Niño Southern Oscillation (ENSO) or the North Atlantic Oscillation (NAO), and their atmospheric teleconnections strongly affect weather patterns at smaller scales (Doblas-Reyes et al., 2013). The pioneering work of (Wood et al., 2002) showed the applicability of dynamical seasonal prediction systems in principle. The recently released Climate Forecast System version 2 (Saha et al., 2014), fulfilling all requirements for seasonal initial value based predictions as claimed by (Kanamitsu et al., 2002), provides seasonal forecasts every 6 h operationally. Data is available online and its capabilities for hydrological applications are considered as promising (Yuan et al., 2011). The physical basis as well as the availability of data makes CFSv2 an interesting source of information about anomalies of meteorological variables for the next few months even if the spatial resolution is comparatively coarse and the predictive accuracy decreases with increasing lead times. In the framework of this paper, first results of CFSv2 forecasts for a small catchment in Tyrol/Austria (Gepatschalm) are presented. A simple statistical downscaling approach is applied in order to scale gridded monthly averages of temperature and precipitation to monthly station recordings. Then, a water balance model with snow and ice melt capabilities that has been designed for this task is calibrated and validated for different periods of time using reanalysis data of the climate model. Finally, first results of the melting season 2015 and the following accumulation season are presented.

## 2 Study area

In order to test the applicability of seasonal predictions for small meso-scale catchments relevant for hydropower generation, the Gepatschalm catchment (55 km$^2$) has been selected. It represents a typical high altitude Alpine catchment with an altitudinal range between 1895 and 3504 m a.s.l. The latter value is close to the altitude of the highest summit of the Ötztal Alps, the Wildspitze (3770 m a.s.l.). Approximately 37 % of the area is covered by glaciers, according to the 3rd Austrian Glacier Inventory updated in 2006 (Fischer et al., 2015). Moreover, downstream of the Gepatschalm gauge a reservoir was built and completed in 1964. The reservoir capacity is 139 million m$^3$; it has been designed for additional inflow of several catchments in the vicinity that are diverted to the reservoir. Here, the natural catchment area is investigated (Fig. 1).

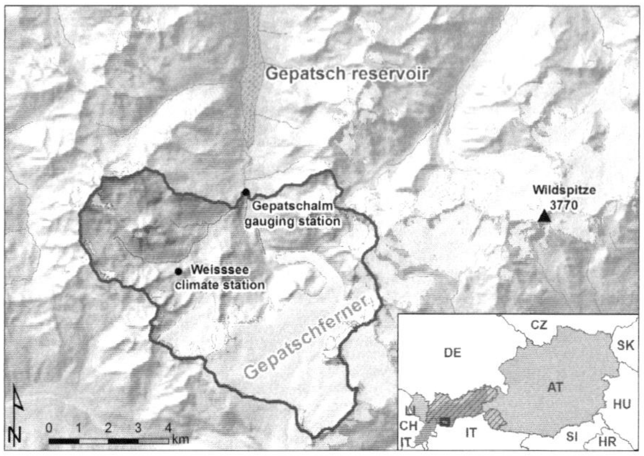


**Figure 1.** Map showing the study area located in Tyrol/Austria (hatched area in the small map) including the Gepatschalm catchment upstream of the Gepatsch reservoir. The Gepatschferner is the largest glacier in Western Austria (glaciers in light grey). The Inn river catchment is highlighted in green in the small map.

The runoff as recorded by the Gepatschalm gauge since 1983 is strongly influenced by seasonal effects as snow and ice melt runoff dominates in late spring and summer resulting in mean monthly runoff depths greater 500 mm whereas the winter runoff is very low (less than 10 mm per month). According to the Austrian Glacier Inventory, the Gepatschferner – the largest glacier of the study region – covered 16.6 km$^2$ in 2006. At present, glaciers undergo a rapid decline both in the study area and worldwide (Fischer et al., 2015; Marzeion et al., 2014). For long-term investigation, the areal glacier extent has to be continuously updated in order to incorporate the variable surface ice melt areas (Schueller et al., 2015).

## 3 Set-up of seasonal forecasts and water balance simulations

### 3.1 The CFSv2 model and linear regression downscaling

The NCEP Climate Forecast System version 2 (CFSv2) has been developed at the Environmental Modeling Center at the National Centers for Environmental Prediction (NCEP). Since 2011, the model has been made operational. The model software, all relevant data and the seasonal forecasts are available to the public (see, e.g., http://cfs.ncep.noaa.gov/). Computations are performed on a horizontal grid with a resolution of roughly 100 km and 64 vertical levels of the atmosphere (Saha et al., 2014). Besides the atmospheric model, CFSv2 also includes a land surface, cryosphere and ocean model. These model components are coupled in order to adequately describe interactions among the systems described by each of these sub-models. A detailed description of the model dynamics and physics is provided by Saha et al. (2014). For practical purposes in the framework

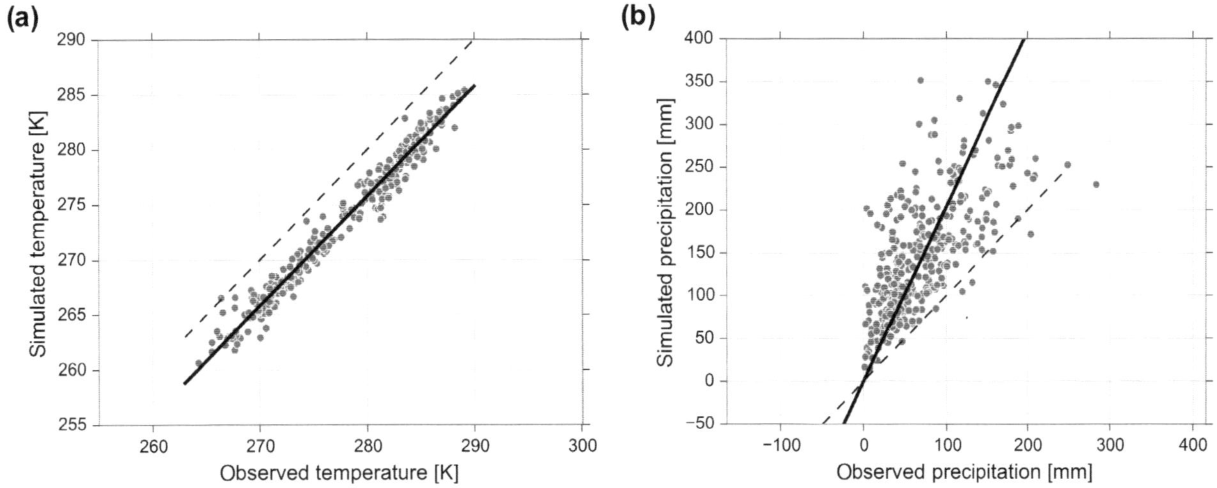


**Figure 2.** Scatter plots of modelled vs. observed temperature **(a)** and precipitation **(b)**. Each point represents the mean temperature and the precipitation depth of one month in the period 1986–2009, respectively. Besides the diagonal line (dashed), the regression line is also added to each plot (solid).

of this study, we focus on some features that are relevant for hydrological modelling.

CFSv2 is run 4 times a day performing simulations for the next 9 months. Additionally, a re-analysis dataset ranging from 1979–2014 is available. In order to assess the applicability of CFSv2 data for the study area, monthly time series of temperature and precipitation extracted from the re-analysis dataset have been compared to observations recorded by a meteorological station. For investigating the Gepatschalm catchment in this context, the nearest meteorological station that provides long-term recording is located in the vicinity of the dam of the Gepatsch Reservoir (approx. 7 km north of the gauge). Long-term time series are needed as a sound database for model evaluations, also with respect to sub-sequent bias corrections.

Figure 2 shows scatter plots for temperature and precipitation with modelled values plotted against observations. Each point represents one month within the period 1986–2009. Mean monthly temperature coincides well with observation. A cold bias yielding approximately −3.4 K can be observed from the plot in Fig. 2a. For the subsequent simulations, the respective slope and intercept parameter derived from regression analyses are applied in order to linearly adjust the underestimation of temperature for the study area. This procedure can be seen as a simple bias correction method ($R^2 = 0.97$). The modelled monthly precipitation depth is on average twice the value obtained through long-term records. Instead of using a linear regression, a pure multiplicative correction (Kuentz et al., 2015) is used instead without assuming an intercept parameter that might yield negative values in the transformation procedure ($R^2 = 0.47$, Fig. 2b). Time series of shortwave radiation are also derived from the CFSv2 data. Other than temperature and precipitation, shortwave radiation recordings are only available as of 2011. Since reanalysis data are available until 2014, at least three years of both observed and modelled values are directly comparable (see Fig. 3). CFSv2 tracks the seasonal characteristics very well. However, shortwave radiation is overestimated during the summer months. Due to the short overlapping period that covers both observed and modelled values, monthly mean shortwave radiation time series are used "as is" in this study as the number of values is too small for statistical corrections. Instead, overestimation of shortwave radiation is compensated through calibration of the water balance model as described in the next section. This approach is seen as feasible method to compensate for model biases since the melt computations include an adjustable radiation dependent melt factor. Thus, the radiation biases are corrected through model calibration in this study.

### 3.2 The water balance model AWARE and its application

Nowadays, hydrological models typically operate at one hour or one day time steps. For flood forecasting, hourly time steps are typical, whereas in urban hydrology, even smaller time steps are employed. For monthly time series of input data, these models cannot be applied without utilising further temporal disaggregation techniques, which scale monthly values to smaller time steps. For instance, (Wood et al., 2002) propose a scaling of historical daily time series so that these time series match the respective monthly means of the forecast. Another possibility is to revisit water balance models that operate at one month time steps (McCabe and Markstrom, 2007), which might be seen as feasible approach given that

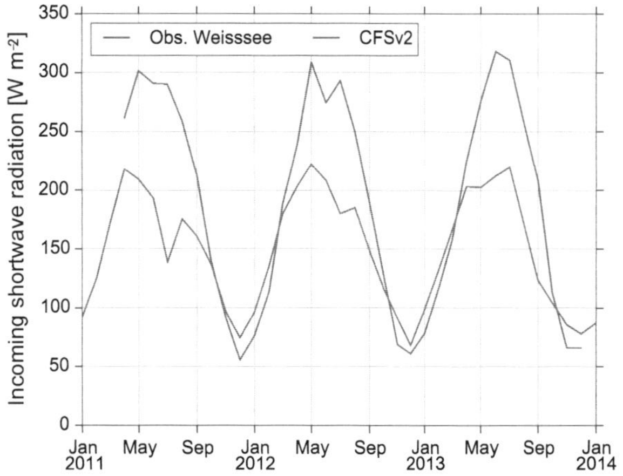


**Figure 3.** Modelled and observed shortwave radiation. Observations have been recorded at the Weisssee climate station (2480 m a.s.l., see Fig. 1), whereas modelled values are derived from CFSv2 without any modification.

**Table 1.** Model performance measures for the water balance model AWARE forced by CFSv2 re-analysis data. NSE = Nash–Sutcliffe model efficiency, PBIAS = Percent Bias, RMSE = Root Mean Square Error, RSR = RMSE observations standard deviation ratio, $R$ = Pearson's correlation coefficient.

| Performance measure | Calibration (1995–2009) | Validation (1985–1992) |
|---|---|---|
| Number of months [–] | 180 | 96 |
| NSE [–] | 0.78 | 0.76 |
| PBIAS [%] | −23.67 | −5.83 |
| RMSE [mm month$^{-1}$] | 79.43 | 90.36 |
| RSR [–] | 0.47 | 0.49 |
| $R$ [–] | 0.91 | 0.87 |

seasonal forecasts address anomalies rather than predicting the weather of specific days. In this study, the latter methodology is pursued for practical reasons.

Here, a simple water balance model ("AWARE": Alpine Water balance And Runoff Estimation model) has been developed including all relevant processes for the study area. The model design also addresses initialisation of system states through other models operating operationally. The physically-based hydro-climatological model AMUNDSEN (Strasser, 2008; Strasser et al., 2011), running at hourly time step, is used to derive snow water equivalent (SWE) maps using 50 m grid spacing in operational mode for the Inn River catchment in Tyrol and adjacent regions (green area in the small map of Fig. 1). In order to employ areal SWE maps as initial conditions for the water balance model with meteorological forcing using CFSv2, the water balance model is a distributed model and has been designed to be capable of running on the same grids as AMUNDSEN.

The water balance model includes 5 different process modules: (i) meteorological pre-processing, (ii) snow and ice melt, (iii) evapotranspiration, (iv) soil water balance, and (v) groundwater. In order to account for lower temperatures and higher precipitation totals at higher altitudes, empirical gradients for temperature and precipitation are applied to the meteorological input downscaled for the meteorological station situated close to the Gepatsch reservoir. The methodology and parameters are similar to the ones in AMUNDSEN. Snow and ice melt are calculated separately assuming different parameterisations of the enhanced degree day method which also incorporates radiation (Hock, 1999), which is crucial for snow and ice melt. In contrast to the cited study that considers hourly melt rates, the proposed methodology is applied to monthly averages of temperature and radiation. This is viewed as an adequate adaptation since the original degree day method can be also reliably applied using average monthly degree day values (Rango and Martinec, 1995). Evapotranspiration is computed according to the (Thornthwaite, 1948) approach. Water fluxes associated with the soil water balance are approximated according to (McCabe and Markstrom, 2007). A linear storage is implemented to account for the groundwater storage and low flow recession.

A split sample test was applied in order to calibrate and validate the model using different periods of time (Klemeš, 1986) as shown in Fig. 4. The period 1995–2009 was used for calibration, while the period 1985–1992 was used for validation (due to some missing values in the CFSv2 re-analyses data, the years 1993–1994 were not considered). Table 1 summarises the calibration and validation period for the water balance model driven by CFSv2 re-analyses. To quantify the deviations between model and observations, a set of performance measures were calculated (Moriasi et al., 2007): Nash–Sutcliffe model efficiency (NSE) values between 0.65 and 0.75 are viewed as "good" model performance rating for monthly time steps. Hence, the NSE value achieved for the calibration period is within the range of "good" values whereas the respective value of the validation period is even "very good" according to this categorisation. These findings also hold true for the RMSE observations standard deviation ratio (RSR). For the percent bias the achieved results fall into the category "satisfactory" (calibration period) and "very good" (validation period) indicating smaller differences between modelled and observed runoff volume in the validation period. However, the modelled peak melt runoff is underestimated in some cases even though modelled runoff is on average slightly higher compared to the observations. In contrast, the model overestimates low flow during the winter months. Notwithstanding these uncertainties in model skill, the modelled time series tracks the observations well. Thus, it is assumed that the model is capable of transforming seasonal predictions of meteorological fields to corresponding runoff and SWE predictions for the study area.

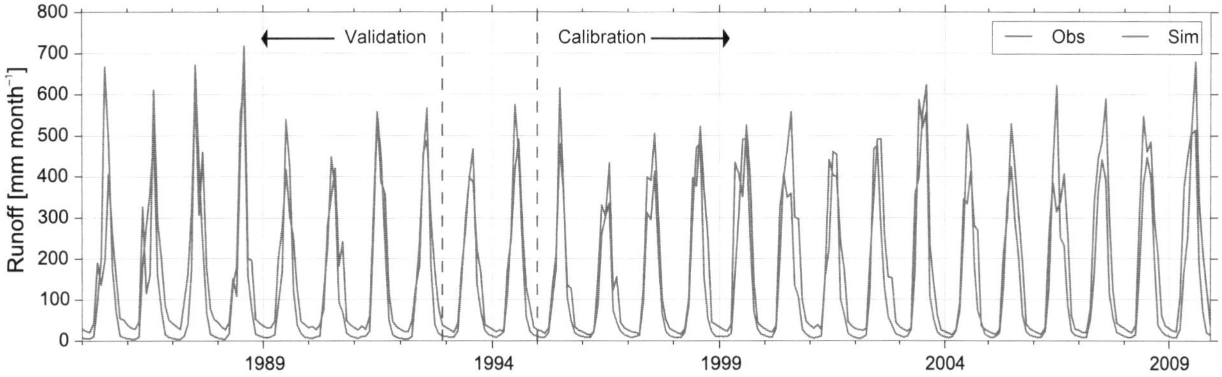


**Figure 4.** Time series of observed (blue line) and modelled runoff (red line) calculated using the water balance model AWARE forced by CFSv2 data.

## 4 Results and discussion

Test operation of the water balance model for seasonal prediction using CFSv2 data has been started in early summer 2015. As initial conditions are crucial for seasonal predictions, appropriate initial states for the most important hydrological storages need to be prepared prior to the model runs. For mountainous and sub-arctic catchments, SWE is the most important variable that needs to be initialised in order to accurately predict snowmelt. Operational snow maps obtained by AMUNDSEN are used as initial conditions in each seasonal forecast run of the water balance model (for an example see Fig. 5). SWE reaches up to several hundred millimetres at the beginning of June 2015. Snow free conditions can be observed in the lower elevation bands, whereas SWE is highest on the glaciers.

Figure 6 shows the first results of the water balance model in forecast mode. Seasonal forecasts performed by CFSv2 were prepared for one day each in June, July, and August 2015. The correction methods derived by evaluating the reanalysis data were applied in order to account for systematic errors in the forecasts likewise. The range of catchment-scale SWE values (average of all grid cells) derived by different ensemble members is subjected to small variations in autumn. Beyond mid-winter, the results indicate a large variety of possible SWE evolutions. A peak accumulation in May 2016 yielding a similar areal SWE is just as possible as a total melt. The first case indicates average conditions whereas the latter case can be viewed as a winter with little snowfall.

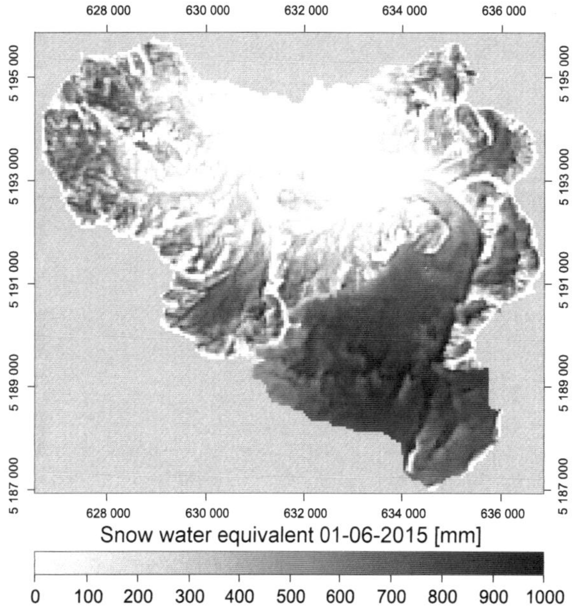


**Figure 5.** Areal initial values of snow water equivalent (SWE) provided by AMUNDSEN for 1 June 2015.

The runoff depth as modelled by the water balance model is of the same order of magnitude as the values recently observed in July, August and September 2015. For instance, the runoff time series of the run initialised in June distinctively overestimate observations in the first month of the simulation (precipitation input 240 mm $month^{-1}$) whereas the value predicted for July 2015 is smaller than the corresponding observation. From August 2015, simulated runoff depths are overestimated compared to observations. This finding also holds true for the runs initialised in July and August. However, the range of values derived by four independent initialisations of the CFSv2 is large as it covers values from approx. 200 to 1000 mm $month^{-1}$. In contrast, the ensemble mean represents typical values observed for August. The bandwidth of values derived by ensemble runs decreases in autumn and winter as melt diminishes and precipitation falls mainly as snow merely contributing to an increase in SWE. In June and August 2015, the overestimation of runoff depth in some runs can be addressed to exceptionally high precipitation depths computed by CFSv2.

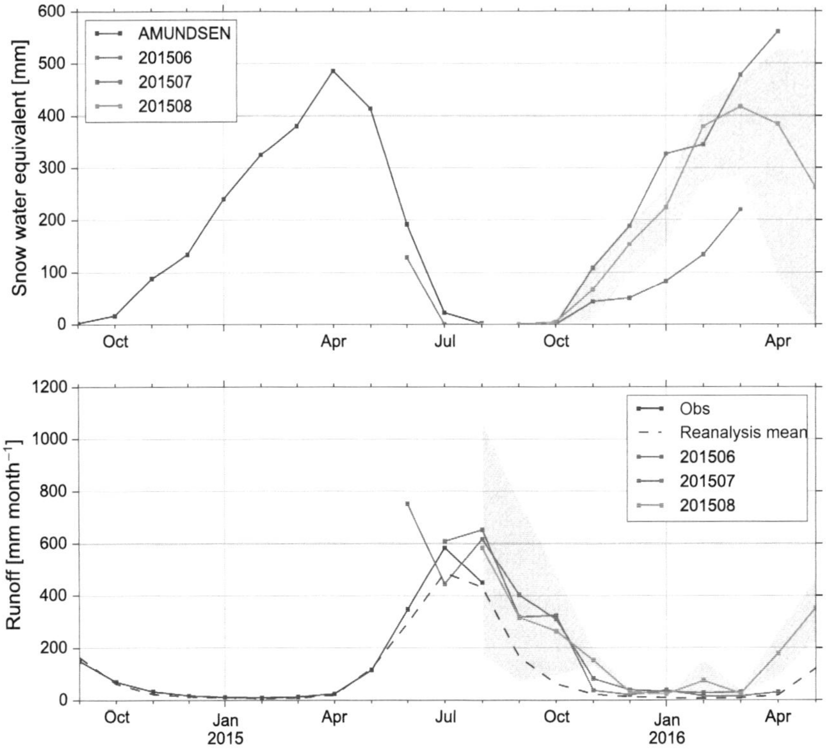


**Figure 6.** Seasonal forecasts of SWE (top) and runoff (bottom) using the water balance model AWARE, initialised by AMUNDSEN and forced by CFSv2. In addition to the ensemble mean, the range of results achieved by four CFSv2 runs in August 2015 is shown as shaded area. The dashed green line indicates average runoff conditions computed using re-analysis data (1985–2009).

When considering the entire simulation period, differences in total precipitation depth among the ensemble members are relevant for seasonal snow accumulations as well. For instance, a total precipitation depth of 710 mm is predicted for the period August 2015–May 2016 if the run initialised on 10 August 2015 at 06:00 z (06:00 UTC) is considered. In contrast, the same evaluation applied to the run initialised at 18:00 z on the same day yields 980 mm (see Fig. 7).

Apart from the run initialised in June, runs that are initialised in consecutive months coincide reasonably with recent observations, at least for one or two months. The differences between two consecutive runs are small for up to three months in advance. However, if single ensemble runs are compared with each other, a large range of forecasted runoff depths in summer and SWE in winter and spring is obvious.

## 5 Conclusions and outlook

The present study analyses available CFSv2 forecasts for runoff and SWE in a meso-scale Alpine catchment and hence, represents a first assessment of feasibility. The results obtained by the water balance model, which is forced using CFSv2 data, are promising with respect to future applications of seasonal forecasts in the study area. The presented results indicate that the model system as yet applied to a glaciated catchment – provides useful results for lead times up to three months. In this snow accumulation and melt dominated catchment, increased uncertainties regarding SWE and runoff appear as the superposition of the uncertain precipitation and temperature inputs for lead times larger than three months. However, if the positive trend of results for lead times up to approximately 100 days could be verified in further model tests, such a model system could support the day-by-day operation as water management tool for hydropower companies. Therefore, an operational use of the model chain CFSv2 → AWARE is pursued to provide estimates of runoff for the upcoming one to three months. In this way, seasonal predictions complement the already existing flood forecasting system for the Inn River called HoPI – "Hochwasserprognose für den Tiroler Inn" (Achleitner et al., 2012; Huttenlau et al., 2016).

Further evaluations of the forecast data and subsequent water balance simulations for the study area are still needed to understand possible uncertainties in seasonal forecasts. For instance, the relatively high precipitation depth values achieved by runs in June and August 2015 resulted in model overestimation with respect to runoff. Moreover, SWE evo-

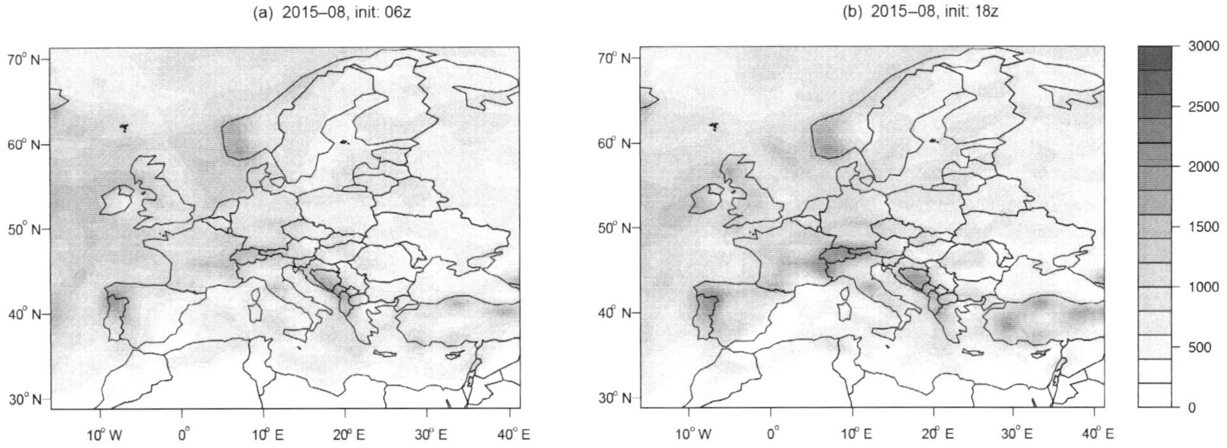


**Figure 7.** Areal distribution of precipitation depth in mm for the period August 2015–May 2016 as predicted on 10 August 2015 by two CFSv2 runs initialised at (a) 06:00 z and (b) 18:00 z, respectively.

lution in the subsequent winter season strongly depends on seasonal precipitation depth. To address these issues, a systematic evaluation of forecast data is foreseen in order to achieve quantitative information about the accuracy of the forecasts in the study area, i.e. by gathering forecasts (ensembles) throughout several months (years) in order to relate uncertainties in predictions to observations. Then, Model Output Statistics (MOS) (Warner, 2011) might be also beneficial to improve forecasts by means of correction approaches that rely on previous forecasts and observations. Therefore, the application of ensemble runs need to be performed more rigorously to gain insight into uncertainties and to provide probability frameworks. Besides methodological enhancements, extensions to the study area are also foreseen to incorporate catchment areas that are artificially drained to the Gepatsch Reservoir.

## 6 Data availability

Both re-analysis data as well as operational forecasts of the CFSv2 model are available for free. CFSv2 re-analysis data (Saha et al., 2010): http://cfs.ncep.noaa.gov/pub/raid0/cfsv2/climo_cfsr_month/flxf06/

Operational CFSv2 7 Day Rotating Archive (Saha et al., 2014): http://nomads.ncep.noaa.gov/pub/data/nccf/com/cfs/prod/cfs/

**Acknowledgements.** This work was carried out as part of the project "W01 MUSICALS II – Multiscale Snow/Ice Melt Discharge Simulation for Alpine Reservoirs" at alpS – Centre for Climate Change Adaptation in Innsbruck, Austria. The K1-Centre alpS is funded through the Federal Ministry of Transport, Innovation and Technology (BMVIT), the Federal Ministry of Science, Research and Economy (BMWFW), as well as the Austrian Provinces of the Tyrol and Vorarlberg within the scope of COMET – Competence Centers for Excellent Technologies. The Programme COMET is managed by the Austrian Research Promotion Agency (FFG). We want to thank TIWAG – Tiroler Wasserkraft AG for the collaboration and co-funding the project, as well as the Editor W. Grabs for his valuable comments.

## References

Achleitner, S., Schöber, J., Rinderer, M., Leonhardt, G., Schöberl, F., Kirnbauer, R., and Schönlaub, H.: Analyzing the operational performance of the hydrological models in an alpine flood forecasting system, J. Hydrol., 412–413, 90–100, doi:10.1016/j.jhydrol.2011.07.047, 2012.

Doblas-Reyes, F. J., García-Serrano, J., Lienert, F., Biescas, A. P., and Rodrigues, L. R. L.: Seasonal climate predictability and forecasting: Status and prospects, WIREs Clim Change, 4, 245–268, doi:10.1002/wcc.217, 2013.

Fischer, A., Seiser, B., Stocker Waldhuber, M., Mitterer, C., and Abermann, J.: Tracing glacier changes in Austria from the Little Ice Age to the present using a lidar-based high-resolution glacier inventory in Austria, The Cryosphere, 9, 753–766, doi:10.5194/tc-9-753-2015, 2015.

Hock, R.: A distributed temperature-index ice- and snowmelt model including potential direct solar radiation, J. Glaciol., 45, 101–111, 1999.

Huttenlau, M., Bellinger, J., Schattan, P., Förster, K., Oesterle, F., Schneider, K., Achleitner, S., Schöber, J., Raffeiner, G., and Kirnbauer, R.: Flood forecasting system for the Tyrolean Inn River (Austria): current state and further enhancements of a modular forecasting system for alpine catchments, in: Living with natural risks, 13th Congress INTERPRAEVENT 2016, Lucerne, 909–916, International Research Society INTERPRAEVENT, Klagenfurt, Austria, 2016.

Kanamitsu, M., Kumar, A., Juang, H.-M. H., Schemm, J.-K., Wang, W., Yang, F., Hong, S.-Y., Peng, P., Chen, W., Moorthi, S., and Ji, M.: NCEP Dynamical Seasonal Forecast System

2000, B. Am. Meteorol. Soc., 83, 1019–1037, doi:10.1175/1520-0477(2002)083<1019:NDSFS>2.3.CO;2, 2002.

Klemeš, V.: Operational testing of hydrological simulation models, Hydrolog. Sci. J., 31, 13–24, doi:10.1080/02626668609491024, 1986.

Klemeš, V.: The modelling of mountain hydrology: the ultimate challenge, in: Hydrology of mountainous areas, edited by: Molnár, L., 29–43, International Association of Hydrological Sciences, Wallingford, 1990.

Koboltschnig, G. R. and Schöner, W.: The relevance of glacier melt in the water cycle of the Alps: the example of Austria, Hydrol. Earth Syst. Sci., 15, 2039–2048, doi:10.5194/hess-15-2039-2011, 2011.

Kuentz, A., Mathevet, T., Gailhard, J., and Hingray, B.: Building long-term and high spatio-temporal resolution precipitation and air temperature reanalyses by mixing local observations and global atmospheric reanalyses: the ANATEM model, Hydrol. Earth Syst. Sci., 19, 2717–2736, doi:10.5194/hess-19-2717-2015, 2015.

Marzeion, B., Cogley, J. G., Richter, K., and Parkes, D.: Glaciers. Attribution of global glacier mass loss to anthropogenic and natural causes, Science, 345, 919–921, doi:10.1126/science.1254702, 2014.

McCabe, G. J. and Markstrom, S. L.: A Monthly Water-Balance Model Driven By a Graphical User Interface, U.S. Geological Survey Open-File report, 2007.

Moriasi, D. N., Arnold, J. G., van Liew, M. W., Bingner, R. L., Harmel, R. D., and Veith, T. L.: Model evaluation guidelines for systematic quantification of accuracy in watershed simulations, Trans. ASABE, 50, 885–900, 2007.

Rango, A. and Martinec, J.: Revisting the degree-day method for snowmelt computations, J. Am. Water Resour. Assoc., 31, 657–669, doi:10.1111/j.1752-1688.1995.tb03392.x, 1995.

Saha, S., Moorthi, S., Pan, H.-L., Wu, X., Wang, J., Nadiga, S., Tripp, P., Kistler, R., Woollen, J., Behringer, D., Liu, H., Stokes, D., Grumbine, R., Gayno, G., Wang, J., Hou, Y.-T., Chuang, H.-Y., Juang, H.-M. H., Sela, J., Iredell, M., Treadon, R., Kleist, D., Van Delst, P., Keyser, D., Derber, J., Ek, M., Meng, J., Wei, H., Yang, R., Lord, S., Van Den Dool, H., Kumar, A., Wang, W., Long, C., Chelliah, M., Xue, Y., Huang, B., Schemm, J.-K., Ebisuzaki, W., Lin, R., Xie, P., Chen, M., Zhou, S., Higgins, W., Zou, C.-Z., Liu, Q., Chen, Y., Han, Y., Cucurull, L., Reynolds, R. W., Rutledge, G., and Goldberg, M.: The NCEP Climate Forecast System Reanalysis, B. Am. Meteorol. Soc., 91, 1015–1057, doi:10.1175/2010BAMS3001.1, 2010.

Saha, S., Moorthi, S., Wu, X., Wang, J., Nadiga, S., Tripp, P., Behringer, D., Hou, Y.-T., Chuang, H.-Y., Iredell, M., Ek, M., Meng, J., Yang, R., Mendez, M. P., van den Dool, H., Zhang, Q., Wang, W., Chen, M., and Becker, E.: The NCEP Climate Forecast System Version 2, J. Climate, 27, 2185–2208, doi:10.1175/JCLI-D-12-00823.1, 2014.

Schaefli, B.: Projecting hydropower production under future climates: a guide for decision-makers and modelers to interpret and design climate change impact assessments, WIREs Water, 2, 271–289, doi:10.1002/wat2.1083, 2015.

Schueller, F., Förster, K., Hanzer, F., Huttenlau, M., Marzeion, B., Strasser, U., Achleitner, S., and Kirnbauer, R.: Multiscale Snow/Icemelt Discharge Simulations into Alpine Reservoirs: adding Glacier Dynamics to a Hydrological Model, Geophysical Research Abstracts, p. 3244, 2015.

Strasser, U.: Modelling of the mountain snow cover in the Berchtesgaden National Park, vol. 55 of Berchtesgaden National Park research report, Nationalparkverwaltung Berchtesgaden, Berchtesgaden, 2008.

Strasser, U., Warscher, M., and Liston, G. E.: Modeling Snow–Canopy Processes on an Idealized Mountain, J. Hydrometeor., 12, 663–677, doi:10.1175/2011JHM1344.1, 2011.

Thornthwaite, C. W.: An Approach toward a Rational Classification of Climate, Geogr. Rev., 38, 55–94, doi:10.2307/210739, 1948.

Viviroli, D., Dürr, H. H., Messerli, B., Meybeck, M., and Weingartner, R.: Mountains of the world, water towers for humanity, Water Resour. Res., 43, 1–13, W07447, doi:10.1029/2006WR005653, 2007.

Wood, A. W., Maurer, E. P., Kumar, A., and Lettenmaier, D. P.: Long-range experimental hydrologic forecasting for the eastern United States, J. Geophys. Res., 107, 4429 (1–16), doi:10.1029/2001JD000659, 2002.

Yuan, X., Wood, E. F., Luo, L., and Pan, M.: A first look at Climate Forecast System version 2 (CFSv2) for hydrological seasonal prediction, Geophys. Res. Lett., 38, 1–7, L13402, doi:10.1029/2011GL047792, 2011.

Yuan, X., Wood, E. F., and Ma, Z.: A review on climate-model-based seasonal hydrologic forecasting: Physical understanding and system development, WIREs Water, 2, 523–536, doi:10.1002/wat2.1088, 2015.

Proc. IAHS, 374, 151–157, 2016
proc-iahs.net/374/151/2016/
doi:10.5194/piahs-374-151-2016


# Assessment of water resources and seasonal prediction of rainfall in India

**P. R. Rakhecha**[1,a]

[1]independent researcher: D-82, Shastrinagar, Jodhpur-342003, India
[a]formerly at: Indian Institute of Tropical Meteorology, Pune, India

*Correspondence to:* Pukhraj Rakhecha (rakhechapukhraj@gmail.com)

Published: 17 October 2016

**Abstract.** From the analysis of rainfall data available at about 5000 stations, this paper is prepared to provide comprehensive assessment of water availability in different states of India as well as for the country as a whole. Changes in water availability per person are occurring mainly as a consequence of changes in population. The water availability per capita in India was over 6000 $m^3$ per annum in 1951, it now stands at about 1900 $m^3$ per capita. Nevertheless, this amount is nearly twice the water need by the people of developed countries. The developed countries require about 1000 $m^3$ of freshwater per capita per year. The increasing demand for water is in the 4 states of Haryana, UP, Bihar and West Bengal in the face of increasing population and possible changes in climate, the water available in these states is to be used most economically and judiciously so that the vagaries of nature do not put any severe strain.

## 1 Introduction

Freshwater is essential for most human purposes and currently it is used in five major sectors: domestic (including drinking water), agriculture, industry, hydropower generation and ecosystem conservation. In recent years there has been an extraordinary increase of interest in the reliable assessment, optimum utilization, and efficient management of water resources in almost every country of the world including India. This is natural since during the past 2 decades the water requirements of every country have enormously increased as a result of population growth and expansion in various important fields of human activity. Of late some hydrologists have come to believe that the world is fast coming to a stage where there is going to be an actual shortage of water. It is in this context we have to view the assessment and management of water resources of India which are unfortunately unevenly distributed by nature. We know that rainfall brings water to the land and the occurrence of rainfall can be considered to be a process of weather systems.

India with 2.4 % of the world surface area of 135.79 million $km^2$ and one sixth (16 %) of the world's population and diversity of topography, soils, climate, etc., has always been facing challenges in the area of water availability received from the southwest (June–September) and northeast (October–December) monsoons. For example, the water availability per capita in India was over 6000 $m^3$ per annum in 1951, it now stands at about 1900 $m^3$ per capita (see Table 2). Careful use of water resources requires a detailed account of various characteristics of climate systems and the quantitative distribution of rainfall over the country. Moreover, the rainfall in a given area is not the same every year and it may range from half of the normal in one year to twice the normal the next year. This leads to either too much water from high rainfall or too little water from low rainfall resulting in droughts. The rainfall therefore is one of the most irregular climatic variables and this inter-annual variability in rainfall makes rainfall prediction very important for policy makers.

The objective of this paper therefore is to provide comprehensive assessment of water resources that is how much water from rainfall is received by the country as a whole as well as by each state of the country and what the estimates are of surface water and groundwater for utilization so that the information can be useful to manage this important resource in a sustainable manner as also to solve water sharing issues and disputes between the states. India with a total area

Published by Copernicus Publications on behalf of the International Association of Hydrological Sciences.

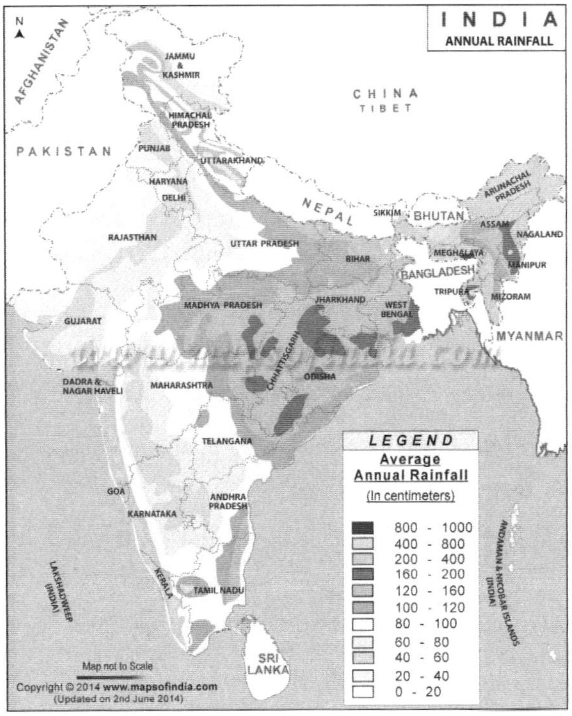


**Figure 1.** Different states of India and the annual rainfall (cm) over India.

of 3 287 263 $km^2$ is divided into 29 states and 7 union territories (see Fig. 1).

## 2 Rain gauge network in India

The assessment of water resources requires measurement of rainfall. Rainfall measurement in India began towards the end of the 18th century when the first rainfall station was set up in 1784 at Calcutta (Kolkata). Thereafter, at Madras (Chennai) in 1792, at Bombay (Mumbai) in 1823 and at Shimla in 1840. With the establishment of the India Meteorological Department (IMD) in 1875, the rainfall station network grew rapidly. At the time of independence in 1947, there were around 3000 rainfall stations. Observations of rainfall from these stations are available extending over more than 100 years. The need for increasing the rainfall stations for agriculture and water resources development work was felt after 1947. The rain gauge network at present comprises of about 5000 stations whose data are published, processed and analyzed by the IMD.

## 3 Climate systems

From north to south, the Indian climate shifts from subtropical to tropical. For the most part of the country the year can be divided into the four seasons viz., winter (January to February), summer (March to May), southwest monsoon or rainy (June to September) and northeast monsoon (October to December). The winter season is characterized by low temperatures averaging around 10–15 °C in the north to around 20–25 °C in the south. The month of March marks the beginning of summer and the temperature starts rising progressively from April through June, averaging around 30–40 °C. The southwest monsoon months constitute the major rainy season which brings water to the Indian landscape. Most parts of the country receive about 75–80 % of their annual rainfall during this season. From time to time monsoon weather systems (depressions and tropical storms) from the Bay of Bengal and the Arabian sea move across the country and bring heavy rainfall and floods into the Indian rivers. The northeast monsoon period is comparatively drier throughout the country but serves as a major source of rainwater for the states of Andhra Pradesh, Telangana and Tamil Nadu.

## 4 Distribution and assessment of annual rainwater resources over India

The source of all freshwater in India is rainfall from the southwest and northeast monsoons. The mean annual rainwater distribution over the country is shown in Fig. 1 which apparently is shaped by the country's physical geography. Figure 1 shows that rainwater is distributed unevenly across the country. In the high reaches of Western Ghats there are places which receive rainwater of the order of 600 cm. However, on the lee side the rainwater decreases sharply to as low a figure as 60 cm. Again in the southern slopes of Khasi-Jaintia Hills in Meghalaya the influence of topography is predominant and there are stations where the normal annual rainwater is of the order of 1000 cm. Whereas on the northern side the influence of Himalayan ranges is apparent in the rainwater pattern in the sub-montane regions, on the northwestern side the rainwater rapidly decreases and over the west Rajasthan desert the annual amounts decreases to about 10 cm.

Dhar and Rakhecha (1975, 1979) have estimated the average monthly, seasonal and annual rainwater of the contiguous Indian area using 60 years (1901–1960) rainfall data of about 3000 stations (see Table 1). The mean annual rainwater for the country as a whole is about 117 cm with a CV of 10 %. This table shows that each of the monsoon months contributes more than 15 % of annual rainwater. The CV of these four months varies from 13 to 22 % while other months have high variability. The average seasonal rainwater amounts are 2.8, 12.6, 89.7 and 12.3 cm for winter, summer, southwest monsoon and northeast monsoon contributing 2.4, 10.7, 76.4 and 10.5 % respectively to the annual rainwater amount. A CV value of 10 % demonstrates that in approximately 68 out of every 100 years, the rainfall in the country will range from an excess of 10 % over the mean annual rainfall to a deficiency of 10 %. The deviation will be greater than 10 % in the remaining 32 years: a deficit greater than 10 % in approximately 16 of those years and an excess greater than 10 % in the other 16 years.

The average annual rainfall of the country is about 117 cm which is almost one and half times larger than the world average of 80 cm year$^{-1}$. The total volume of annual rainwater calculated from the product of the rainfall and the land area of India (3 287 263 km$^2$) is approximately 4000 km$^3$ with a standard deviation of 385 km$^3$. Of the total rainwater received by the country, about 1400 km$^3$ of water is lost by evaporation and transpiration and approximately 730 km$^3$ of water goes into the soil. After deduction for evapotranspiration and infiltration, according to CWC (2002) the average annual surface runoff in the river systems of the country is 1869 km$^3$. This shows that roughly 47 % of the annual rainfall is converted into surface runoff. The ground water recharge from rainfall in India made by the Central Ground Water Board, New Delhi is about 432 km$^3$ (CGWB, 1991; Chadha, 2006). Thus the total water resource of India is 2301 km$^3$. The present water availability for a population of 1210 million people in India stands approximately 1902 m$^3$ per capita per year. Developed countries require approximately 1000 m$^3$ of freshwater per capita per year. In this sense, water resources of India are relatively high.

The series of annual rainwater volume over the country is Gaussian. The values of annual rainwater volume which can be exceeded at probability of 5, 10, 20, 30, 40, 50, 60, 70, 80, 90 and 95 % levels have been calculated as 4635, 4492, 4325, 4203, 4098, 4000, 3902, 3798, 3677, 3507 and 3365 km$^3$ respectively.

## 5 Population and freshwater availability trends in India

India's finite and fragile water resources are stressed and depleting while population has been growing rapidly. Population growth is the key to the whole equation of water availability and its use because of the fact that India's population has grown four fold over the past 70 years. The population of India which at the turn of the twentieth century was around 238.4 million increased to reach 1210 million at the dawn of the twenty first century. The population of India as recorded at each decennial census from 1901 has grown steadily except for a decrease during 1911–1921. Decadal growth of population from 1901 is shown in Table 2. This table shows that per capita water availability is reducing with population growth. The water per capita in India was over 6000 m$^3$ per annum in 1951, it is now stands at about 1902 m$^3$ per capita. With every increase in population there is corresponding decline in per capita availability of water (Fig. 2). It is estimated that the population of India would be around 1.5 to 1.8 billion in 2050. The per capita availability of water will then be around 1200 m$^3$ year$^{-1}$, consequently effective water management in India is of greater urgency than in most of other countries.

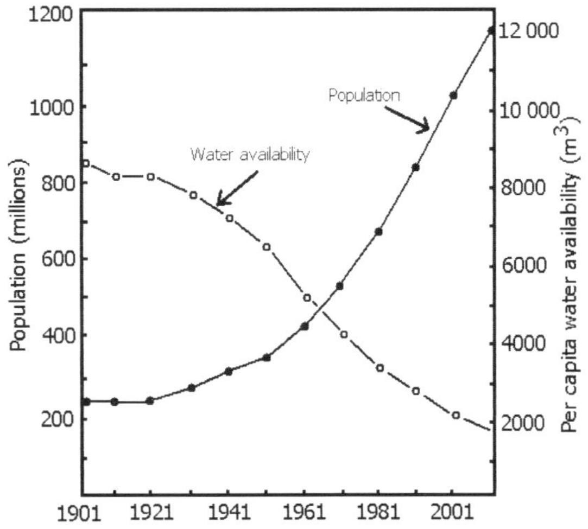


**Figure 2.** Population and water availability trends 1901–2011.

## 6 Assessment of annual water resources of different states of India

It is important to know how much water from rainfall is received by each state of the country, and determine estimates of surface water and groundwater for utilization. Table 3 gives area average annual rainfall values and the coefficient of variability of annual rainfall for each of the 29 states, as well as for the country as a whole estimated from point rainfall values (IMD, 1962, 1981). Table 3 shows that the rainfall in India varies widely. The highest average annual rainfall is 300 cm in Kerala and the lowest annual rainfall is 55 cm in Haryana. About 10 % of India has an annual rainwater greater than 200 cm while about 40 % has less than 100 cm. These proportions play an important part in the relative water availability statuses of the 29 states.

The rainfall-runoff ratio has been determined to be approximately 47 % for the country as a whole. Average annual rainfall values are available for each state of India in Table 3. By assuming a rainfall-runoff ratio of 47 %, estimates of annual runoff water available in each state of the country can be made. A summary of the population, the average annual rainfall and runoff, the average annual amount of groundwater and per capita annual water availability in each state of India are given in Table 3. Figures suggest that the disparity among states in per capita water availability is wide. The highest water availability is 65 071 m$^3$ year$^{-1}$ per person in Arunachal Pradesh, and the lowest is 831 m$^3$ year$^{-1}$ per person in Bihar. According to Table 3, four contiguous states (Haryana, Uttar Pradesh, Bihar and West Bengal) have total water availability less than 1000 m$^3$ year$^{-1}$ per person, 3 between 1000 and 1500 m$^3$ year$^{-1}$ per person, eight between 1500 and 3000 m$^3$ year$^{-1}$ per person, five between 3000 and 5000 m$^3$ year$^{-1}$ per person and 9 have over 5000 m$^3$ year$^{-1}$ per person. Areas with water resources in the

**Table 1.** Rainwater statistics of the contiguous Indian area-1901 to 1960 (Dhar and Rakhecha, 1975, 1979).

| Months | Mean rainfall (cm) | % of annual | SD (cm) | CV (%) | Highest rainfall (cm) | Lowest rainfall (cm) |
|---|---|---|---|---|---|---|
| January | 1.1 | 0.9 | 1.1 | 100 | 5.3 | 0.4 |
| February | 1.7 | 1.7 | 1.3 | 77 | 5.1 | 0.3 |
| March | 2.5 | 2.1 | 1.1 | 44 | 5.7 | 0.7 |
| April | 3.8 | 3.2 | 0.9 | 24 | 6.9 | 1.8 |
| May | 6.3 | 5.4 | 1.6 | 25 | 10.3 | 3.4 |
| June | 16.6 | 14.1 | 3.5 | 21 | 22.5 | 9.7 |
| July | 29.7 | 25.3 | 3.8 | 13 | 35.9 | 17.1 |
| August | 26.0 | 22.3 | 3.7 | 14 | 33.9 | 17.8 |
| September | 17.4 | 14.8 | 3.9 | 22 | 27.1 | 10.3 |
| October | 7.8 | 6.7 | 3.0 | 39 | 16.0 | 2.1 |
| November | 3.1 | 2.6 | 1.7 | 55 | 7.5 | 0.3 |
| December | 1.4 | 1.2 | 0.8 | 57 | 3.7 | 0.1 |
| Winter (Jan–Feb) | 2.8 | 2.4 | 1.2 | 43 | 6.1 | 0.7 |
| Summer (Mar–May) | 12.6 | 10.7 | 2.1 | 17 | 14.8 | 5.6 |
| Southwest (Jun–Sep) | 89.7 | 76.4 | 8.5 | 9 | 100.5 | 65.1 |
| Northeast (Oct–Nov) | 12.3 | 10.5 | 3.5 | 29 | 21.0 | 5.0 |
| Annual | 117.4 | 100 | 11.7 | 10 | 144.6 (1917) | 96.1 (1918) |

**Table 2.** Population and water availability trends in India (M.I&B, 2012).

| Census years | Population (million) | Decadal growth (%) | Progressive growth rate over 1901 (%) | Per capita water availability ($m^3$) |
|---|---|---|---|---|
| 1901 | 238.4 | – | – | 9652 |
| 1911 | 252.1 | 5.8 | 5.8 | 9127 |
| 1921 | 251.3 | −0.31 | 5.4 | 9156 |
| 1931 | 279.0 | 11.0 | 17.0 | 8247 |
| 1941 | 318.7 | 14.2 | 33.7 | 7220 |
| 1951 | 361.1 | 13.3 | 51.5 | 6372 |
| 1961 | 439.2 | 21.6 | 84.3 | 5239 |
| 1971 | 548.2 | 24.8 | 129.9 | 4197 |
| 1981 | 683.3 | 24.7 | 186.7 | 3368 |
| 1991 | 846.4 | 23.9 | 255.0 | 2719 |
| 2001 | 1028.7 | 21.5 | 331.5 | 2236 |
| 2011 | 1210.0 | 17.6 | 407.5 | 1902 |

range of 1000 to 1500 $m^3$ $year^{-1}$ per capita begin to experience water stressed condition. Increasing shortages are felt at local levels which can spread to the regional level as the population continues to grow. Jain et al. (2007) have written a comprehensive book which covers the water resources of India in sufficient detail for the benefit of hydrologists.

At the time of India's independence in 1947, the country's population was less than 360 million and water available per capita was over 6000 $m^3$ $year^{-1}$. Now, after 68 years of independence, India's population has increased to approximately 1.2 billion and per capita water availability has fallen to approximately 1902 $m^3$ $year^{-1}$. This per capita water availability will further fall to approximately 1500 $m^3$ $year^{-1}$ by the year 2025 due to increasing population which will exert further pressure on water availability. This necessitates great changes to be made with a national perspective in this regard.

## 7 Seasonal forecasting of Indian monsoon rainfall

Rainfall forecasts are required in planning and management of rainwater resources for agriculture and other human activities. Hence, in recent years the need for improved forecasting of monsoon rainfall in India has become increasingly important. We shall study some of the methods of weather forecasting for the benefit of hydrologists.

Generally there are three main approaches used for long-range forecast of the southwest monsoon in India. The first is the statistical method which uses the historical relationship between the southwest monsoon and various global weather

**Table 3.** Land, Population and Water Resources of 29 states in India (Source: IMD, 1962 and M.I&B, 2012).

| Sr No. | State | Area (km$^2$) | People (million) (2011) | Avg. annual rainfall (cm) | CV (%) | Total water (km$^3$ year$^{-1}$) | Runoff (km$^3$ year$^{-1}$) | Ground Water (km$^3$ year$^{-1}$) | Water Resource (km$^3$ year$^{-1}$) | Per capita water (m$^3$ year$^{-1}$) |
|---|---|---|---|---|---|---|---|---|---|---|
| 1 | Andhra Pradesh | 160 229 | 49.4 | 88 | 20 | 141.0 | 66.3 | 5.3 | 71.6 | 1449 |
| 2 | Arunachal Pradesh | 83 743 | 1.4 | 228 | 11 | 190.9 | 89.7 | 1.4 | 91.1 | 65 071 |
| 3 | Assam | 78 438 | 31.2 | 252 | 11 | 196.0 | 92.9 | 22.5 | 115.4 | 3699 |
| 4 | Bihar | 94 161 | 103.8 | 134 | 13 | 126.2 | 59.3 | 27.0 | 86.3 | 831 |
| 5 | Chattisgarh | 136 034 | 25.6 | 135 | 16 | 183.6 | 86.3 | 16.1 | 102.4 | 4000 |
| 6 | Goa | 3702 | 1.5 | 254 | 20 | 9.4 | 4.4 | 0.2 | 4.6 | 3066 |
| 7 | Gujarat | 196 024 | 60.4 | 83 | 30 | 162.7 | 76.5 | 20.4 | 96.9 | 1604 |
| 8 | Haryana | 44 212 | 25.4 | 55 | 28 | 24.3 | 11.4 | 11.2 | 22.6 | 890 |
| 9 | Himachal Pradesh | 55 673 | 6.9 | 175 | 21 | 97.4 | 45.8 | 0.3 | 46.1 | 6681 |
| 10 | Jammu & Kashmir | 222 236 | 12.6 | 100 | 22 | 222.2 | 104.4 | 4.4 | 108.8 | 8635 |
| 11 | Jharkhand | 79 714 | 33.0 | 119 | 16 | 94.9 | 44.6 | 6.6 | 51.2 | 1551 |
| 12 | Karnataka | 191 791 | 61.2 | 136 | 18 | 260.8 | 122.6 | 16.2 | 138.8 | 2268 |
| 13 | Kerala | 38 863 | 33.4 | 300 | 14 | 116.6 | 54.8 | 7.9 | 62.7 | 1877 |
| 14 | Madhya Pradesh | 308 144 | 72.6 | 122 | 20 | 375.9 | 176.7 | 34.8 | 211.5 | 2913 |
| 15 | Maharashtra | 307 713 | 112.4 | 132 | 21 | 406.2 | 190.9 | 37.4 | 228.3 | 2031 |
| 16 | Manipur | 22 327 | 2.7 | 252 | 11 | 56.3 | 26.5 | 3.2 | 29.7 | 11 000 |
| 17 | Meghalaya | 22 429 | 3.0 | 283 | 11 | 63.5 | 29.8 | 0.5 | 30.3 | 10 100 |
| 18 | Mizora | 21 081 | 1.1 | 283 | 11 | 59.0 | 28.1 | – | 28.1 | 25 545 |
| 19 | Nagaland | 16 579 | 2.0 | 283 | 11 | 46.9 | 22.1 | 0.7 | 22.8 | 11 400 |
| 20 | Orissa | 155 707 | 42.0 | 149 | 14 | 232.0 | 109.1 | 20.1 | 129.2 | 3076 |
| 21 | Punjab | 50 362 | 27.7 | 63 | 34 | 31.7 | 14.9 | 18.2 | 33.1 | 1195 |
| 22 | Rajasthan | 342 239 | 68.6 | 59 | 31 | 201.9 | 94.9 | 12.6 | 107.5 | 1567 |
| 23 | Sikkam | 7096 | 0.6 | 274 | 12 | 19.5 | 9.2 | – | 9.2 | 15 333 |
| 24 | Tamil Nadu | 130 058 | 72.2 | 101 | 14 | 131.4 | 61.8 | 26.4 | 88.2 | 1222 |
| 25 | Telangana | 114 840 | 35.3 | 93 | 20 | 106.8 | 50.2 | 3.8 | 54.0 | 1530 |
| 26 | Tripura | 10 491 | 3.7 | 252 | 11 | 26.4 | 12.4 | 0.7 | 13.1 | 3540 |
| 27 | Uttaranchal | 53 483 | 10.2 | 103 | 19 | 55.1 | 25.9 | 28.4 | 54.3 | 5234 |
| 28 | Uttar Pradesh | 238 566 | 199.3 | 99 | 20 | 236.2 | 110.0 | 82.6 | 192.6 | 966 |
| 29 | West Bengal | 88 752 | 91.4 | 165 | 15 | 146.5 | 68.9 | 23.1 | 92.0 | 1007 |
| – | India | 3 287 263 | 1210 | 117 | 10 | 4000 | 1869 | 432 | 2301 | 1902 |

parameters. The second approach is the empirical method which uses time series analysis of past rainfall data to predict the monsoon. The third is the dynamical method, which uses general circulation models of the atmosphere and ocean to predict the southwest monsoon. Prediction models based on the statistical approach have so far yielded the most satisfactory results for the Indian monsoon. However, none of the models can claim 100 % accuracy because there are several factors like the correlations between the parameters and changing predictability of the model over a period of time because of weather anomalies.

### 7.1 Statistical method used for predicting the southwest monsoon rainfall

Prior to 2002, the IMD used to issue the annual forecast using a model based on 16 parameters. The 16 parameters are broadly classified into 3 groups, namely (1) pressure, (2) upper winds and temperatures and, (3) snow cover and atmospheric oscillations. But the model failed in 2002. Since 2003, two new models were introduced which instead of 16 used 8 and 10 parameters to forecast the southwest monsoon in India.

Weather forecasts generally have uncertainty due to small errors in the initial conditions and model approximations. Together they limit the skill of a deterministic forecast system. Ensemble forecasting has emerged as the practical way of estimating the forecast uncertainty and making probabilistic forecasts. Ensemble forecasts are generated based on multiple perturbed initial conditions which sample the errors in the initial conditions to estimate the forecast uncertainty. The skill of the ensemble forecast shows marked improvement over the deterministic forecast. Since 2007, an ensemble forecasting technique is being used to forecast the monsoon.

At present, the predictors used to forecast the monsoon include the surface temperature of the north Atlantic, equatorial southeast Indian Ocean and Central Pacific Ocean along with the air temperature of northwest Europe. Apart from this, the warm water volume of the equatorial Pacific, the air pressure over the north Atlantic and East Asia as well as the wind patterns over the north central Pacific Ocean are the parameters used to predict the southwest monsoon.

### 7.2 Forecasting Northeast monsoon rainfall over Tamil Nadu

Tamil Nadu (see Fig. 1) is the only state of the Indian union which receives more rainfall in the northeast monsoon (October–December) season than in the southwest monsoon (June–September) season. Thus the rainfall received in Tamil Nadu during the northeast monsoon season is of great economic value. Major agricultural operations are normally undertaken during that season. It has been noted that the rainfall during northeast monsoon is highly variable. Therefore, if its behavior could be predicted in advance, it would go a long way toward helping the agricultural and industrial activities of the state.

Dhar and Rakhecha (1983) examined the association between the southwest and northeast monsoon rainfall over Tamil Nadu for the 100 year period from 1871 to 1978. A correlation analysis between the two rainfall series revealed that the southwest monsoon rainfall is negatively correlated with that of the northeast monsoon rainfall. That is an excess or deficit of southwest monsoon rainfall over Tamil Nadu is generally followed by an opposite tendency to the northeast monsoon rainfall. A Chi-square test of dependence further indicated that the rainfalls in the two monsoons are not independent of each other. The negative rainfall relationship can be useful tool in forecasting the northeast monsoon rainfall over Tamil Nadu. The observed southwest monsoon rainfall over Tamil Nadu can be used to provide a forecast of the behavior of the following northeast monsoon season rainfall.

## 8 Water use in India

According to the Ministry of Water Resources, Government of India (2002), about 690 $km^3$ from surface water and 432 $km^3$ from groundwater are considered to be suitable water for public use. It is found that the five major sectors (agriculture, domestic, industry, hydro power generation and environment) use nearly 60 % of the total water estimated at about 1122 $km^3$ in 2010 (Rakesh Kumar et al., 2005). Agriculture is the mainstay of India's economy. As such the first consumer is agriculture which uses about 543 $km^3$ of the total volume. The second largest consumption of water is domestic water of nearly 42 $km^3$. The industry, power generation and environment use about 37, 18 and 5 $km^3$ per year respectively. The population in India is steadily increasing which means an increase in the water demand. With an average growth rate of 3.5 % per year in population, Rakesh Kumar et al. (2005) estimated the total water requirements for agriculture, domestic use, industry, power generation and environment would be about 561, 55, 67, 31 and 10 $km^3$ per year respectively in the year 2025.

## 9 Climate change impacts on India's rainfall

According to the Intergovernmental Panel on Climate Change (IPPC, 2007) the world's climate may be changing in response to land use change and from the increase in carbon dioxide and other green house gases. Emissions of carbon dioxide grew twelve fold between 1900 and 2000 from 534 million metric tons per year in 1900 to 6.59 billion metric tons in 1997. In the same period, the human population has nearly quadrupled from 1.6 billion to 6.1 billion progressively consuming greater quantities of fossil fuels. The IPCC note that greenhouse gases are likely to warm the atmosphere by 2 °C by 2100. Therefore, it appears natural that higher temperatures will intensify the global hydrological cycle and in turn considerably influence the variability and strength of the Indian monsoon rainfall. The current question is to see whether or not a climate change exercises any significant influence on the rainfall of India. A large number of studies pertaining to trends and periodicities in either the annual or seasonal rainfall for individual stations (Koteswaram and Alvi, 1969; Jagannathan and Parthasarthy, 1973; Rupa Kumar et al., 1992), for different subdivisions (Dhar et al., 1982; Jain et al., 2013) as well for India as a whole (Mooley and Parthasarathy, 1984; Kumar et al., 2010) showed that there is no statistically significant trend of increase or decrease in Indian rainfall. In addition Rakhecha and Soman (1994) examined annual extreme rainfall data of 1 to 3 days for the period 1901–1980 obtained from 316 stations distributed across the country and discovered that the annual extreme rainfall records of most stations are free from trend and persistence.

## 10 Conclusions

It is a known fact that rainwater in a geographic sense is unevenly distributed in India. The assessment of water resources has shown that the amount of available freshwater has not changed and there is enough fresh water from rainfall in India as a whole to everyone's satisfaction. Nevertheless, four contiguous states of Haryana, Uttar Pradesh, Bihar and West Bengal are characterized as the hot spots of water stressed where urgent action is needed within water sector. However, times have changed as demographic pressure in the last 5 decades (for example 360 million in 1951, 680 million in 1961 and 1210 million at the end of 2011) demonstrates that access to freshwater is a challenge for the current population. The management of water therefore needs cooperation among communities.

## 11 Data availability

The data on rainfall and groundwater play a vital role in the assessment of water resources of any region. Such data in India are available in various publications of India Meteorological Department (IMD) and in the professional journals of meteorology and hydrology. Monthly and annual averages of rainfall and of rainy days for the period 1901–1950 for nearly 3000 rain gauge stations in India are available in the Memoirs of the India Meteorological Department published in 1962 (IMD, 1962) and in the Climatological Atlas of India part A (rainfall) published in 1981. The Groundwater data in respect of different states of India are available in the Bhu-Jal News (special issue on groundwater statistics), a quarterly journal of Central Ground Water Board (CGWB, 1991) published in 1991 by the Ministry of Water Resources, Government of India. These data have been used systematically to study the assessment of water resources of India.

**Acknowledgements.** The author thanks Colin Clark, Charldon Hill Research Station, Bruton, England for comments on an earlier draft of the paper. Also thanks to the reviewers for the constructive review of the paper.

## References

Central Ground Water Board (CGWB): Bhu-Jal News (special issue on groundwater statistics), J. CGWB, Ministry of Water Resources Publ. Government. of India, 6, 1–162, 1991.

Central Water Commission (CWC): National Water Policy 2002, Min. of Water Resources Government of India, New Delhi, 2002.

Chadha, D. K.: Development of groundwater resources of India. A overview, groundwater modeling & management, Capital Publishing Company, New Delhi, 2006.

Dhar, O. N. and Rakhecha, P. R.: A review of hydrometeorological studies of Indian rainfall, Proc Second World Congress on Water Resources organized by International Water Resources Association, held at New Delhi, 449–462, 3, 1975.

Dhar, O. N. and Rakhecha, P. R.: A review of rainfall relationship based upon Indian data, Water Resources Journal ESCAP, Bangkok, 16–25, 1979.

Dhar, O. N. and Rakhecha, P. R.: Foreshadowing northeast monsoon rainfall over Tamil Nadu, India, Mon. Weather Rev., 3, 109–112, 1983.

Dhar, O. N., Rakhecha, P. R., and Kulkarni, A. K.: Fluctuations in northeast monsoon rainfall of Tamil Nadu, J. Climatol., 2, 339–345, 1982.

India Meteorological Department (IMD): Monthly and annual averages of rainfall and of rainy days, Memoirs of IMD, 31, 65–208, 1962.

India Meteorological Department (IMD): Climatological Atlas of India, part A (rainfall), 1981.

Intergovernmental Panel on Climate Change (IPCC): Fourth assessment Report on Climate change, Synthesis report, Cambridge Univ, Press, Cambridge, 2007.

Jagnnathan, P. and Parthasarthy, B.: Trends and peridiocities of rainfall over India, Mon. Weather Rev., 101, 371–375, 1973.

Jain, S. K., Agarwal, P. K., and Singh, V. P.: Hydrology & Water Resources of India, Springer Verlag, Netherlands, 2007.

Jain, S. K., Kumar, V., and Saharia, M.: Analysis of rainfall & temperature in northeast India, Int. J. Climatol., 33, 968–978, 2013.

Koteswaram, P. and Alvi, S. M. A.: Secular trends and periodicities in rainfall at west coast stations in India, Current Sci., 38, 229–231, 1969.

Kumar, V., Jain, S. K., and Singh, Y.: Analysis of long term rainfall trends in India, Hydrolog. Sci., 55, 484–496, 2010.

Ministry of Information & Broadcasting, India 2012: A reference Annual, M. I & B publ. Government of India, New Delhi, 2012.

Ministry of Water Resources: National water policy, Government of India, 1–9, 2002.

Mooley, D. A. and Parthasarathy, B.: Fluctuations in all India summer monsoon rainfall during 1871–1978, Climatic Change, 6, 287–301, 1984.

Rakesh Kumar, Singh, R. D., and Sharma, K. D.: Water resources of India, Curr. Sci. India, 89, 794–811, 2005.

Rakhecha, P. R. and Soman, M. K.: Trends in the annual extreme rainfall events of 1 to 3 days duration over India, Theor. Appl. Climatol., 48, 227–237, 1994.

Rupa Kumar, K., Pant, G. B., Parthasarthy, B., and Sontakke, N. A.: spatial and subseasonal patterns of the long term trends of Indian summer monsoon rainfall, Int. J. Climatol., 12, 257–268, 1992.

Proc. IAHS, 374, 159–163, 2016
proc-iahs.net/374/159/2016/
doi:10.5194/piahs-374-159-2016


Open Access PIAHS

# Comparison of cross-validation and bootstrap aggregating for building a seasonal streamflow forecast model

**Simon Schick[1,2], Ole Rössler[1,2], and Rolf Weingartner[1,2]**

[1]Institute of Geography, University of Bern, Bern, Switzerland
[2]Oeschger Centre for Climate Change Research, Bern, Switzerland

*Correspondence to:* Simon Schick (simon.schick@giub.unibe.ch)

Published: 17 October 2016

**Abstract.** Based on a hindcast experiment for the period 1982–2013 in 66 sub-catchments of the Swiss Rhine, the present study compares two approaches of building a regression model for seasonal streamflow forecasting. The first approach selects a single "best guess" model, which is tested by leave-one-out cross-validation. The second approach implements the idea of bootstrap aggregating, where bootstrap replicates are employed to select several models, and out-of-bag predictions provide model testing. The target value is mean streamflow for durations of 30, 60 and 90 days, starting with the 1st and 16th day of every month. Compared to the best guess model, bootstrap aggregating reduces the mean squared error of the streamflow forecast by seven percent on average. Thus, if resampling is anyway part of the model building procedure, bootstrap aggregating seems to be a useful strategy in statistical seasonal streamflow forecasting. Since the improved accuracy comes at the cost of a less interpretable model, the approach might be best suited for pure prediction tasks, e.g. as in operational applications.

## 1 Introduction

Small sample sizes challenge the application of statistical models for seasonal streamflow forecasting. For example, a daily hydrometeorological time series of length 30 years can be considered as a long record. However, at seasonal time scales the series provides 30 cases (e.g. summer means). Following the nomenclature described by Hastie et al. (2009), the model building procedure then has to cope with these 30 cases for:

1. model training, i.e. fit models with varying complexity or different predictors;
2. model selection, i.e. validate the models and choose the best one(s); and
3. model testing, i.e. estimate the final models prediction error (possibly by combining several models).

To overcome small sample sizes, resampling is commonly used for model selection and testing. In addition, seasonal streamflow forecasting often encounters weak predictor-predictand relationships, introduced by missing or noisy predictors – e.g. precipitation and temperature of the target season. Models out of any resampling thus can differ markedly, which leads us to the following question: Are there any benefits if the models resulting from resampling are combined in a systematic way? To address this question, we compare (1) the selection of a single "best guess" model along with leave-one-out cross-validation against (2) bootstrap aggregating along with out-of-bag prediction error estimates. Bootstrap aggregating was introduced by Breiman (1996a, "bagging" or "bagged model" for short) and aims to reduce the variance of a statistical model by applying it to bootstrap replicates of the data set and combining the corresponding predictions afterwards.

Below Sect. 2 briefly presents the data set, Sect. 3 outlines the methodology, and in Sects. 4 and 5 the results are presented and discussed, respectively.

Published by Copernicus Publications on behalf of the International Association of Hydrological Sciences.

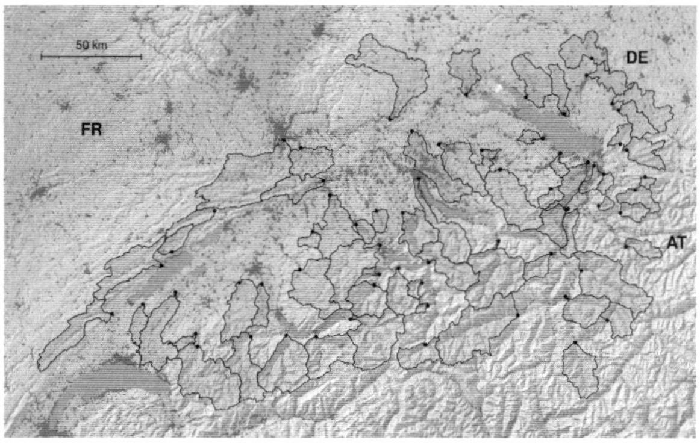


**Figure 1.** The study region comprises parts of Germany, Austria, and Switzerland. Grey shaded regions indicate urban areas, points mark gauging stations.

## 2 Data

The hindcast experiment is applied to 66 sub-catchments (no nesting) of the Swiss Rhine at Basel, ranging in mean elevation from 500 to 2300 m and in area from 20 to 900 $km^2$ (Fig. 1). Streamflow is regulated and routed for the purpose of hydro power, flood protection, water supply, and ecological conservation. Up to 10 catchments can be considered as heavily regulated; for the remaining catchments we assume that anthropogenic effects on the catchments hydrology do not have any impacts at seasonal time scales. Daily mean streamflow in $m^3\,s^{-1}$ for the period 1982–2013 is provided by public authorities of Germany, Austria, and Switzerland, whereas daily precipitation and temperature series are catchment averages derived from the E-OBS gridded data set version 12.0 in 0.25° resolution (approximately 19 and 28 km in longitude and latitude; Haylock et al., 2008).

## 3 Methodology

The comparison of the two model building procedures relies itself on the principle of cross-validation, i.e. some cases are in turn excluded from the data set and the complete model building procedures are conducted by using the remaining cases. Section 3.1 first introduces the regression model, which is common to both model building procedures. Section 3.2 then describes the two resampling approaches; in case of the best guess model, resampling is solely used to estimate the prediction error, whereas in case of the bagged model resampling is at the heart of the model building procedure. Section 3.3 finally states the cross-validation implementation and the statistical test in order to contrast the two procedures.

### 3.1 Regression model

The regression model follows closely the approach of Garen (1992), i.e. initial conditions are considered only. The predictand $\boldsymbol{y}_{i,j}$ is in turn mean streamflow of duration $i = 30, 60, 90$ d, starting at the 1st and 16th day of every month (date of prediction $j = 1, \ldots, 24$). For a particular choice of $i$ and $j$, the regression equation is given by

$$\boldsymbol{y}_{i,j} = b_0 \boldsymbol{1} + \mathbf{X}\boldsymbol{b} + \boldsymbol{\varepsilon} \tag{1}$$

where $b_0$ denotes the intercept, $\boldsymbol{1}$ a vector of ones, $\boldsymbol{b}$ the vector of regression coefficients, and $\boldsymbol{\varepsilon}$ the errors. The $n \times p$ matrix $\mathbf{X}$ has in its $p = 3$ columns antecedent streamflow, antecedent precipitation, and antecedent temperature as predictors; $n$ equals the number of years. The time aggregation is individually selected for each predictor according to Spearman's rank correlation, but has to be one of $10, 20, \ldots, 720$ d. Since these predictors can be highly correlated, the regression coefficients $\boldsymbol{b}$ are estimated using partial least squares (Mevik and Wehrens, 2007). Partial least squares is related to principal components regression, but decomposes the cross-covariance matrix $\mathbf{X}^T\boldsymbol{y}$ instead of the predictors covariance matrix. Regarding model selection, we decide to select at least the first partial least squares direction, as otherwise the regression model shrinks to $b_0$ in Eq. (1). Please note that we do not make any distributional assumptions about $\boldsymbol{\varepsilon}$.

### 3.2 Resampling approaches

The regression model from Eq. (1) is applied twofold for a particular catchment and predictand $\boldsymbol{y}_{i,j}$:

1. A single best guess model is selected and the mean squared error of prediction $E_{\mathrm{MSP}}$ is estimated according to leave-one-out cross-validation.

2. For each of 100 bootstrap replicates of the data set, one model is selected. These models are then combined by simply averaging their predictions (bootstrap aggregating; Breiman, 1996a). Here, out-of-bag predictions (Breiman, 1996b) are used to estimate $E_{\mathrm{MSP}}$.

Breiman (1996a) showed that the aggregation of unstable models can help to decrease the prediction error. Instability (or high model variance) refers to the case when small changes in the data set lead to large changes in the final estimated model. A simple linear model fitted by ordinary least squares can be considered as an example for a stable model whereas neural networks or regression trees generally are examples for unstable models. The present model from Sect. 3.1 is in our view a stable model – it is linear, consists of three predictors (where only the time aggregations are allowed to vary), and partial least squares further tries to reduce the dimensionality of the predictor space.

The out-of-bag approach is closely related to the leave-one-out procedure in that one case is left out at a time, i.e. the model averaging considers only those models for which the "left-out" case was not included in the corresponding replicates. Since prediction error estimates of out-of-bag and leave-one-out approximately converge with an increasing number of bootstrap replicates, the out-of-bag estimate provides a convenient alternative – testing a bagged model via leave-one-out can be computationally expensive due to the involved bootstrap. Finally, the choice of 100 replicates is based on the recommendation by Hastie et al. (2009) that model training can be stopped as soon as the out-of-bag error has stabilised.

Hereafter, the two approaches are named BGS/LOO (best guess model BGS in combination with leave-one-out LOO) and BAG/OOB (bagged model BAG in combination with out-of-bag OOB), respectively.

### 3.3 Hindcast experiment

In order to contrast BGS/LOO with BAG/OOB, the 32 year period of investigation is used for an additional leave-one-out cross-validation. Doing so, we get an estimate of $E_{\mathrm{MSP}}$ independently of LOO and OOB. Here, also the three adjacent years of the left out case are omitted to avoid spurious skill due to catchment memory (hence $n = 25$ in Eq. 1). Since BGS/LOO and BAG/OOB are nested inside this "buffered" leave-one-out cross-validation, we refer to the latter as the outer cross-validation. Considering a particular catchment and predictand $\boldsymbol{y}_{i,j}$, three steps are applied:

1. $\boldsymbol{y}_{i,j}$ is centred to mean 0 and scaled to standard deviation 1 with respect to the period 1982–2013.

2. Each year (together with its three adjacent years) is left out once, while the remaining years are used for the application of the model building procedures BGS/LOO and BAG/OOB.

3. The mean value of $\boldsymbol{y}_{i,j}$ serves as a competing model (hereafter named the seasonal regime, SRG). Analogue to BGS, $E_{\mathrm{MSP}}$ is estimated by LOO as well as the outer cross-validation.

Paired differences of $E_{\mathrm{MSP}}$ are then used for inference. Here, paired differences are calculated such that $E_{\mathrm{MSP}}$ of the more complex model is subtracted from $E_{\mathrm{MSP}}$ of the less complex model (always per catchment). The mean difference $\mu$ is used for a right-sided $t$ test (alternative hypothesis $\mu > 0$). Also a nonparametric bootstrap is applied to estimate the probability $P\{\mu > 0\}$, since the differences not necessarily follow a Gaussian distribution.

## 4 Results

The results are arranged in three sections: Firstly, we contrast BGS with BAG in order to see whether bagging improves the predictions. Secondly, model skill is evaluated by comparing

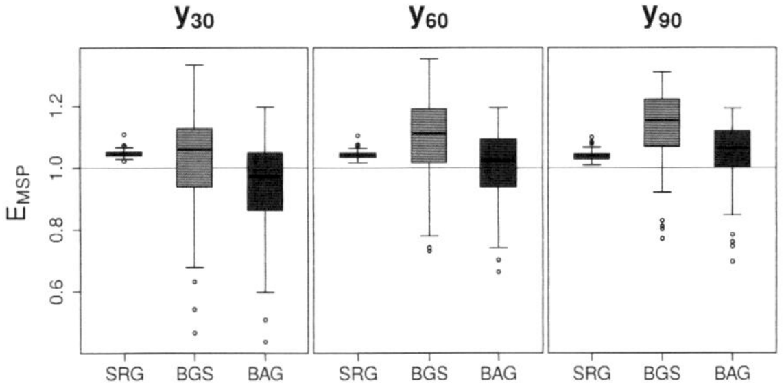


**Figure 2.** $E_{\mathrm{MSP}}$ for all catchments and predictands as obtained from the outer cross-validation; $n = 66$.

BGS and BAG against SRG. Thirdly, we analyse the accuracy of $E_{\mathrm{MSP}}$ estimates from LOO and OOB. In the following subscript $j$ is dropped when error statistics are averaged over $j$.

### 4.1 Comparison of prediction error

The comparison of BGS against BAG focuses on $E_{\mathrm{MSP}}$ from the outer cross-validation: Fig. 2 suggests that BAG scores on average the smaller $E_{\mathrm{MSP}}$. Also the $p$ values indicate that BAG is most likely able to reduce $E_{\mathrm{MSP}}$ (third row in Table 1). Table 2 lists additionally $E_{\mathrm{MSP}}$ of BGS and BAG for $\boldsymbol{y}_{30}$, $\boldsymbol{y}_{60}$, and $\boldsymbol{y}_{90}$, averaged over all catchments (i.e. the mean value of the corresponding whisker boxes in Fig. 2). Independently of the predictand, reduction of $E_{\mathrm{MSP}}$ by using BAG instead of BGS amounts to 7 to 8 %.

### 4.2 Comparison of model skill

For the evaluation of model skill we focus again on the $E_{\mathrm{MSP}}$ estimates from the outer cross-validation (Fig. 2). Due to standardisation of $\boldsymbol{y}_{i,j}$, the benchmark model SRG shows an $E_{\mathrm{MSP}}$ near 1 for all catchments (a perfectly estimated mean value would yield an $E_{\mathrm{MSP}}$ of 1). On average SRG is a serious competitor and outperforms BGS and BAG in several catchments. Reduction of $E_{\mathrm{MSP}}$ by using BGS and BAG instead of SRG is strongest for $\boldsymbol{y}_{30}$ and weakest for $\boldsymbol{y}_{90}$. These findings are also supported by Table 1, which reports the $p$ values of the $t$ test and the bootstrap: It is questionable to unlikely that BGS reduces $E_{\mathrm{MSP}}$ on average, whereas BAG very likely does for $\boldsymbol{y}_{30}$ and $\boldsymbol{y}_{60}$, but not for $\boldsymbol{y}_{90}$.

### 4.3 Comparison of prediction error estimation

Figure 3 shows the differences in $E_{\mathrm{MSP}}$, when LOO and OOB estimates are subtracted from the estimates obtained in the outer cross-validation, which are here considered to be the reference. Thus, a positive difference can be attributed as an underestimation and a negative difference as an overestimation of the prediction error. Apart from a few outliers, the differences lie in the interval $[-0.1, 0.1]$ and are symmetri-

**Table 1.** $p$ values for the null hypothesis "the simple model outperforms the complex model", estimated by a right-sided $t$ test (paired differences with mean difference $\mu$ and alternative hypothesis $\mu > 0$). Paired differences here follow the rule that $E_{\mathrm{MSP}}$ of the more complex model is subtracted from $E_{\mathrm{MSP}}$ of the less complex model, as specified in the first column. In parentheses also the probabilities $P\{\mu > 0\}$ according to a nonparametric bootstrap with 10 000 replicates are listed. Only $E_{\mathrm{MSP}}$ from the outer cross-validation is considered; $n = 66$.

| | $y_{30}$ | $y_{60}$ | $y_{90}$ |
|---|---|---|---|
| SRG-BGS | 0.1 (0.09) | 0.99 (0.99) | 0.99 (0.99) |
| SRG-BAG | <0.01 (<0.01) | <0.01 (<0.01) | 0.44 (0.45) |
| BGS-BAG | <0.01 (<0.01) | <0.01 (<0.01) | <0.01 (<0.01) |

**Table 2.** $E_{\mathrm{MSP}}$ of BGS and BAG from the outer cross-validation, averaged over all catchments; $E_{\mathrm{MSP}}$ is based on centred and standardised $y_{i,j}$. The last row indicates the reduction in $E_{\mathrm{MSP}}$ when BAG is used instead of BGS.

| | $y_{30}$ | $y_{60}$ | $y_{90}$ |
|---|---|---|---|
| BGS | 1.02 | 1.09 | 1.13 |
| BAG | 0.95 | 1.00 | 1.04 |
| 1 – BAG/BGS | 0.07 | 0.08 | 0.08 |

cally centred around zero – on average neither LOO nor OOB tend to optimism or pessimism. The heavy negative outliers correspond to the same catchment, which turns out to be regulated due to hydro power.

## 5 Discussion

In the present study, a hindcast experiment was conducted that mimics the operational use of a simple forecasting system. The objective was the comparison of two model building procedures, which both rely on the same regression model, but use different resampling strategies: A single best guess model, which is tested by leave-one-out cross-validation, and a bagged model, which employs the bootstrap technique in order to build an ensemble of models. An useful byproduct of bagging is the out-of-bag prediction error estimate, which in theory can replace an additional resampling. Regarding the methodology, several points need some attention:

- Catchments were selected without a priori reasoning about their adequacy for seasonal streamflow forecasting. Strictly speaking, none of these catchments exhibits natural streamflow, though some anthropogenic effects might be averaged out due to the seasonal time scale. However, most of these effects are hardly quantifiable and it is not clear whether or not they favour model skill.
- The standardisation of $y_{i,j}$ attaches all seasons and catchments equal weights for the analysis. Doing so, model skill in seasons/catchments with large streamflow variability is masked, e.g. in spring when snow melting occurs and the models perform best (not shown).
- In order to compare the model building procedures, $E_{\mathrm{MSP}}$ estimates from the outer cross-validation are considered as the "true" values. This assumption is indeed critical, but unavoidable in the present context – otherwise the real-world data set has to be replaced with a synthetic one.
- The residual analysis (not shown) reveals that the prediction errors are not independent and identically distributed. High flow is commonly underestimated, whereas low flows are often overestimated. Technically, the model can be considered as misspecified, since it lacks relevant predictors (most likely precipitation and temperature during the season to predict). Therefore, common techniques to estimate prediction intervals are not applicable. It remains to be tested whether a substitution of the missing predictors by climate indices or seasonal climate predictions mitigates model misspecification.

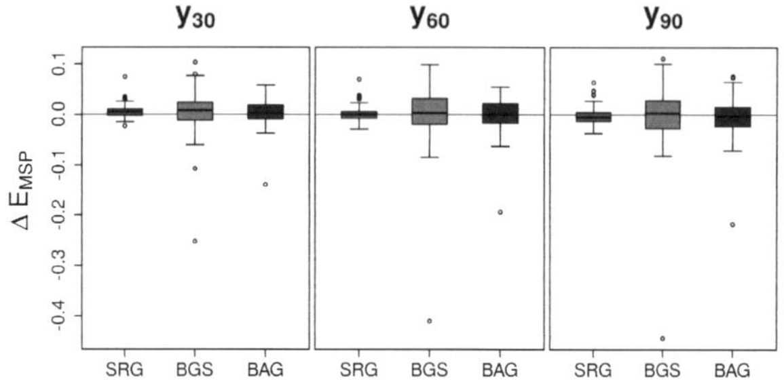


**Figure 3.** $E_{\mathrm{MSP}}$ estimates of LOO (in case of SRG and BGS) and OOB (in case of BAG) subtracted from $E_{\mathrm{MSP}}$ estimates of the outer cross-validation for all catchments and predictands; $n = 66$.

## 6 Conclusions

The results are valid only for the present data set, though the sample size of 66 catchments in combination with 72 predictands might permit more general conclusions:

- BAG scores on average the lower $E_{\mathrm{MSP}}$ than BGS. Bagging is useful if the model is unstable (Breiman, 1996a). Since we consider the applied model as rather simple and stable, we argue that instability is introduced by weak predictor-predictand relationships in combination with small sample sizes. These weak (and sometimes spurious) relationships propagate through the screening of the time aggregation, the selection of partial least squares directions, and the final regression coefficients. Small changes in the data set thus often cause that completely different models are identified as the correct one.

- For 30 and 60 day mean streamflow, BAG outperforms in the majority of catchments a naive forecasting strategy, which relies on long-term averages only (SRG). Otherwise it is either questionable (30 day mean streamflow in case of BGS and 90 day mean streamflow in case of BAG) or very unlikely that BGS and BAG provide on average a smaller $E_{\mathrm{MSP}}$ than SRG.
- LOO and OOB estimates of $E_{\mathrm{MSP}}$ are for most catchments close to $E_{\mathrm{MSP}}$ from the outer cross-validation. Neither LOO nor OOB tend to optimistic or pessimistic estimates. Thus, instead of testing the bagged model via the outer cross-validation, also the OOB estimates had been quite accurate.

In practice, statistical seasonal streamflow forecasting is commonly confronted with small sample sizes and weak predictor-predictand relationships due to missing or noisy predictors. The results of the present study indicate that bagging is also able to reduce a pseudo model variance, introduced by weak relationships and intensified by small sample sizes. If resampling is anyway part of the model building procedure and weak relationships come along with small sample sizes, we propose to prefer bagging to the best guess model approach – the computational costs are nearly the same, out-of-bag predictions provide model testing, and prediction errors are likely to decrease. This benefit however comes at the cost of a hardly interpretable model. We thus argue that bagging is most useful when prediction alone is the goal, i.e. in operational forecasting, be it seasonal streamflow or another environmental variable.

## 7 Data availability

The streamflow series are provided by federal offices and were manually compiled. The corresponding data policies do not allow data dissemination, though for Bayern (http://www.gkd.bayern.de/fluesse/abfluss/karten/index.php?thema=gkd&rubrik=fluesse&produkt=abfluss&gknr=0, GKDB, 2016) and Austria (http://ehyd.gv.at/, BMLFUW, 2016) the series can be accessed online. The E-OBS data set is publicly available at http://www.ecad.eu/ (E-OBS, 2016), the Corine Land Cover at http://www.eea.europa.eu/data-and-maps/data/corine-land-cover-2006-raster-2 (CORINE, 2016), and the EU-DEM at http://www.eea.europa.eu/data-and-maps/data/eu-dem (EU-DEM, 2016).

**Acknowledgements.** Runoff series and catchment boundaries are provided by the following authorities: Landesanstalt für Umwelt, Messungen und Naturschutz Baden-Württemberg; Bayerisches Landesamt für Umwelt; Land Vorarlberg (data.vorarlberg.gv.at); Bundesministerium für Land- und Forstwirtschaft, Umwelt und Wasserwirtschaft Österreich; and Schweizerisches Bundesamt für Umwelt. We also acknowledge the E-OBS data set from the EU-FP6 project ENSEMBLES (ensembles-eu.metoffice.com) and the data providers in the ECA&D project (www.ecad.eu). Figure 1 is produced by using Copernicus data and information funded by the European Union (EU-DEM layers) as well as the Corine Land Cover 2006 raster data (version 17) of the European Environment Agency (EEA). Bernhard Wehren made available additional streamflow data for the river Kander at Hondrich. We also thank David M. Hannah for the careful review. The study was funded by the Group of Hydrology, which is part of the Institute of Geography at the University of Bern, Switzerland.

## References

BMLFUW: Streamflow monitoring Austria, Bundesministerium für Land- und Forstwirtschaft, Umwelt und Wasserwirtschaft, http://ehyd.gv.at/, last access: 11 August 2016.

Breiman, L.: Bagging Predictors, Mach. Learn., 24, 123–140, doi:10.1023/A:1018054314350, 1996a.

Breiman, L.: Out-of-bag estimation, https://www.stat.berkeley.edu/~breiman/OOBestimation.pdf (10 August 2016), 1996b.

CORINE: Corine Land Cover 2006 raster data, European Environment Agency, http://www.eea.europa.eu/data-and-maps/data/corine-land-cover-2006-raster-3, last access: 11 August 2016.

E-OBS: Daily temperature and precipitation fields in Europe, ECA&D – European Climate Assessment & Dataset, http://www.ecad.eu/download/ensembles/ensembles.php, last access: 11 August 2016.

EU-DEM: Digital Elevation Model over Europe, European Environment Agency, http://www.eea.europa.eu/data-and-maps/data/eu-dem, last access: 11 August 2016.

Garen, D. C.: Improved techniques in regression-based streamflow volume forecasting, J. Water Res. Pl.-ASCE, 118, 654–670, doi:10.1061/(ASCE)0733-9496(1992)118:6(654), 1992.

GKDB: Streamflow monitoring Bayern, Gewässerkundlicher Dienst Bayern, http://www.gkd.bayern.de/fluesse/abfluss/karten/index.php?thema=gkd&rubrik=fluesse&produkt=abfluss&gknr=0, last access: 11 August 2016.

Hastie, T., Tibshirani, R., and Friedman, J.: The Elements of Statistical Learning, Springer New York Inc., 2. edn., doi:10.1007/978-0-387-84858-7, 2009.

Haylock, M. R., Hofstra, N., Klein Tank, A. M. G., Klok, E. J., Jones, P. D., and New, M.: A European daily high-resolution gridded data set of surface temperature and precipitation for 1950–2006, J. Geophys. Res.-Atmos., 113, d20119, doi:10.1029/2008JD010201, 2008.

Mevik, B.-H. and Wehrens, R.: The pls Package: Principal Component and Partial Least Squares Regression in R, 18, 1–23, doi:10.18637/jss.v018.i02, 2007.

Proc. IAHS, 374, 165–173, 2016
proc-iahs.net/374/165/2016/
doi:10.5194/piahs-374-165-2016


Open Access
PIAHS

# Investigation of hydrological variability in the Korean Peninsula with the ENSO teleconnections

**Sunkwon Yoon**[1] **and Taesam Lee**[2]

[1]Climate Application Department, APEC Climate Center, Busan, 48058, Republic of Korea
[2]Department of Civil Engineering, ERI, Gyeongsang National University, 52828, Republic of Korea

*Correspondence to:* Sunkwon Yoon (skyoon@apcc21.org)

Published: 17 October 2016

**Abstract.** This study analyzes nonlinear behavior links with atmospheric teleconnections between hydrologic variables and climate indices using statistical models during warm season (June to September) over the Korean Peninsula (KP). The ocean-related major climate factor, which is the El Niño-Southern Oscillation (ENSO) was used to analyze the atmospheric teleconnections by principal component analysis (PCA) and a singular spectrum analysis (SSA). The nonlinear lag time correlations between climate indices and hydrologic variables are calculated by Mutual Information (MI) technique. The nonlinear correlation coefficients (CCs) by MI were higher than linear CCs, and ENSO shows a few months of lag time correlation. The warm season hydrologic variables in KP shows a significant increasing tendency during the warm pool (WP), and the cold tongue (CT) El Niño decaying years shows a significant decreasing tendency, while the La Niña year shows slightly above normal conditions, respectively. A better understanding of the relationship between climate indices and streamflow, and their local impacts can help to prepare for the river discharge management by water managers and scientists. Furthermore, these results provide useful data for policy makers and end-users to support long-range water resources prediction and water-related policy.

## 1 Introduction

The changing of global climate characteristics has induced more frequent extreme climate events, which are closely related to the seasonal variations of regional water resource variabilities (Wang et al., 2000; Pizarro and Lall, 2002; Yoon, 2015). Many climate scientists believe that extreme climate characteristics have changed with the increase in the Earth's temperature, and the frequency of occurrence and intensity of extreme climactic events is expected to gradually increase (IPCC, 2007). In particular, seasonal variations of hydrometeorological variables are closely related to the characteristic changes in the global climate. For the efficient management and prediction of water resources, it is necessary to investigate the relationship between climate factors and hydrological indicators (Kim et al., 2012; Yoon et al., 2013).

The effect of local hydro-meteorological long-term changes do not occur independently and their correlations appear to be based on large-scale circulation pattern changes. It can be predicted a teleconnection based abnormal sea surface temperature (SST), and its nonlinear climate systems behavior in ocean-related climate indices such as the El Niño-Southern Oscillation (ENSO) in the Tropical Pacific Ocean (TPO) region (Piechota and Dracup, 1996; Piechota et al., 1998; Klein et al., 1999). Based on many studies on ENSO, which is major sources of atmospheric large-scale environmental changes, it is well known that they are closely correlated to seasonal variations of local hydrological pattern changes such as precipitation and streamflow (McPhaden et al., 2006).

The different phases of ENSO, which manifests as abnormally warm SST conditions in the central Pacific Ocean region, and its physical mechanism have been actively examined by the TPO coastal region countries scientists. These phases have been called El Niño Modoki, Central Pacific (CP) El Niño, and warm-pool (WP) El Niño etc. Several studies have shown this abnormal SST conditions will occur more frequently due to current conditions of global warming and changing climate. As a representative study, Feng et al. (2010) revealed that the El Niño Modoki phenomenon is affecting areas across of the world through significant changes of the large-scale climate and hydrological circulation patterns. The conventional climate indices Niño3 and Niño4 in the TPO region were used to identify the two types of ENSO (Ashok and Yamagata, 2009; Kug et al., 2009; Yeh et al., 2009). However, Ren and Jin (2011) argued for a distinction in the different types of El Niño, cold-tongue (CT) El Niño and WP El Niño, by using the Niño3 and Niño4 SST observation data.

In this study, we analyzed the influence of the long-term hydro-meteorological variability over the East Asia (EA) and the Korean Peninsula (KP) by looking at the conventional strong CT El Niño, different types of WP El Niño, and strong La Niña events as defined by Ren and Jin (2011). Specifically, this study had three main objectives: (1) To analyze the significant large-scale pattern changes and long-term hydrological variability in KP region by the ocean-related abnormal climate phenomenon in accordance with the ENSO; (2) to investigate linkages of the atmospheric teleconnections using a statistical correlation analysis between climatic indices and hydro-meteorological variables over KP; (3) to perform a diagnostic study on the possibility of seasonal prediction for water resources management using climate indices.

## 2 Stud basin and data used

### 2.1 Study basin

The five major river basins (Han River basin, Nakdong River basin, Geum River basin, Seomjin River basin, and Youngsan River basin) and their sub-watersheds are located in the Korean peninsula (126°24′–129°02′ E, 36°30′–38°55′ N). Most of the eastern part of the KP is composed of mountains and valleys with steep slopes, and the western part of the KP is composed of wide plains. 70 % of the annual rainfall occurs during the summer season (June–September) and is characterized by localized rain resulting primarily from tropical cyclones and heavy rainfall during the rainy season. In 2013 survey dataset from the Korean government's Water Resources Management Information System (WAMIS) for the five major river basins showed a total flooded area of about 1.86 million ha and the total flood damage at over 1,673 million with steady increases in the damage scale and loss with every new flood season.

### 2.2 Data used

ENSO data used in this research was obtained from the National Oceanic and Atmospheric Administration (NOAA), which provides monthly ENSO data from the four sites (Niño1+2, Niño3, Niño3.4, and Niño4). Among these sites, observation data from Niño3 (5° S–5° N, 150–90° W) and Niño4 (5° S–5° N, 160° E–150° W) sites are known to have strong correlation with El Niño (Trenberth, 1997). The area-averaged sea surface temperature (SST) anomaly over this region is a well-known ENSO index. SST data provided by Hadley Centre, called HadISST, was used to investigate the correlation between a number of hydrologic parameters for each sub-basin and large-scale atmospheric circulation patterns. This dataset has a spatial resolution of 1° × 1° and is revised monthly (Rayner et al., 2003).

The 1966 to 2012 precipitation and streamflow data was obtained from WAMIS in Korea. The average precipitation was calculated by the Thiessen polygon network method from the 125 precipitation gauge stations in the five major river basins of the KP. The streamflow data was calculated using the Precipitation-Runoff Modeling System (PRMS), a widely applied semi-distributed long-term runoff simulation model developed by the US Geological Survey (Leavesley et al., 1983; Dressler et al., 2006). By using PRMS to calculate the natural streamflow, the effect of artificial dams was eliminated from the streamflow data.

## 3 Methodology

### 3.1 Teleconnections

ENSO phenomenon in the TPO region induced large-scale atmospheric climate pattern changes. To understand their local hydro-environmental impacts in the KP, it is necessary to analyze the linear and non-linear correlations and its lag time correlations between climate indices and hydro-meteorological variables. In general, abnormal SST in the TPO region has triggering effects for the troposphere temperature rise. A teleconnection, enabled by the jet stream, connects the abnormal SST variation in these subtropical regions to hydrological variations worldwide.

To explore this, we performed a teleconnection based time series analysis and analyzed its hydrological local impacts in the KP region, shown in Fig. 1. First, we analyzed stationary processes by the logarithmic transformation in each time series data set, such as ocean-related climate indices and hydrological data. Second, we analyzed the low frequency for the time series data sets by PCA using a SSA. Third, the joint probability distribution density function was derived from the SSA results using a kernel density method. The linear and non-linear lag time correlation of the joint probability distribution density function, derived using the MI technique, was used to analyze the atmospheric teleconnections. Finally, we assessed the impact of this teleconnection on hydro-meteorological variability and the resulting implications for local water resource management.

### 3.2 Classification of ENSO events

The emerging issue of the new type of El Niño phenomenon has caused significant changes in the hydro-meteorological patterns throughout the world. However, quantitative studies on features of CT/WP El Niño were found to be relatively insufficient for the Korean watersheds. The effect of different phases of ENSO on the summer precipitation and streamflow characteristics was analyzed by adopting the classification criteria proposed by Ren and Jin (2011) for hydrologic variables from the five major river basins in Korea and their 113 sub-basins. We calculated these variables and divided into the two different indices ($N_{CT}$ and $N_{WP}$), which are indicated by the cold tongue and warm pool El Niño categorization proposed by Ren and Jin (2011) that used a simple nonlinear transformation to distinguish a CT El Niño from a WP El Niño. The equation is defined as follows:

$$N_{CT} = N_3 - \alpha N_4 \qquad \alpha = \begin{cases} 2/5, N_3 N_4 \\ 0, \text{otherwise} \end{cases}. \qquad N_{WP} = N_4 - \alpha N_3 \tag{1}$$

where, $N_3$ and $N_4$ mean indices of Niño3 and Niño4, respectively. $N_{CT}$ and $N_{WP}$ are new indices based on a simple transformation from Niño3 and Niño4 indices, and $\alpha$ indicate the correction constant.

### 3.3 Principal component analysis using SSA

Singular spectrum analysis (SSA) is a non-parametric spectral estimation method that reduces dispersion by changing the coordinates in time-series in derived techniques from principal component analysis (PCA), which enables the extraction of information from noisy time series. By removing the non-harmonic components from the original time series data, one can more easily understand its long-term frequency and trend (Moon and Lall, 1996). SSA embeds a time series data $X_i\,(1 < i < N)$ in a vector space of dimension $M$ and applies the empirical orthogonal function (EOF) method.

To separate the non-harmonic components, the size of the eigenvalues is defined by the orthogonal process. The orthogonal function is calculated between $\alpha_i^1$ and $\alpha_i^2$, the orthogonal coefficients of the PC1 and PC2 time series, to correspond with the harmonic components in original coordinate $X$ and $Y$. Finally, it converted to the reconstruction component (RC) to $R_i^1$ and $R_i^2$. It is possible to configure the estimation forecasting model to reflect specific characteristics (such as frequency and trend) in original data. By reconstructing the data using Eq. (2), it is possible to replace the original data with a new time series, which has constant frequency and less noise.

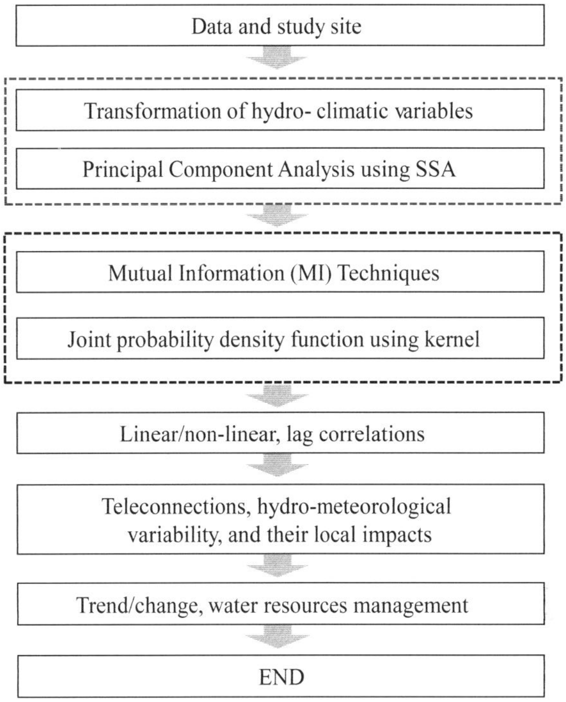


**Figure 1.** Flow chart of teleconnection based seasonal prediction study between climate indices and hydro-meteorological variables.

$$(R_A X)_i = \frac{1}{i}\sum_{j=1}^{i}\sum_{k\in A}^{i}\alpha_{i-j}^{k}E_j^k,\ 1 \le i \le M-1 \tag{2}$$

$$(R_A X)_i = \frac{1}{M}\sum_{j=1}^{M}\sum_{k\in A}^{i}\alpha_{i-j}^{k}E_j^k,\ M \le i \le N-M+1$$

$$(R_A X)_i = \frac{1}{N-i+1}\sum_{j=1-N+M}^{i}\ \sum_{k\in A}^{i}\alpha_{i-j}^{k}E_j^k,$$

$$N-M+2 \le i \le N$$

Here, $\alpha_i^k$ means an orthogonal coefficient, $E_j^k$ is the empirical orthogonal function $(1 \le k \le M)$, $M$ indicates a dimension, and $\tau$ is the sampling rate.

### 3.4 Mutual information techniques

Mutual information (MI) is a quantitative measurement of how much one random variable ($Y$) tells us about another random variable ($X$) and can also be thought of as the reduction in uncertainty about one random variable given knowledge of another. This is a useful tool for determining non-linear correlation between the two different data sets. Here, we use it to extract information about the non-linear correlation between climate indices and hydrological variables in the KP.

If we have two kinds of time series data sets such as $(s_1, s_2, s_3, \cdots, s_n, q_1, q_2, q_3, \cdots, q_n)$, where $n$ is the observed periods, then the MI value between observation $s_i$ and $q_j$ is defined by Moon et al. (1995) as follows:

$$\mathrm{MI}_{s,q}\left(s_i, q_j\right) = \log_2\left(\frac{P_{s,q}\left(s_i, q_j\right)}{P_s\left(s_i\right) P_q\left(q_j\right)}\right). \quad (3)$$

Here, $P_{s,q}\left(s_i, q_j\right)$ indicates the joint probability density function between $s$ and $q$ calculated by a time series of $\left(s_i, q_j\right)$. $P_s\left(s_i\right)$ and $P_q\left(q_j\right)$ are the marginal probability densities calculated from $s_i$ and $q_j$, respectively. The overall concern is the correlation between the two series of data. The average mutual information $(I_{s,q})$ of two discrete random variables s and $q$ can be defined as Eq. (4):

$$I_{s,q} = \sum_{i,j} P_{s,q}\left(s_i, q_j\right) \log_2\left(\frac{P_{s,q}\left(s_i, q_j\right)}{P_s\left(s_i\right) P_q\left(q_j\right)}\right) \quad (4)$$

where $P_{s,q}\left(s_i, q_j\right)$ is the joint probability distribution function of $X$ and $Y$, and $P_s(s_i)$ and $P_q\left(q_j\right)$ are the marginal probability distribution functions of $s$ and $q$, respectively. This equation is useful to search for the components in multivariate sampling whether independent and dependent.

In this study, we performed a linear regression (LR) approach with the estimated average MI value and a nonlinear regression using the method of Kernel density estimation (KDE). The 90 % confidence bounds were estimated using 1000 bootstrap resampling replications, allowing for a more accurate calculation of the confidence bounds given the limited data.

## 4 Analysis and results

### 4.1 Strong ENSO events and composite anomalies of SST

A comparison of Niño3 and Niño4 anomalies using a scatter plot revealed a correlation of 0.77 ($R^2 = 0.593$) between the two time series. However, a comparison between CT El Niño ($N_{\mathrm{CT}}$) and WP El Niño ($N_{\mathrm{WP}}$) indices yielded a correlation of only 0.19 ($R^2 = 0.034$). Therefore, we confirmed the independence between $N_{\mathrm{CT}}$ and $N_{\mathrm{WP}}$ indices that showed a smaller relative correlation. To further examine local hydrologic responses within the context of seasonal variability, the six strongest events of CT El Niño years (1965/66, 1972/73, 1982/83, 1987/88, 1991/92, and 1997/98) were identified using the new $N_{\mathrm{CT}}$ index. The six strongest events of WP El Niño years (1968/69, 1990/91, 1994/95, 2002/03, 2004/05, and 2009/10) were identified using the Nwp index. The five strongest events of La Niña years (1973/74, 1975/76, 1988/89, 1999/00, and 2010/11) were selected using the classification criteria provided by the NOAA climate prediction center (CPC) during the SST observation period 1950–2013.

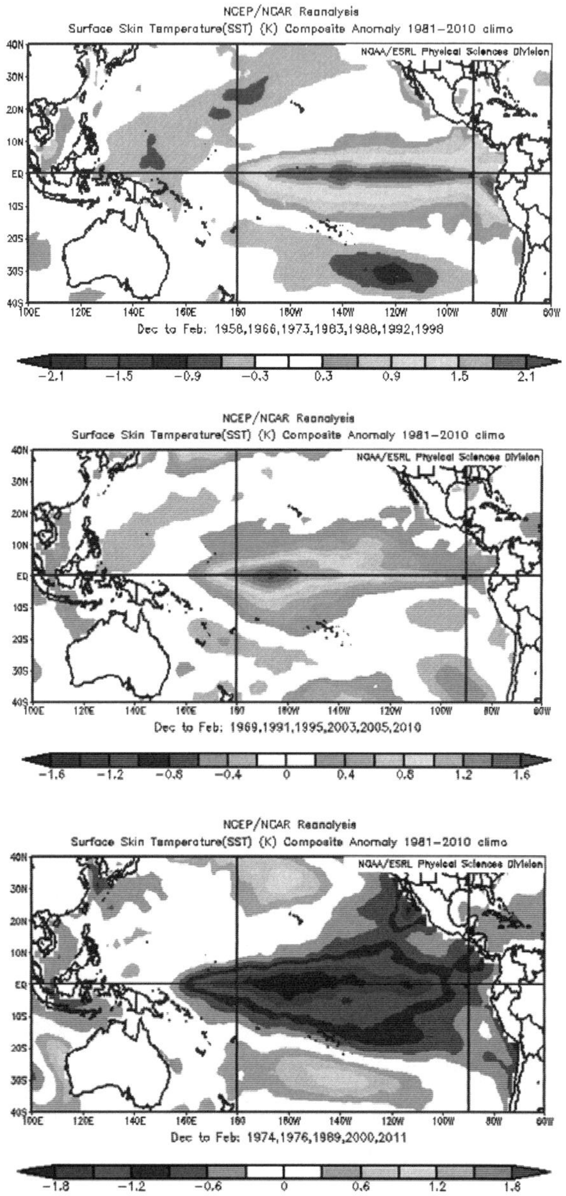


**Figure 2.** Composite sea surface temperature anomalies (SSTA) in the TPO region in developing phases of two types of El Niño and La Niña during December–January. **(a)** to **(c)** indicate SSTA during CT/WP El Niño and La Niña years, respectively.

Through this classification research, we confirmed that the WP type El Niño is occurring more frequently than the CT type El Niño in the Tropical Pacific Ocean in recent years. From the changes in frequency of WP to CT, we can see that using the subdivision proposed by Ren and Jin is necessary for more accurate analysis of the impacts of these major climatic events.

Figure 2 shows the composite sea surface temperature anomalies of the mean SST in the TPO region in developing phases of two types of El Niño and La Niña during December–January. The SST data was obtained from the National Centers for Environmental Prediction (NCEP) and the National Center for Atmospheric Research (NCAR) reanalysis version 2 and the 30-year climatology for the period of 1981–2010. In the case of CT El Niño years, the above-normal anomalies occurred in the central Pacific and eastern Pacific regions while below-normal anomalies occurred throughout the western Pacific region (Fig. 2a). This warm anomaly grew stronger and extended to the west during the fall. In the winter, a positive SST anomaly pattern strengthened, but as spring arrived the CT El Niño weakened and disappeared. However, the case of WP El Niño pattern shows a slight yet clear difference when compared to the CT El Niño. In the WP El Niño years, above-normal SST anomalies clearly appeared in the central Pacific region and a slightly above-normal SST anomaly was observed in the eastern Pacific regions (Fig. 2b). This central Pacific warm anomaly grew stronger and extended to the east during the fall. In the wintertime, a positive SST anomaly pattern strengthened, but as spring arrived, the WP El Niño weakened and disappeared. In the case of La Niña years, a cold SST anomaly pattern was located in the eastern Pacific and central Pacific Ocean (Fig. 2c). This cold anomaly grew stronger and extended to the west during the fall. In the wintertime a negative SST anomaly pattern strengthened but, as spring arrived, the La Niña weakened and disappeared. Many recent studies (Kug et al., 2009; Feng et al., 2010) show these different SST anomaly patterns may affect air circulation and they are findings significant changes in different regions in seasonal precipitation over the Pacific region countries.

## 4.2 Composite analysis for GPCC

Figure 3 shows the composite anomalies (1981–2010 climatology) of Global Precipitation Climatology Center (GPCC) precipitation from June to September during strong ENSO events over the North East Asia domain. We extracted monthly precipitation for 1901–2013 from the GPCC full quality control data reanalysis version 6.0 at 0.5° × 0.5° spatial resolution from Deutscher Wetterdienst in Germany. The data set is the most accurate in situ precipitation reanalysis data set of GPCC (Schneider et al., 2011).

Data for two different CT/WP El Niño years, and La Niña years allowed for a characteristic analysis of GPCC rainfall over the Korean Peninsula. Figure 3a to c shows the regional scale variability of Korean summer (June–September) precipitation during CT/WP El Niño and La Niña years. During CT El Niño years (Fig. 3a), the GPCC precipitation shows anomalously dry conditions (10–50 mm lower than long-term normal) on the KP, while WP El Niño years show anomalously wet conditions (20–50 mm higher than long-term normal) (Fig. 3b). As well, La Niña years showed

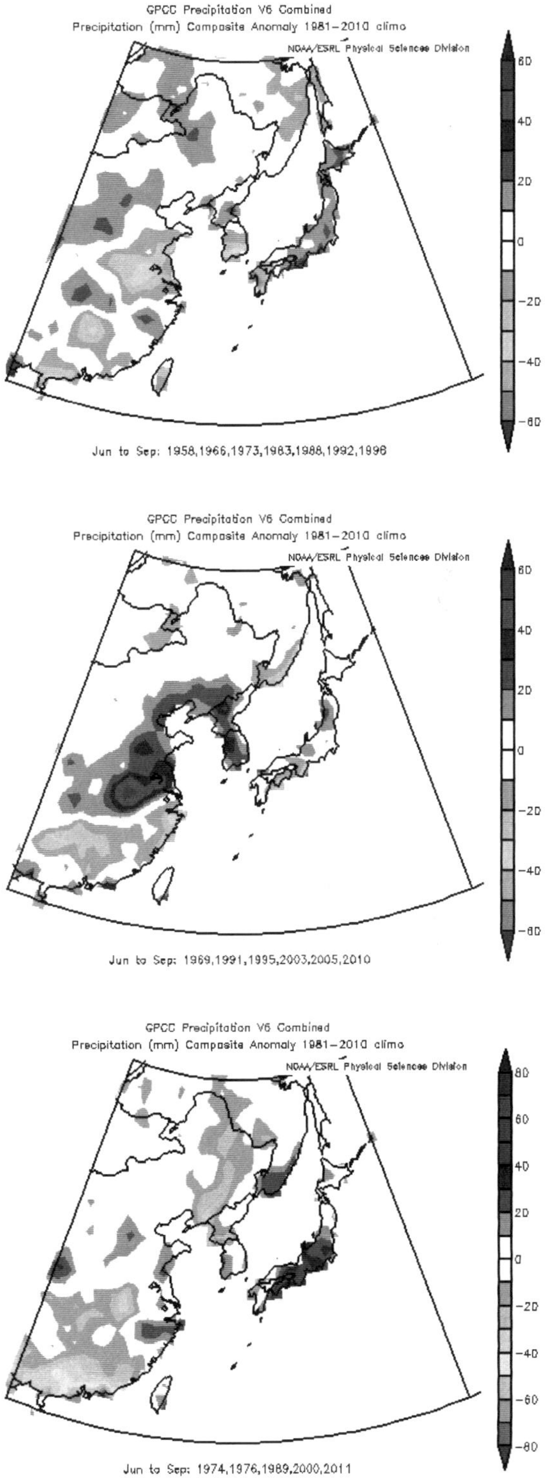


**Figure 3.** Composite anomalies (1981–2010 climatology) of GPCC Full V6 combined precipitation (JJAS: June to September) during ENSO events over the North East Asia domain.

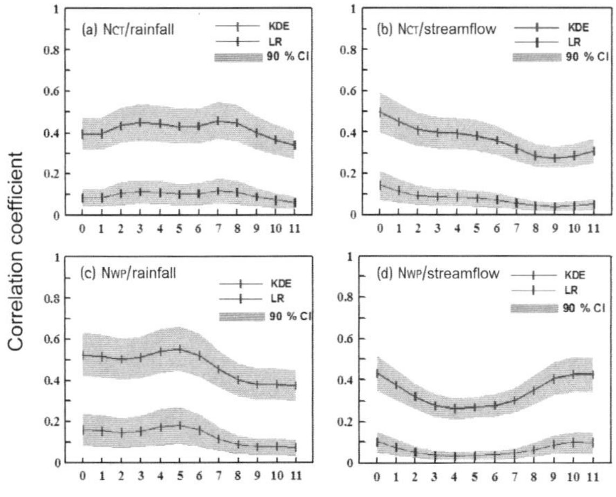


**Figure 4.** Nonlinear and linear CCs with their 90 % confidence bounds between climate index and hydro-meteorological variable in the Korean Peninsula, using KDE and LR approaches, respectively. The confidence bounds are given by 5 and 95 % quantiles of 1000 bootstrap replications.

slightly wet conditions: 0–10 mm higher than the long-term normal. While the results of this study rely on short observations and relatively limited data set, these findings are important to seasonal prediction for streamflow in basin scale and establishing adaptive solutions to changing climate related issues.

### 4.3 Teleconnection results

Figure 4 shows nonlinear and linear correlation coefficients (CCs) with their 90 % confidence bounds between climate index and hydro-meteorological variable in the KP, using KDE and LR approaches respectively. As described in the methodology, we obtained the non-linear behaviors and their correlations between climate indices and hydro-meteorological variable from a MI analysis. The linear and non-linear correlation of the $N_{CT}$ index and KP rainfall (LR: 0.118, KDE: 0.459) showed a 7-month lag time (Fig. 4a), while the correlation between the $N_{CT}$ index and the KP streamflow (LR: 0.182, KDE: 0.553) showed direct influence (Fig. 4b). Moreover, an analysis of the correlation between the $N_{WP}$ index with KP rainfall (LR: 0.118, KDE: 0.459) and streamflow (LR: 0.104, KDE: 0.433) showed a 5 and 10 months lag time respectively (Fig. 4c, d). These time-lags indicate the existence on a non-linear correlation between climate indices and hydrological variables, potentially enabling the use of large scale climate indices in the seasonal prediction of water resources in KP.

### 4.4 Seasonal variability during ENSO

Figures 5 and 6 show composite anomalies of annual and JJAS (June to September) rainfall and river discharge during CT/WP El Niño and La Niña years in the basins of the KP. The hatched polygons indicate statistically significant changes in annual rainfall based on the 10 % significance level by Student's $t$ test. As shown in Fig. 5, the annual rainfall anomalies reflect the sub-basin scale variability of Korean summer rainfall during CT/WP El Niño and La Niña events. Looking at the detailed variations in the sub-watershed areas, we can capture the spatial and seasonal variability caused by ENSO.

During CT El Niño events (Fig. 5a, d), the annual/JJAS rainfall on the KP was $-7.06$ %/$-11.18$ % lower than the long-term norm (1971–2000). The annual/JJAS streamflow during CT El Niño events (Fig. 6a, d) were decreased to $-10.04$ %/$-18.28$ % in the basins of the KP. During WP El Niño events (Fig. 5b, e), the annual/JJAS rainfall on the KP increased to 10.94 %/20.09 %. The annual/JJAS streamflow during WP El Niño events (Fig. 6b, e) substantially increased to 15.58 %/24.52 % in the basins of the KP. And during La Niña events (Fig. 5c, f), the annual/JJAS rainfall on the KP was 6.99 %/12.74 % higher than long-term norm. The annual/JJAS streamflow during La Niña events (Fig. 6c, f) increased to 9.92 %/17.35 % in the basins of the KP.

As a result of this composite analysis, a substantial increasing tendency of annual/JJAS rainfall and streamflow was found during WP El Niño events, while a significant decreasing trend was found during CT El Niño events, as compared to long-term normal years. In addition, the annual/JJAS rainfall and streamflow during La Niña event shows a slight tendency above normal.

Figure 7 shows a scatter plot of percentage change anomaly (departures from the 1971–2000 normal) changes between precipitation and streamflow during annual and JJAS season in different climate conditions of the ENSO events. ENSO conditions clearly show that the differences during the WP El Niño year (+17.93 %, +26.99 %), the CT El Niño year ($-8.20$ %, $-15.73$ %), and the La Niña year (+8.89 %, +15.85 %), respectively.

Throughout the results, different phase of ENSO events clearly show influence on hydrologic seasonal variability in the KP and its sub-watersheds. This shows a linkage of possible teleconnections and characteristic changes between major climactic events and hydrological indicators.

## 5 Summary and conclusions

The mechanism of major climate phenomena such as El Niño and La Niña has not yet been fully understood. It is currently impossible to completely simulate these large-scale climate phenomena using the physical climate model, which would require perfect knowledge of the global environment and its weather conditions. Furthermore, analyzing climatic

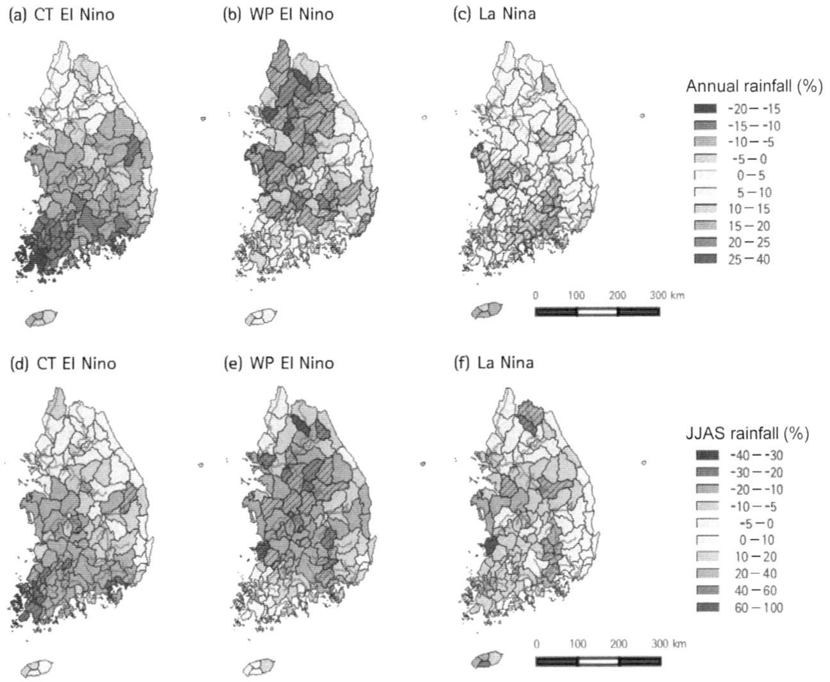


**Figure 5.** Composite anomalies of annual and JJAS (June to September) rainfall during CT/WP El Niño and La Niña years. The hatched polygons indicate statistically significant changes in annual rainfall based on the 10 % significance level.

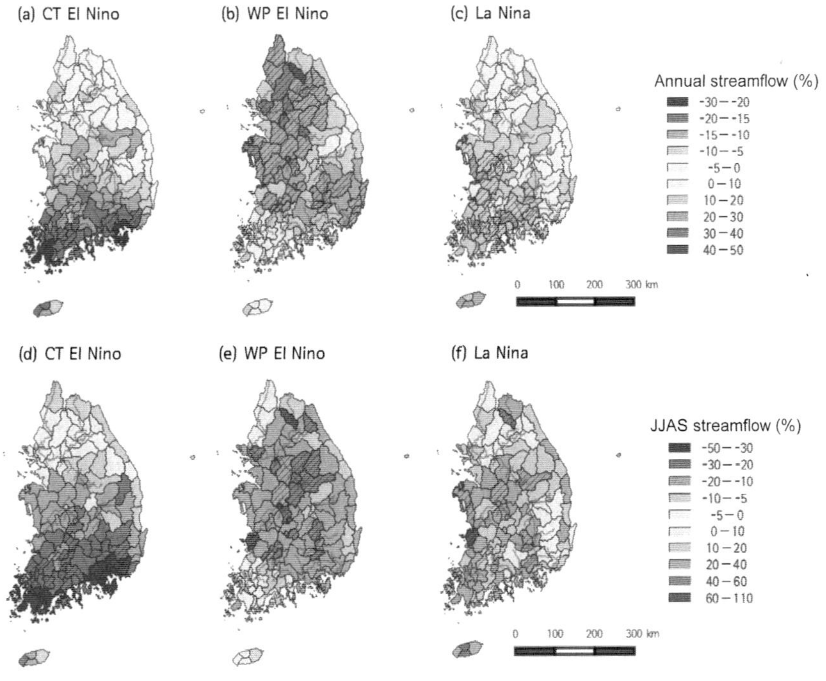


**Figure 6.** Same as Fig. 5, but for streamflow.

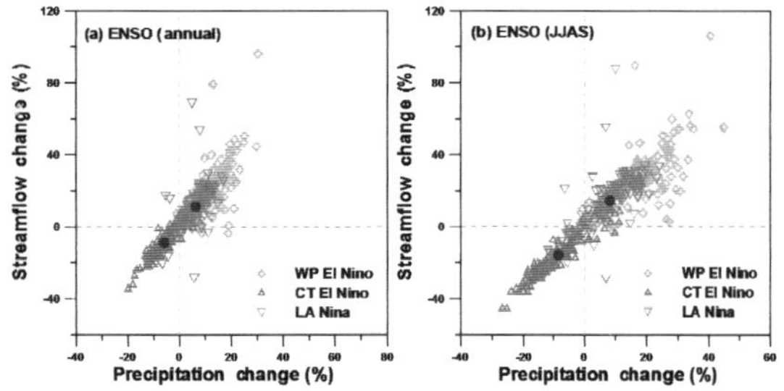


**Figure 7.** Scatter plot of percentage change anomaly (departures from the 1971–2000 normal) changes between precipitation and streamflow in different climate conditions. **(a)** and **(b)** show during annual and JJAS season in strong ENSO events in the 113 subwatershed in KP.

phenomena through the physical model is only possible in limited geographical areas, as predicting involves considerable difficulties. With these considerations, this study analyzed non-linear behavior links of the atmospheric teleconnections between hydrologic variable and climate indices using statistical models. One disadvantage of the statistical approach is that the limited number of observation reduces the significance level of the results. This is why the statistical model is utilized to complement to the prediction result of the physical model, resulting in more significant results.

Results from composite analysis in JJAS seasonal rainfall and river discharge during the CT/WP El Niño and the La Niña events in the KP and its sub-watershed show that the WP El Niño years show a substantially increasing tendency than normal years. The CT El Niño decaying years show a significantly decreasing tendency compared to the normal years. In addition, the La Niña years shows a slightly increasing tendency and normal condition. Also the MI based nonlinear and linear teleconnections using KDE and LR between climate indices and hydrologic variables show that the nonlinear CCs were higher than linear correlations, and the ENSO has a few months lag time correlation on rainfall and streamflow anomalies over the KP region.

This result shows a methodology of the teleconnection based long-range water resources seasonal prediction for reducing a changing climate uncertainty, when abnormal SSTA (such as, El Niño, La Niña) phenomenon occurs in the TPO regions. Furthermore, these results are useful water managers and end-users to support long-range water resources prediction and water-related policy making. A better understanding on hydrologic natural and anthropogenic drivers of this relationship between ENSO and streamflow could help to scientists to prepare river discharge control in the major river basins in KP. In addition, this research will facilitate well-informed decision-making in minimizing climate-related disasters such as flood and drought on agriculture sector, and supports optimal development of strategies ensuring use of water resources in best management under a changing climate.

## 6 Data availability

The data set of precipitation and streamflow over Korean Peninsula are respectively available at the Korean government's Water Resources Management Information System (WAMIS), http://www.wamis.go.kr/eng/WKW_ABGRF_LST.aspx and http://www.wamis.go.kr/eng/WKW_CMS_LST.aspx. Sea surface temperature data set from the National Oceanic and Atmospheric Administration (NOAA) is accessible at the Earth System Research Laboratory (ESRL), http://www.esrl.noaa.gov/psd/data/gridded/tables/sst.html. The composite Global Precipitation Climatology Center (GPCC) Precipitation V6 Combined data is accessible at the Physical Sciences Division at NOAA's ESRL, http://www.esrl.noaa.gov/psd/cgi-bin/data/composites/printpage.pl/.

**Acknowledgements.** This research was supported by the APEC Climate Center and we appreciate the reviewers for their valuable comments of this article.

## References

Ashok, K. and Yamagata, T.: Climate change: The El Niño with a difference, Nature, 461, 481–484, doi:10.1038/461481a, 2009.

Dressler, K. A., Leavesley, G. H., Bales, R. C., and Fassnacht, S. R.: Evaluation of girdded snow water equivalent and satellite snow cover products for mountain basins in a hydrologic model, Hydrol. Process., 20, 673–688, 2006.

Feng, J., Chen, W., Tam, C. Y., and Zhou, W.: Different impacts of El Niño and El Niño Modoki on China rainfall in the decaying phases, Int. J. Climatol., 31, 2091–2101, doi:10.1002/joc.2217, 2010.

GPCC (Global Precipitation Climatology Center): http://www.esrl.noaa.gov/psd/cgi-bin/data/composites/printpage.pl, last access: November 2014.

IPCC (Intergovernmental Panel on Climate Change): Managing the Risks of Extreme Events and Disasters to Advance Climate Change Adaptation (SREX) – Special Report of the Intergovernmental Panel on Climate Change, Cambridge University Press, 1–594, 2007.

Kim, J. S., Jain, S., and Yoon, S. K.: Warm season streamflow variability in the Korean Han River Basin: Links with atmospheric teleconnections, Int. J. Climatol., 32, 635–640, doi:10.1002/joc.2290, 2012.

Klein, S. A., Soden, B. J., and Lau, N. C.: Remote sea surface temperature variations during ENSO: Evidence for a tropical atmospheric bridge, J. Climate, 12, 917–932, 1999.

Kug, J. S., Jin, F.-F., and An, S.-I.: Two types of El Niño events: Cold tongue El Niño and warm pool El Niño, J. Climate, 22, 1499–1515, doi:10.1175/2008JCLI2624.1, 2009.

Leavesley, G. H., Lichty, R. W., Troutman, B. M., and Saindon, L. G.: Precipitation-Runoff Modeling System: User's Manual, Water Resources Investigations: 83-4238, U.S. Geological Survey, Denver, Colorado, 1983.

McPhaden, M. J., Zebiak, S. E., and Glantz, M. H.: ENSO as an integrating concept in Earth Science, Science, 314, 1740–1745, 2006.

Moon, Y. I. and Lall, U.: Atmospheric flow indices and interannual Great Salt Lake variability, J. Hydrol. Eng., 1, 55–61, 1996.

Moon, Y., Rajagopalan, B., and Lall, U.: Estimation of mutual information using kernel density estimators, Phys. Rev. E, 52, 2318–2321, 1995.

NOAA (National Oceanic and Atmospheric Administration), National Weather Service Climate Prediction Center, 2014: http://www.cpc.ncep.noaa.gov/data/indices/, last access: November 2014.

Piechota, T. C. and Dracup, J. A.: Drought and regional hydrologic variation in the United States: Associations with the El Niño-Southern Oscillation, Water Resour. Res., 32, 1359–1373, 1996.

Piechota, T. C., Chiew Francis, H. S., Dracup, J. A., and McMachon, T. A.: Seasonal streamflow forecasting in eastern Australia and the El Niño-Southern Oscillation, Water Resour. Res., 34, 3035–3044, 1998.

Pizarro, G. and Lall, U.: El Niño-induced flooding in the U.S. West: What can we expect?, Eos Trans. Amer. Geophys. Union, 83, 349–352, 2002.

Rayner, N. A., Parker, D. E., Horton, E. B., Folland, C. K., Alexander, L. V., and Rowell, D. P.: Global analyses of sea surface temperature, sea ice, and night marine air temperature since the late nineteenth century, J. Geophys. Res., 108, doi:10.1029/2002JD002670, 2003.

Ren, H. L. and Jin, F. F.: Niño indices for two types of ENSO, Geophys. Res. Lett., 38, L04704, doi:10.1029/2010GL046031, 2011.

Schneider, U., Becker, A., Finger, P., Meyer-Christoffer, A., Rudolf, B., and Ziese, M.: GPCC Full Data Reanalysis Version 6.0 at 0.5°: Monthly Land-Surface Precipitation from Rain-Gauge built on GTS-based and Historic Data, doi:10.5676/DWD_GPCC/FD_M_V6_050, 2011.

Trenberth, K. E.: The definition of El Niño, B. Am. Meteorol. Soc., 78, 2771–2777, 1997.

WAMIS (Water Management Information System): http://wamis.go.kr/eng/, last access: September 2015.

Wang, B., Wu, R., and Fu, X.: Pacific-East Asia Teleconnection: How Does ENSO Affect East Asian Climate?, J. Climate, 13, 1517–1536, 2000.

Yeh, S. W., Kug, J. S., Dewitte, B., Kwon, M. H., Kirtman, B. P., and Jin, F.-F.: El Niño in a changing climate, Nature, 461, 511–514, doi:10.1038/nature08316, 2009.

Yoon, S. K.: Extreme climate patterns and their hydrologic impacts over the Korean Peninsula: Links with atmospheric teleconnections, APEC Climate Center Research Report, 2015-12, 1–181, 2015.

Yoon, S. K., Kim, J. S., Lee, J. H., and Moon, Y. I.: Hydrometeorological variability in the Korean Han River Basin and its subwatersheds during different El Niño phases, Stoch. Environ. Res. Risk Assess., 27, 1465–1477, 2013.

Proc. IAHS, 374, 175–185, 2016
proc-iahs.net/374/175/2016/
doi:10.5194/piahs-374-175-2016


# Development of an integrated method for long-term water quality prediction using seasonal climate forecast

**Jaepil Cho[1], Chang-Min Shin[2], Hwan-Kyu Choi[2], Kyong-Hyeon Kim[2], and Ji-Yong Choi[3]**

[1]Research Department, APEC Climate Center, Busan, 48058, Korea
[2]National Institute of Environmental Research, Incheon, 22689, Korea
[3]Seoul National University, Pyeongchang, 25354, Korea

*Correspondence to:* Jaepil Cho (jpcho89@gmail.com)

Published: 17 October 2016

**Abstract.** The APEC Climate Center (APCC) produces climate prediction information utilizing a multi-climate model ensemble (MME) technique. In this study, four different downscaling methods, in accordance with the degree of utilizing the seasonal climate prediction information, were developed in order to improve predictability and to refine the spatial scale. These methods include: (1) the Simple Bias Correction (SBC) method, which directly uses APCC's dynamic prediction data with a 3 to 6 month lead time; (2) the Moving Window Regression (MWR) method, which indirectly utilizes dynamic prediction data; (3) the Climate Index Regression (CIR) method, which predominantly uses observation-based climate indices; and (4) the Integrated Time Regression (ITR) method, which uses predictors selected from both CIR and MWR. Then, a sampling-based temporal downscaling was conducted using the Mahalanobis distance method in order to create daily weather inputs to the Soil and Water Assessment Tool (SWAT) model. Long-term predictability of water quality within the Wecheon watershed of the Nakdong River Basin was evaluated. According to the Korean Ministry of Environment's Provisions of Water Quality Prediction and Response Measures, modeling-based predictability was evaluated by using 3-month lead prediction data issued in February, May, August, and November as model input of SWAT. Finally, an integrated approach, which takes into account various climate information and downscaling methods for water quality prediction, was presented. This integrated approach can be used to prevent potential problems caused by extreme climate in advance.

## 1 Introduction

Demand from water resources managers for seasonal climate prediction information with a lead-time of several months is increasing as this information can provide key knowledge on issues like long-term dam inflow and water quality prediction information. Long-term water quality forecasts are particularly important in watershed management because they allow for these managers to implement proactive water quality control management techniques. The importance of utilizing long-term forecasts for proactive management of water quality is becoming more important, particularly in non-point source pollution cases. Non-point source pollution flows into the water bodies during rainfall events and gradually induces the water quality problems such as algae (e.g. Meng et al., 2010; Xu et al., 2012). However, seasonal forecast information has yet to be widely utilized in this manner mainly due to its high uncertainty levels. In addition to the issue of high uncertainty, there are large differences in the spatio-temporal scale between the forecast data that is necessary for water quality management and the data that is currently provided. Therefore, in order to address these concerns, the development of spatiotemporal downscaling techniques are imperative in order to fully apply long-term climate prediction information to water quality management in a large region.

There are two approaches to downscaling: (1) dynamic, and (2) statistical. Dynamic downscaling approach is not commonly used mainly due to its more time- and cost-intensive nature compared to that of the statistical approach. For this reason, the statistical downscaling approach has been used to attempt to enhance the utilization of data created using dynamical models. In previous studies the regression

Published by Copernicus Publications on behalf of the International Association of Hydrological Sciences.

method, which indirectly uses atmosphere-ocean variables from dynamic prediction data as predictors for forecasting local target variables based on high correlation, was applied instead of directly using the target variables in order to improve predictability (Kang et al., 2009, 2014). Other statistical approaches that improve predictability, use only those lagged climate indices that have high correlation with local climate variables. The lag time based climate index approach has been used in regions that have strong tele-connection between global climate indices and local climate variables (Bridhikitti, 2013; Hamlet and Lettenmaier, 1999; Schepen et al., 2012; Räsänen and Kummu, 2013). In Korea, there has been research on seasonal prediction using lagged climate indices, which have high correlation with precipitation and temperature on the Korea peninsula (Kim et al., 2007; Kim and Kim, 2010). In addition, various studies have been conducted using climate indices in connection with the precipitation of the East Asian region, including the Korea Peninsula (Lee et al., 2008; Wang et al., 2008). Recently, the Hybrid approach, which combines dynamically and statistically predicted climate information, has been applied to improve the predictability of the seasonal forecast (Robertson and Wang, 2012; Schepen et al., 2012).

Therefore, the objectives of this study are: (1) to develop a hybrid downscaling technique for predicting long-term precipitation and temperature on the Korea peninsula, by considering both the multi-model based prediction data provided by the APEC Climate Center (APCC) and the statistical prediction information based on teleconnection for water resources management; and (2) to evaluate the applicability of the seasonal forecast information in long-term future water quality predictions by using the predicted climate information as input to watershed modeling.

## 2 Methodology

Figure 1 shows a flow chart of the overall study: (1) Seasonal Climate Forecasting (steps j–n in the figure); (2) Temporal Downscaling (steps o and p); and Long-Term Water Quality Forecasting (steps q and r). Steps j–n include the development and application of four downscaling methods to predict the monthly precipitation and temperature depending on the available weather information. Steps o and p include converting monthly forecast information to daily data at the weather station levels using temporal downscaling based on the average, minimum (MIN), and maximum (MAX) of the multi-model ensemble (MME). Steps q and r include the assessment of the predictability in simulating monthly water quality by aggregation of daily water quality output, using predicted daily weather data as input to the watershed modeling.

### 2.1 Seasonal climate forecasting

The purpose of seasonal climate forecasting is to generate monthly precipitation and temperature prediction data on the target area while improving predictability by using different source of global climate information. The overall seasonal climate forecasting technique combines four different downscaling methods according to the degree of using dynamic prediction data produced by global climate models (GCMs). These methods include: (1) the Simple Bias Correction (SBC) method, which directly uses APCC's climate prediction data with 3 to 6 month lead time; (2) the Moving Window Regression (MWR) method, which indirectly utilizes the dynamic prediction data; (3) the Climate Index Regression (CIR) method, which predominantly uses the observation-based climate indices without using any prediction data; and (4) Integrated Time Regression (ITR) method, which uses predictors selected from both CIR and MWR. Since predictability on the Korean peninsula may differ depending on the target month and selected method, predictability was evaluated using the simple average of all available forecast information.

Simple Bias Correction (SBC) is a forecast-based direct downscaling method which uses GCM's prediction data to adjust the monthly mean of predicted precipitation and temperature through a simple bias-correction. For example, if the precipitation and temperature prediction data on Korean peninsula is needed, SBC directly uses the grid values of precipitation and temperature variables, which are produced from GCMs over the corresponding area. The systematic bias is adjusted for precipitation and temperature by using the ratio and addition, respectively, in order to make the monthly average of prediction same to the average of observation for the same period.

Moving Window Regression (MWR), which is similar to Kang et al. (2009) in the concepts and methodology, is a forecast-based indirect statistical downscaling method, which uses the proxy variables, produced by GCMs as predictors of regression model when high correlation exists between proxy variables and regional target variables. If there are limitations in directly predicting target variables such as precipitation (PREC) and temperature at 2 m (T2M) in the target area, the MWR method uses the oceanic and atmospheric circulation variables as predictors to improve the seasonal prediction predictability in the target region. Available proxy variables provided by APCC include temperature at 850 hPa (T850), zonal wind at 200 hPa (U200), meridional wind at 200 hPa (V200), zonal wind at 850 hPa (U850), meridional wind at 850 hPa (V850), geopotential height at 500 hPa (Z500), sea level pressure (SLP), and sea surface temperature (SST). In this study, only climate information from the latitudinal range of −40–40° (centered on the equator) was used for the predictor selection procedures.

Climate Index Regression (CIR) is an observation-based indirect statistical downscaling method that can be used

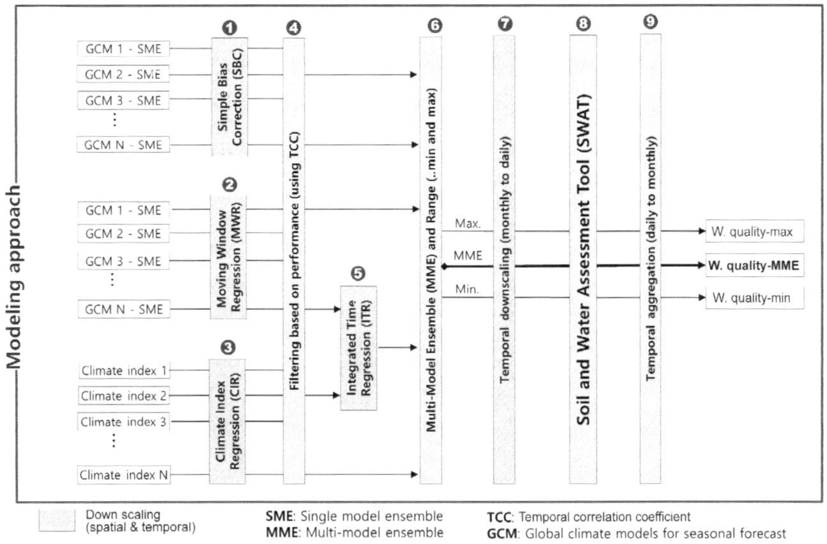


**Figure 1.** Flow chart of integrated seasonal climate and water quality forecasting based on a modeling approach.

when there is a high correlation between global climate indices and regional target variables with lag time. For real time operation of CIR in predicting monthly precipitation and temperature using climate indices, the lag time between the monthly precipitation/temperature and indices should be larger than the lead-time. The CIR method is similar to the MWR method in that both methods indirectly utilize the correlation between regional target variables and global scale climate variables related to oceanic and atmospheric circulation. There is a difference between the CIR and MWR methods when selecting predictors to forecast future seasonal target variable values. While the MWR method uses simultaneous proxy predictors that are predicted by GCMs, the CIR method uses the observed climate information from a few months ago by taking into account the lag time. A 2 to 6 month range of lag time was used in this study for 3 month lead forecasting.

Integrated Time Regression (ITR) is an indirect statistical downscaling method that uses both forecast and observation based predictors from the MWR and CIR methods, respectively. As a result, it can be used only when the MWR and CIR methods simultaneously select predictors for a particular target period. From the best predictors determined by the MWR and CIR methods, a selection of final predictors for the multivariate regression model are finally selected through the Akaike Information Criterion (AIC) analysis.

The statistical downscaling model was constructed separately for each month and target variable. The concepts of both cross-validation and split-validation were applied in developing statistical downscaling methods such as MWR and CIR in order to prevent overfitting problems, which can occur when constructing statistical forecasting models. The Leave-one-out cross-validation (LOOCV) technique was applied to the observation period (1983–2013). In other words, when predicting target variables for a specific target period (year/month), all predictors for the same target period are removed from the model construction procedure in order to reproduce the same conditions as the real time forecasting. For example, when predicting for January 1983, only predictors from January 1984 to 2013 are utilized in constructing the regression model. Predictions are made in the same way for the rest of simulation period. For each cross-validation process, the split validation approach was applied, and then the best predictors that showed consistent performance for both training and verification periods were finally selected.

## 2.2 Temporal downscaling

In addition to predicted monthly precipitation and temperature, additional long-term climate variables including wind speed, solar radiation, and relative humidity, are necessary at the daily time-scale in order to use the seasonal forecast information as inputs to watershed modeling. From a geographical standpoint, considering the spatial correlation among weather stations is very important in the temporal downscaling procedure. In this study, a sampling approach that extracts daily weather variables from the past observations within the target region was selected. First, the year/month from previous observed data that is most similar to the regional average in predicted climate patterns is determined, considering both precipitation and temperature simultaneously. This was done using the Mahalanobis distance method (Mahalanobis, 1936) by considering the covariance between precipitation and temperature. Then, the daily weather variables for each weather station are extracted from the selected year/month.

### 2.3 Evaluating Predictability

It is necessary to conduct calibration and validation procedures first in order to evaluate the modeling-based water quality predictability using watershed-scale model. In the case of the flow rate, the total error (Err) and Nash-Sutcliffe Efficiency index (NSE) were used to evaluate the performance of the model based on simulated and monitored data. For water quality components such as Total Nitrogen (TN) and Total Phosphorus (TP), graphical approach based on trial-and-error methods was used to evaluate model performance.

$$\mathrm{Err} = \frac{\left|\sum_{i=1}^{n} O_i - \sum_{i=1}^{n} P_i\right|}{\sum_{i=1}^{n} O_i} \times 100 \qquad (1)$$

$$\mathrm{NSE} = 1 - \frac{\sum_{i=1}^{n} (O_i - P_i)^2}{\sum_{i=1}^{n} \left(O_i - \bar{O}\right)^2} \qquad (2)$$

where $O_i$ = observed value, $P_i$ = predicted value, $\bar{O}$ = observed mean, and $n$ = number of simulations.

According to the Provisions of Water Quality Prediction and Response Measures of the Korean Ministry of Environment, modeling-based predictability was evaluated by using 3-month lead prediction data forecasted in February, May, August, and November as model inputs in the Soil and Water Assessment Tool (SWAT). The ultimate goal of the modeling-based predictability evaluation is to compare the generated predicted water quality output to the observed water quality, using long-term weather prediction information as watershed model inputs. However, the generated water quality outputs that used observed weather data as model inputs, was also compared with the predicted water quality in order to distinguish the uncertainties caused by parameterization of watershed models and long-term climate prediction.

## 3 Materials

### 3.1 Study watershed and watershed model

The Wecheon watershed within the Nakdong River basin was selected for this study mainly due to the relatively few changes in land-use within the watershed for a long period of time. The watershed has a high percentage of rural areas, and low percentage of waste water treatment plants. As a result, in 2011, it was found that non-point source pollution caused around 80 % of the total Biological Oxygen Demand (BOD) pollutant load within the watershed, which is sensitive to rainfall characteristics. Therefore, the Wecheon watershed is suitable to evaluate the applicability of seasonal forecast information for watershed management purposes when considering higher impacts by climate factors compared to human factors.

Figure 2 shows the locations of the weather, water level, and water quality monitoring station network. Younggok and Wecheon-B stations were selected to calibrate the flow and water quality related parameters of the selected watershed model, respectively, because their measurement points are located relatively close to each other.

The Soil and Water Assessment Tool (SWAT) model was selected in this study to simulate the TN and TP movements from upland areas to streams. SWAT, which was developed by the US Department of Agriculture to predict the long-term behavior of hydrology and contaminants at large watershed scale, divides the watershed into multiple sub-watersheds in order to capture the different spatial characteristics (Neitsch et al., 2005). Each sub-watershed is subsequently subdivided into multiple Hydrologic Response Units (HRUs) by grouping them according to their similar properties, such as land use, soil, and slope. In addition to the hydrological processes, using the Modified Universal Soil Loss Equation (MUSLE) can simulate soil erosion. The transport mechanisms of organic substances like nitrogen, phosphorus, and pesticides, can also be simulated. Water, sediments, and nutrients that are introduced from the HRUs into the water body are simulated through a reaction mechanism within the water body (Neitsch et al., 2005).

In this study, Wecheon watershed was divided into 29 sub-watersheds and 1866 HRUs by combining land use, soil, and the Digital Elevation Model (DEM). Two years of warm-up simulations were conducted for the initialization of the model parameters. Then, the model was calibrated and validated using measured data for 2007–2008 and 2010–2011, respectively. The trial-and-error method was used for the model calibration procedure. Therefore, manual calibration of the flow rate was conducted first and then calibration of the sediment and nutrient related parameters were conducted sequentially.

### 3.2 Climate information

APCC has been collecting monthly dynamic prediction data produced by 16 institutions and has been producing 3-month and 6-month lead Multi-Model Ensemble (MME) climate forecasts every month. In this study, 3-month lead seasonal forecast data which were regrided with 2.5° × 2.5° resolution based on 10 individual Global Climate Models (GCM) were used for the SBC and MWR downscaling methods. Table 1 shows the description of 10 GCMs used.

When using the CIR method, we used the real time 25 climate indices as predictors: 16 climate indices that are updated on a monthly basis by NOAA, through the webpage (http://www.esrl.noaa.gov/psd/data/climateindices/list/), and 9 indices which are extracted monthly at APCC using the NCEP/NCAR Reanalysis 1 data (Table 2).

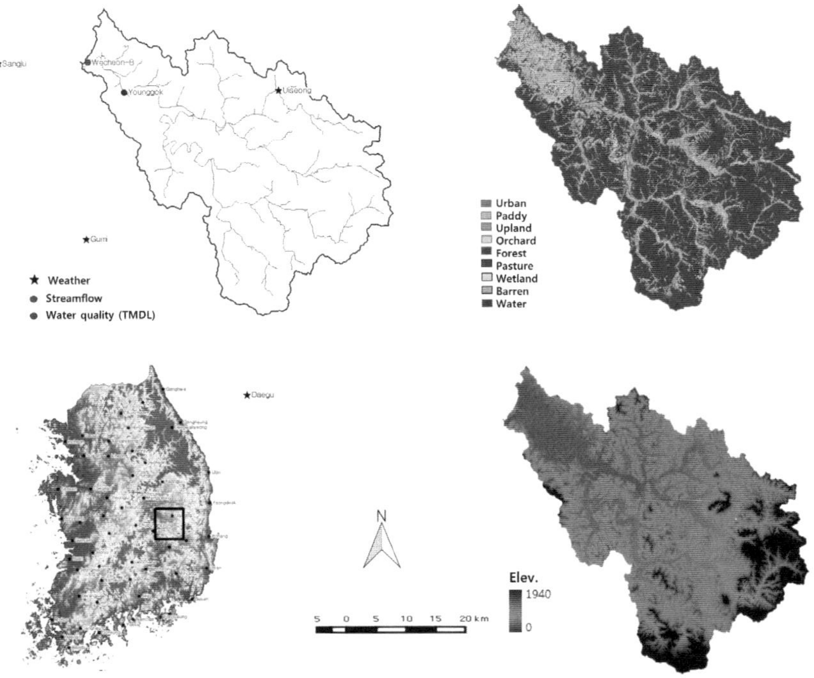


**Figure 2.** **(a)** Location of the monitoring network; **(b)** land use; and **(c)** elevation of the Wecheon watershed.

**Table 1.** Description of ten GCMs used as predictors.

| Model acronym | Institution(country) | Model resolution | Ensemble size |
|---|---|---|---|
| CWB | Central Weather Bureau (Taipei) | T42L18 | 10 |
| GDAPS_F | Korea Meteorological Administration (Korea) | T106L21 | 20 |
| HMC | Hydrometeorological Centre of Russia (Russia) | $1.125° \times 1.40625°$ | 20 |
| JMA | Japan Meteorological Agency (Japan) | T95L40 | 31 |
| MSC_CANCM3 | Meteorological Service of Canada (Canada) | $1.41° \times 0.94°$ | 10 |
| MSC_CANCM4 | Meteorological Service of Canada (Canada) | $1.41° \times 0.94°$ | 10 |
| NASA | National Aeronautics and Space Administration (USA) | $288 \times 181$ | 9 |
| NCEP | NCEP Climate Prediction Center (USA) | T62L64 | 15 |
| PNU | Pusan National University (Korea) | T42L18 | 5 |
| POAMA | Centre for Australian Weather and Climate Research/Bureau of Meteorology (Australia) | T47L17 | 10 |

**Table 2.** Monthly updated climate indices used for seasonal prediction.

| Abbreviation | Full name |
|---|---|
| PNA* | Pacific North American Index |
| EP* | East Pacific/North Pacific Oscillation |
| WP* | Western Pacific Index |
| NAO* | North Atlantic Oscillation |
| SOI* | Southern Oscillation Index |
| NINO3* | Eastern Tropical Pacific SST |
| TNA* | Tropical Northern Atlantic Index |
| TSA* | Tropical Southern Atlantic Index |
| WHWP* | Western Hemisphere warm pool |
| ONI | Oceanic Nino Index |
| MEI* | Multivariate ENSO Index |
| NINO12* | Extreme Eastern Tropical Pacific SST |
| NINO4* | Central Tropical Pacific SST |
| NINO34* | East Central Tropical Pacific SST |
| NOI | Northern Oscillation Index |
| NP | North Pacific pattern |
| TNI | Trans-Niño Index |
| AO | Antarctic Oscillation |
| AAO | Antarctic Oscillation |
| PACWARM | Pacific Warm Pool (1st EOF of SST (60–170° E, 15° S–15° N) SST EOF) |
| EOFPAC | Tropical Pacific SST EOF |
| ATLTRI | Atlantic Tripole SST EOF |
| AMO* | Atlantic multi-decadal Oscillation |
| QBO* | Quasi-Biennial Oscillation |
| ESL* | Equatorial Eastern Pacific SLP |

* Data source is NOAA and remaining is APCC.

This study utilizes the observed monthly precipitation and temperature data from the Korean peninsula, based on 57 Korean weather stations (Fig. 1).

## 4 Results

### 4.1 Evaluation of the downscaling method

Table 3 shows the results of prediction models that have been selected for each case (month and variable). When using the Simple Bias Correction (SBC) method, total of 21 and 36 models were selected for precipitation and temperature, respectively. For precipitation forecasting, a similar number of models were selected for different lead times (6, 7, and 8 models for 1, 2, and 3 month lead times, respectively). For temperature forecasting, the number of models varied between different lead times (20, 10, and 8 models for 1, 2, and 3 month lead times, respectively). The MWR method selected 19 models for precipitation forecasting and 9 models for temperature forecasting. For precipitation forecasting, the number of models varied between different lead times (6, 4, and 9 models for 1, 2, and 3 month lead times, respectively). For temperature forecasting, a similar number of models were selected for different lead times. When utilizing the Climate Index Regression (CIR) method, only one index (the Western Pacific Index, "WP" with 5 month lag) was selected to predict precipitation in July, while two indices such as the Pacific Warm Pool (PACWARM) with 6-month lag and the Atlantic Tripole SST EOF (ATLTRI) with 3-month lag were selected to predict temperature in September and October, respectively. As a result, the ITR method was selected to forecast precipitation levels in July and temperature levels in September and October, When MWR model selections are available. Overall, the SBC method, which is based on dynamic prediction data, shows the highest model selection and is followed by statistical downscaling methods such as MWR, and CIR/ITR. The SBC method shows the highest selection of models for 1-month lead temperature prediction for September with 6 models, while the MWR method shows the highest selection of models for 1-month lead precipitation prediction for September with 3 models. Figure 3 shows an example of spatial distribution of the three predictors that have been selected by the MWR method for 1-month lead precipitation prediction for September.

An evaluation of predictability when issuing forecasts every month was conducted, as shown in Fig. 4. For example, when we predict precipitation levels in August during the month of July, all three prediction results (including 1-month lead prediction issued in July, 2-month lead prediction issued in June, and 3-month lead prediction issued in May) can be used. Figure 5 illustrates an evaluation of predictability using a simple average of multi-model predictions.

Figure 5 shows the temporal correlation coefficient for each month according to changes in lead time. For precipitation prediction, there were difficulties when selecting models for January, March, May, and June. The months of February, July, and December show Temporal Correlation Coefficient (TCC) values that are greater than 0.6 for most lead times. In December, when the selected models are based on dynamic model predictions, there is a decreasing in TCC values as the lead times increase. Figure 5 also shows that there are difficulties when selecting prediction models for temperature predictions February, March, April, and June. The greatest TCC values occurred during the month of September, when most of the model selections are based on the SBC method, directly using dynamic prediction data.

### 4.2 Long-term water quality forecast

Table 4 shows the monthly Total Error (Err) and Nash-Sutcliffe Efficiency Index (NSE) for streamflow during the calibration and validation periods. The Errs were less than 5 % for both calibration and validation periods (−1.6 and −1.1 % Err, respectively). NSE values were greater than 0.9 for both calibration and validation periods (0.95 and 0.92, respectively). This indicates that streamflow simulation using the Soil and Water Assessment Tool (SWAT) is satisfactory. Figures 6 and 7 compare the measured and simu-

**Table 3.** Selected downscaling method and models for each month according to different lead time and variables.

| Month | Var | 1 month lead | 2 month lead | 3 month lead |
|---|---|---|---|---|
| Jan | *P* | B_JMA<br>B_POAMA | | |
| | *T* | B_GDAPS_F<br>B_PNU<br>B_MSC_CANCM3<br>B_MSC_CANCM4<br>M_POAMA | B_POAMA | M_CWB |
| Feb | *P* | | M_CWB | M_GDAPS_F |
| | *T* | | B_JMA | |
| Mar | *P* | B_POAMA | | |
| | *T* | B_GDAPS_F<br>B_JMA | | |
| Apr | *P* | | | B_NASA<br>M_NCEP<br>M_PNU<br>B_HMC |
| | *T* | M_GDAPS_F | B_NASA | |
| May | *P* | | | |
| | *T* | M_CWB | M_PNU | M_MSC_CANCM3 |
| Jun | *P* | M_HMC | M_MSC_CANCM4 | |
| | *T* | | | |
| Jul | *P* | C_Lag | C_Lag | I_PNU |
| | *T* | B_JMA | M_GDAPS_F | M_POAMA |
| Aug | *P* | B_JMA | B_GDAPS_F | |
| | *T* | B_HMC<br>B_PNU | | M_CWB |
| Sep | *P* | M_NCEP<br>M_PNU<br>M_POAMA | B_PNU | B_GDAPS_F<br>B_PNU |
| | *T* | B_GDAPS_F<br>B_HMC<br>B_JMA<br>B_PNU<br>B_NASA<br>B_POAMA<br>I_NASA | B_GDAPS_F<br>B_NASA<br>B_PNU<br>I_MSC_CANCM4<br>I_NCEP | B_GDAPS_F<br>B_HMC<br>B_JMA<br>B_NASA<br>B_PNU<br>C_Lag |
| Oct | *P* | | | M_MSC_CANCM4 |
| | *T* | B_GDAPS_F<br>B_HMC<br>B_PNU<br>I_MSC_CANCM3 | B_GDAPS_F<br>B_JMA<br>B_NASA<br>B_PNU<br>C_Lag | B_GDAPS_F<br>C_Lag |
| Nov | *P* | B_PNU | B_GDAPS_F<br>B_PNU<br>B_JMA | B_PNU<br>B_JMA<br>B_POAMA<br>M_MSC_CANCM3 |
| | *T* | B_POAMA | | B_JMA<br>M_PNU |
| Dec | *P* | B_POAMA<br>M_PNU<br>M_MSC_CANCM4 | B_JMA<br>B_POAMA<br>M_HMC<br>M_POAMA | B_POAMA<br>M_GDAPS_F<br>M_POAMA<br>M_NASA |
| | *T* | B_POAMA | | B_JMA |

*P*: Precipitation, *T*: Temperature, B_: simple bias correction (SBC), M_: Moving Window Regression (MWR), C_: Climate Index Regression (CIR), I_: Integrated Time Regression (ITR)

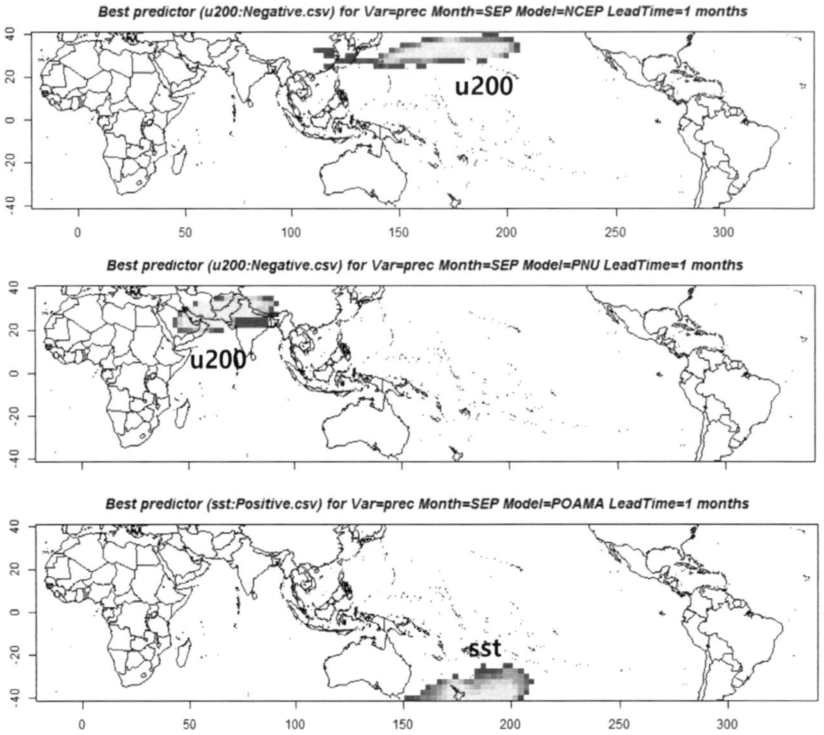


**Figure 3.** Spatial distribution of selected variables by the NCEP, PNU, and POAMA models for 1-month lead precipitation predictions in September (yellow indicates most frequent selection through the cross-validation procedures from 1983 to 2013).

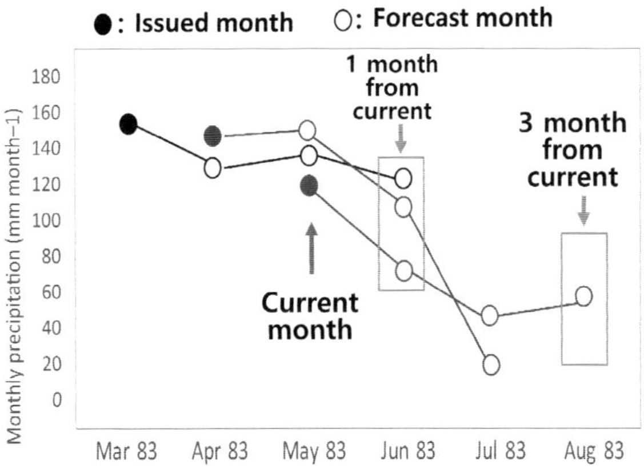


**Figure 4.** Description of the monthly prediction approach and use of different amounts of prediction data according to lead time.

lated Total Nitrogen (TN) and Total Phosphorous (TP) loads at the Wecheon-B station. It was also determined that the SWAT appropriately reflects the seasonal variation of TN and TP loads in the watershed. As a result, the long-term climate forecast data were utilized as weather inputs with fixed SWAT parameters in order to further assess the predictability in long-term water quality predictions.

**Table 4.** Results of SWAT calibration and validation for monthly streamflow.

| Performance measures | Calibration | Validation |
|---|---|---|
| % Error | −1.6 | −1.1 |
| Monthly NSE | 0.95 | 0.92 |

Figure 8 shows the comparison of observed pollutant loads (Observed-WQ), simulated pollutant loads using observed weather data (SWAT-Observed), and simulated pollutant loads using forecasted MME data (SWAT-Forecast). When comparing Observed-WQ and SWAT-Forecast, the temporal correlation coefficient (TCC) was 0.22 for TN. This low TCC value may be due to the overestimation by SWAT-Forecast during the period of 2007–2010. However, TCC increased to 0.58 when comparing SWAT-Forecast to SWAT-Observed due to their good temporal agreement. A comparison of TP load shows a similar trend to TN comparison results by showing TCC values of 0.05 and 0.49 when SWAT-Forecast was compared to Observed-WQ and SWAT-Observed, respectively. We found that uncertainty from the SWAT parameterization was higher than the uncertainty from the climate forecast, illustrated by the big differences in their TCCs. The high uncertainties in model parameterization can

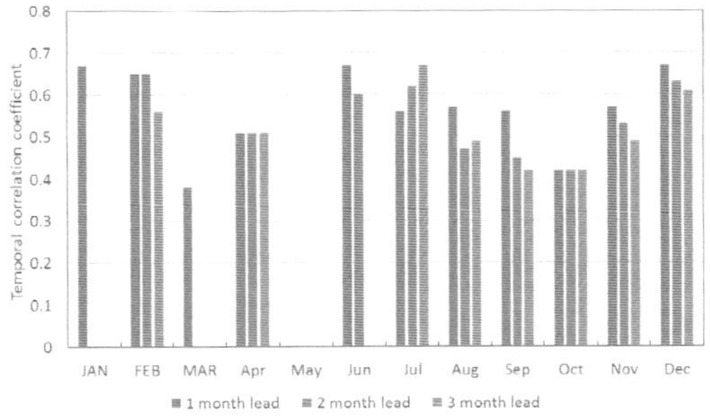


| Month | 1 month lead | 2 month lead | 3 month lead |
|---|---|---|---|
| JAN | 0.67 | | |
| FEB | 0.65 | 0.65 | 0.56 |
| MAR | 0.38 | | |
| APR | 0.51 | 0.51 | 0.51 |
| MAY | | | |
| JUN | 0.67 | 0.6 | |
| JUL | 0.56 | 0.62 | 0.67 |
| AUG | 0.57 | 0.47 | 0.49 |
| SEP | 0.56 | 0.45 | 0.42 |
| OCT | 0.42 | 0.42 | 0.42 |
| NOV | 0.57 | 0.53 | 0.49 |
| DEC | 0.67 | 0.63 | 0.61 |

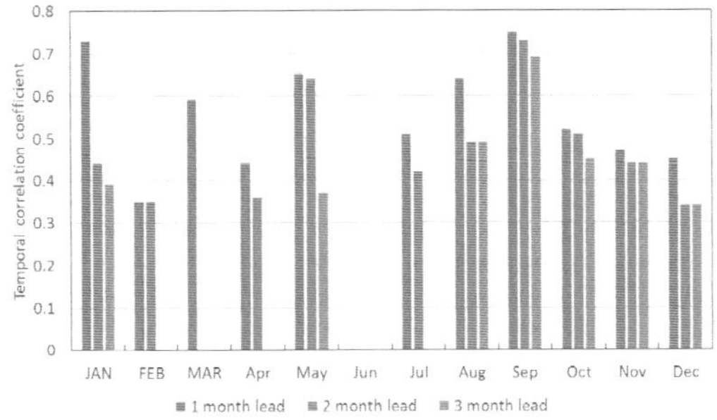


| Month | 1 month lead | 2 month lead | 3 month lead |
|---|---|---|---|
| JAN | 0.73 | 0.44 | 0.39 |
| FEB | 0.35 | 0.35 | |
| MAR | 0.59 | | |
| APR | 0.44 | 0.36 | |
| MAY | 0.65 | 0.64 | 0.37 |
| JUN | | | |
| JUL | 0.51 | 0.42 | |
| AUG | 0.64 | 0.49 | 0.49 |
| SEP | 0.75 | 0.73 | 0.69 |
| OCT | 0.52 | 0.51 | 0.45 |
| NOV | 0.47 | 0.44 | 0.44 |
| DEC | 0.45 | 0.34 | 0.34 |

**Figure 5.** Temporal correlation coefficients (TCC) according to changes in lead time for predicting precipitation (top) and temperature (bottom) using multi-model ensemble (MME) average.

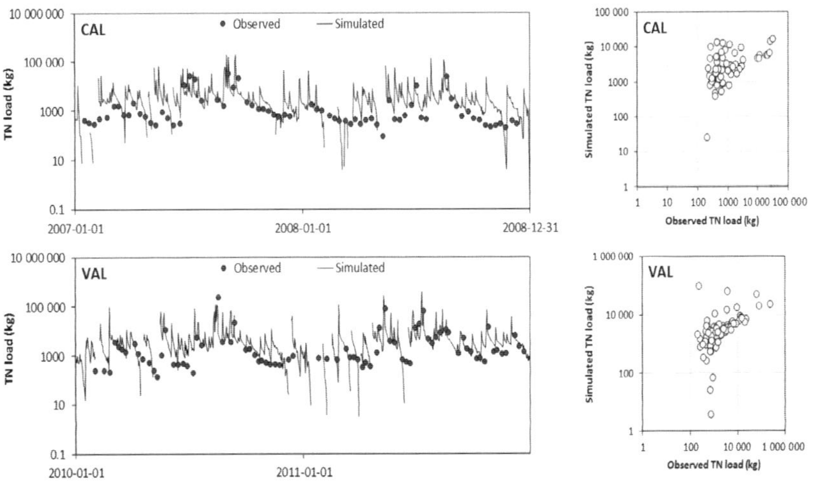


**Figure 6.** Comparison of simulated and observed total nitrogen (TN) load for calibration (top) and validation (bottom) periods.

be caused by the lack of sampled data used to estimate monthly pollutant loads. TN and TP concentrations were measured around four times every month and the monthly pollutant load was estimated by multiplying the concentration and flow rate. As a result, monthly pollutant loads can be affected by flow rate depending on the sampling time.

Table 5 shows the monthly comparison of TCC values according to different references as observation including Observed-WQ and SWAT-Observed. When SWAT-Forecast is compared to Observed-WQ, only March has a TCC value that is greater than 0.5 for TN, and all months have TCC values that are less than 0.5 for TP. However, when SWAT-Forecast is compared to SWAT-Observed, 3 months from March to May have TCC values that are greater than 0.5 for both TN and TP predictions. Even though there were difficulties in selecting models between March and May for both precipitation and temperature, this period had the highest predictability in long-term water quality forecast. On the other hand, water quality predictions in July and December, when predictability in precipitation prediction was high, had

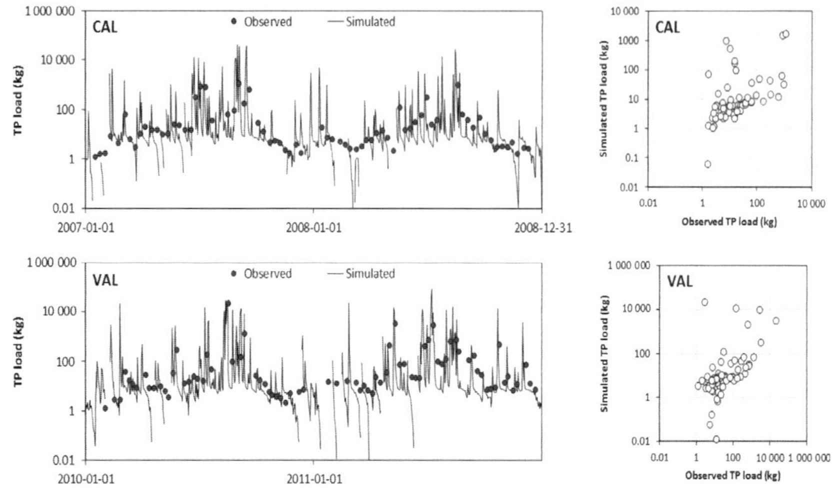


**Figure 7.** Comparison of simulated and observed total phosphorus (TP) load for calibration (top) and validation (bottom) periods.

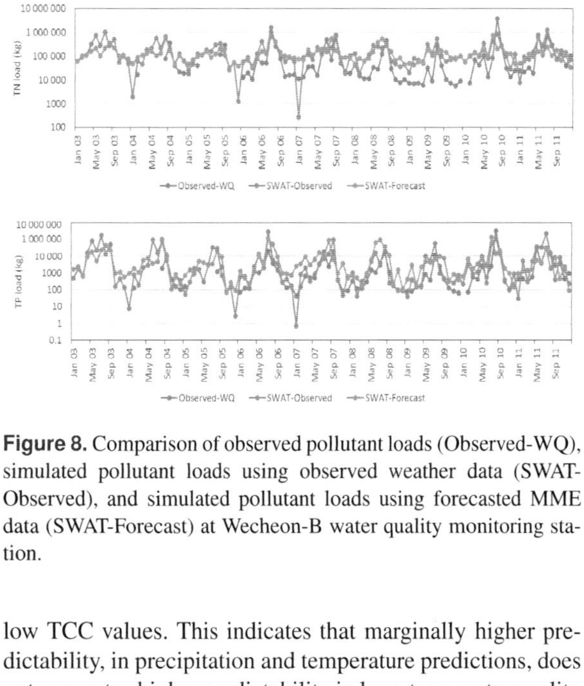


**Figure 8.** Comparison of observed pollutant loads (Observed-WQ), simulated pollutant loads using observed weather data (SWAT-Observed), and simulated pollutant loads using forecasted MME data (SWAT-Forecast) at Wecheon-B water quality monitoring station.

low TCC values. This indicates that marginally higher predictability, in precipitation and temperature predictions, does not guarantee higher predictability in long-term water quality prediction.

## 5 Conclusions

Four different downscaling methods in accordance with the degree of utilizing seasonal climate prediction data were developed in order to improve predictability and refine the spatial scale. These methods include: (1) the Simple Bias Correction (SBC) method, which directly uses the APCC's climate prediction data with 3 month lead; (2) the Moving Window Regression (MWR) method, which indirectly utilizes prediction data; (3) the Climate Index Regression (CIR) method, which predominantly uses observation-based climate indices; and (4) the Integrated Time Regression (ITR) method, which uses predictors selected from both the CIR and MWR methods. Overall, the SBC method, based on dynamic prediction data, shows the highest model selection and is followed by statistical downscaling methods such as MWR, and CIR/ITR.

Then, sampling-based temporal downscaling using the Mahalanobis distance method was conducted in order to create daily weather inputs to the Soil and Water Assessment Tool (SWAT) model. Long-term predictability of water quality within the medium-size Wecheon watershed of the Nakdong River Basin was evaluated. According to the Provisions of Water Quality Prediction and Response Measures of the Korean Ministry of Environment, modeling-based predictability was evaluated by using 3-month lead prediction data forecasted in February, May, August, and November as model inputs of SWAT. The results indicate that marginally higher predictability in precipitation and temperature predictions does not guarantee higher predictability in long-term water quality predictions.

Finally, we presented an integrated approach that takes into account various climate information and downscaling methods for water quality prediction, which can be used to proactively prevent potential problems caused by extreme climate events.

## 6 Data availability

Seasonal forecast data by individual models are available at the APCC Data Service System website (APCC) (http://adss.apcc21.org/). Monthly climate indices data are available at the Climate Indices website (APCC) (http://www.apcc21.

**Table 5.** Monthly comparison of temporal correlation coefficient (TCC) according to different references as observation.

| Month | Total Nitrogen (TN) | | Total Phosphorus (TP) | |
|---|---|---|---|---|
| | SWAT-Forecast vs. Observed-WQ | SWAT-Forecast vs. SWAT-Observed | SWAT-Forecast vs. Observed-WQ | SWAT-Forecast vs. SWAT-Observed |
| Jan | −0.47 | −0.27 | −0.38 | −0.08 |
| Feb | 0.01 | 0.00 | −0.34 | 0.26 |
| Mar | 0.72 | 0.67 | 0.45 | 0.68 |
| Apr | 0.40 | 0.80 | −0.20 | 0.71 |
| May | −0.34 | 0.58 | −0.26 | 0.74 |
| Jun | −0.05 | 0.00 | −0.15 | −0.10 |
| Jul | −0.04 | 0.30 | −0.46 | 0.27 |
| Aug | −0.25 | 0.14 | −0.30 | 0.14 |
| Sep | 0.31 | 0.29 | 0.31 | 0.07 |
| Oct | −0.30 | 0.25 | −0.14 | −0.09 |
| Nov | 0.37 | 0.22 | 0.24 | 0.13 |
| Dec | −0.62 | 0.37 | −0.39 | −0.02 |

org/ser/indic.do?lang=en) and the Monthly Atmospheric and Ocean Time Series website (ESRL) (http://www.esrl.noaa.gov/psd/data/climateindices/list/).

**Acknowledgements.** This research was supported by APEC Climate Center.

## References

APCC: APCC Data Service System, available at: http://adss.apcc21.org/, last access: August 2016.

APCC: Climate Information Service, Climate Indices website available at: http://www.apcc21.org/ser/indic.do?lang=en, last access: August 2016.

Bridhikitti, A.: Connections of ENSO/IOD and aerosols with Thai rainfall anomalies and associated implications for local rainfall forecasts, Int. J. Climatol., 33, 2836–2845, 2013.

ESRL: Climate Indices, Monthly Atmospheric and Ocean Time Series, available at: http://www.esrl.noaa.gov/psd/data/climateindices/list/, last access: August 2016.

Hamlet, A. F. and Lettenmaier, D. P.: Columbia river streamflow forecasting based on ENSO and PDO climate signals, J. Water Resour. Plann. Manage., 125, 333–341, 1999.

Kang, H., Park, C. K., Hameed, S. N., and Ashok, K.: Statistical downscaling of precipitation in Korea using multimodel output variables as predictors, Mon. Weather Rev., 137, 1928–1938, 2009.

Kang, S., Hur, J., and Ahn, J. B.: Statistical downscaling methods based on APCC multi-model ensemble for seasonal prediction over South Korea, Int. J. Climatol., 34, 3801–3810, 2014.

Kang, H., Park, C. K., Hameed, S. N., and Ashok, K.: Statistical downscaling of precipitation in Korea using multimodel output variables as predictors, Mon. Weather Rev., 137, 1928–1938, 2009.

Kim, M. K. and Kim, Y. H.: Seasonal prediction of monthly precipitation in china using large-scale climate indices, Adv. Atmos. Sci., 27, 47–59, 2010.

Kim, M. K., Kim, Y. H., and Lee, W. S.: Seasonal prediction of Korean regional climate from preceding large-scale climate indices, Int. J. Climatol., 27, 925–934, 2007.

Lee, E., Chase, T. N., and Rajagopalan, B.: Seasonal forecasting of East Asian summer monsoon based on oceanic heat sources, Int. J. Climatol., 28, 667–678, 2008.

Mahalanobis, P. C.: On the generalized distance in statistics, Proceedings of the National Institute of Sciences (Calcutta), 2, 49–55, 1936.

Meng, H., Sexton, A. M., Maddox, M. C., Sood, A., Brown, C. W., Ferrara, R. R., and Murtugudde, R.: Modeling Rappahannock River basin using SWAT-Pilot for Chesapeake Bay watershed, Appl. Eng. Agric., 26, 795–805, 2010.

Neitsch, S. L., Arnold, J. G., Kiniry, J. R., Williams, J. R., and King, K. W.: Soil and water assessment tool theoretical documentation and user's manual. Version 2005, GSWR Agricultural Research Service & Agricultural Experiment Station, Temple Texas, 2005.

Räsänen, T. A. and Kummu, M.: Spatiotemporal influences of ENSO on precipitation and flood pulse in the Mekong River Basin, J. Hydrol., 476, 154–168, 2013.

Robertson, D. E. and Wang, Q. J.: A bayesian approach to predictor selection for seasonal streamflow forecasting, J. Hydrometeorol., 13, 155–171, 2012.

Schepen, A., Wang, Q. J., and Robertson, D.: Evidence for using lagged climate indices to forecast Australian seasonal rainfall, J. Clim., 25, 1230–1246, 2012.

Wang, B., Wu, Z., Li, J., Liu, J., Chang, C. P., Ding, Y., and Wu, G.: How to measure the strength of the East Asian summer monsoon, J. Climate, 21, 4449–4463, 2008.

Xu, J., Long, W., Wiggert, J. D., Lanerolle, L. W. J., Brown, C. W., Murtugudde, R., and Hood, R. R.: Climate Forcing and Salinity Variability in Chesapeake Bay, USA, Estuar. Coasts, 35, 237–261, 2012.

Proc. IAHS, 374, 187–192, 2016
proc-iahs.net/374/187/2016/
doi:10.5194/piahs-374-187-2016


Open Access
PIAHS

# River water quality modelling under drought situations – the Turia River case

**Javier Paredes-Arquiola, Javier Macián, María Pedro-Monzonís, Edgar Belda, Andrea Momblanch, and Joaquín Andreu**

Instituto Universitario de Ingeniería del Agua y Medio Ambiente. Universitat Politècnica de València, Valencia, Spain

*Correspondence to:* Javier Paredes-Arquiola (jparedea@hma.upv.es)

Published: 17 October 2016

**Abstract.** Drought and water shortage effects are normally exacerbated due to collateral impacts on water quality, since low streamflow affects water quality in rivers and water uses depend on it. One of the most common problems during drought conditions is maintaining a good water quality while securing the water supply to demands. This research analyses the case of the Turia River Water Resource System located in Eastern Spain. Its main water demand comes as urban demand from Valencia City, which intake is located in the final stretch of the river, where streamflow may become very low during droughts. As a result, during drought conditions concentrations of pathogens and other contaminants increase, compromising the water supply to Valencia City. In order to define possible solutions for the above-mentioned problem, we have developed an integrated model for simulating water management and water quality in the Turia River Basin to propose solutions for water quality problems under water scarcity. For this purpose, the Decision Support System Shell AQUATOOL has been used. The results demonstrate the importance of applying environmental flows as a measure of reducing pollutant's concentration depending on the evolution of a drought event and the state of the water resources system.

## 1 Introduction

Apart from water scarcity (i.e. quantity), a relevant impact of droughts on river basins is the worsening of water quality in river systems. This can render water resources useless for economic activities and human water supply, while seriously damaging the environment (Momblanch et al., 2015). An isolated analysis of water quality limits the action measures that can be proposed. Thus, an integrated evaluation combining water management and water quality models would be desirable.

The use of a water allocation and a water quality model, integrated in the same platform, facilitates the transfer of results and allows interactive simulations, which are useful to perform scenario analysis (Momblanch et al., 2015). Water management models provide indicators about water supply reliabilities and vulnerabilities, which represent the level of satisfaction of water demands. In turn, water quality models generate time series of concentration of pollutants. The purpose is to select a set of target variables to capture the key aspects of the analysed water system, and to assess the effect of different alternatives on them.

Many studies have identified drought impacts, mostly on water scarcity, i.e. lack of water, with regard to human and environmental needs (Chang and Wang, 2013; Sušnik et al., 2012), which is being further associated with the drought hazard (Mishra and Singh, 2011). Besides, there are some other studies concerned about the impacts of droughts and the implicit low-flow conditions on river water quality (Barros et al., 1995; Caruso, 2002; van Vliet and Zwolsman, 2008; Paredes-Arquiola et al., 2010; Paredes et al., 2010; Whitworth et al., 2012; Momblanch et al., 2015).

In this paper we analyse the problems related to nitrates and pathogens in the Turia River System in Spain, which affect the water supply of Valencia City during droughts. Moreover, we take into account the impact of drought on irrigation water demands that represent around 80 % of the total wa-

Published by Copernicus Publications on behalf of the International Association of Hydrological Sciences.

ter use in the basin. To do so, we connect a water allocation model with a water quality model in order to test different scenarios and propose a balanced solution considering water quantity-quality tradeoffs.

## 2 Material and methods

A key objective of this research is to develop a water quality model along the Turia River to analyse the influence of nitrates, and total (TC) and faecal coliforms (FC) in the final stretch of the river, as well as building a water allocation model to assess water supply reliability for agricultural demands. For this purpose, the Decision Support System Shell (DSSS) AQUATOOL (Andreu et al., 1996) will be used. AQUATOOL comprises several modules. Firstly, the SIMGES module (Andreu et al., 1996) allows the evaluation on water resources allocation and secondly, the GESCAL module (Paredes-Arquiola et al., 2010) allows the assessment of water quality. This methodology allows us to analyse different alternatives to cope with problems triggered by droughts in the Turia River Basin. The main features of these modules are described below.

SIMGES is a general simulation model for river basins or complex water resources systems. It includes control or storage elements (surface water and groundwater) and elements for collection, transport, use and/or consumption, and artificial recharge of water. The simulation is done at a monthly scale and it can reproduce the water flow through the system at any spatial scale the user prefers. The SIMGES module also considers the return flows to the surface water system and the recharge to the groundwater system. It also takes into account the evaporation and infiltration losses from reservoirs and the interaction between surface water and groundwater. Water resources management is simulated using operation rules, which try to maintain a similar filling level in reservoirs using reservoir zone curves. It allows definition of environmental flows, as well as setting different water use priorities. The simulation and management of the surface water system are done at the same time using a network flow optimization algorithm. This algorithm determines the flows through the system trying to satisfy the multiple objectives of deficit minimization, and maximum adaptation to reservoir objective volume curves and hydroelectric production objectives. The optimization is improved through an iterative solution process of the network, what allows improving the quality of the simulated non-linear processes, such as infiltration, evaporation and surface and groundwater interaction.

GESCAL was developed to evaluate the water quality at a river basin level. For this, it includes the calculation of all processes or water quality modifications both in rivers and reservoirs. These calculations are done at every element of the river basin scheme defined with SIMGES connecting them according to the calculated flows. Applying it to different decision alternatives in the river basin management it is possible to evaluate the consequences for the water quality in different river stretches. GESCAL allows for the modelling of water temperature, arbitrary constituents, dissolved oxygen, biological oxygen demand (BOD), the nitrogen cycle, phytoplankton and phosphorous. Arbitrary constituents are defined as those for which degradation can be modelled as a first-order kinetic and/or with a sedimentation velocity. Dissolved oxygen can be considered with three possible levels of complexity. The simplest level considers BOD and dissolved oxygen. The intermediate level considers the cycle of nitrogen and the effect on dissolved oxygen, and organic nitrogen, ammonia, and nitrates. The highest level of complexity of BOD models includes the nitrogen cycle, phytoplankton, and the phosphorous cycle, as well as their relationships and the effects on dissolved oxygen. A detailed description of the processes and the model can be found in Paredes et al. (2004).

The first step represents the development of the water management and water quality models, which will be calibrated and validated. At this step, it is very important to include a drought period in the calibration and validation to ensure that the models adequately represent the performance of the system in this specific situation (Momblanch et al., 2015). The results from the calibrated models represent the baseline situation, which will be used as the reference scenario. The analysis of the calibration outcome and the initial results will provide key information to recognize the main problems occurring in the basin.

In the second step, the action measures are designed and combined to generate reasonable and realistic scenarios, in line with the water quantity and quality problems identified in the Turia River Basin during droughts. Thus, the measures that can be simulated are referred either to water quantity or water quality actions. It is important to select the proper target variables that represent the evolution of the addressed problems.

As a final step, it is necessary to analyse the existing tradeoffs by means of the target variables for the selection of the measures or sequence of measures that best facilitate the mitigation of drought effects.

## 3 Case study: the Turia River Basin

### 3.1 Characterization of the study area

The Turia River Basin (TRB) is located in the eastern part of the Iberian Peninsula in Spain (Fig. 1). It is the second most extensive basin (6393 $km^2$) and water resources producer in the Jucar River Basin District. The Turia River has a length of 280 km, traversing the provinces of Teruel, Cuenca and Valencia, and it has its mouth in the city of Valencia at the Mediterranean Sea. Hence, it is a characteristic Mediterranean river, heavily affected by drought events, presenting large differences in flow in different seasons. The main reservoirs are Arquillo de San Blas, Benagéber and Loriguilla.

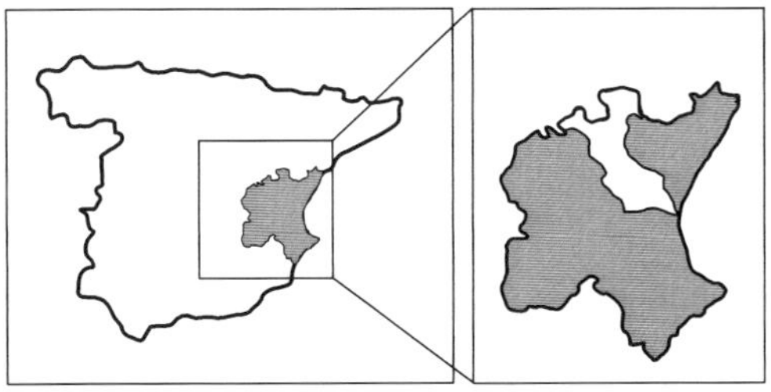

**Figure 1.** Location of Turia River (right) in the Jucar River Basin (Source: http://www.chj.es).

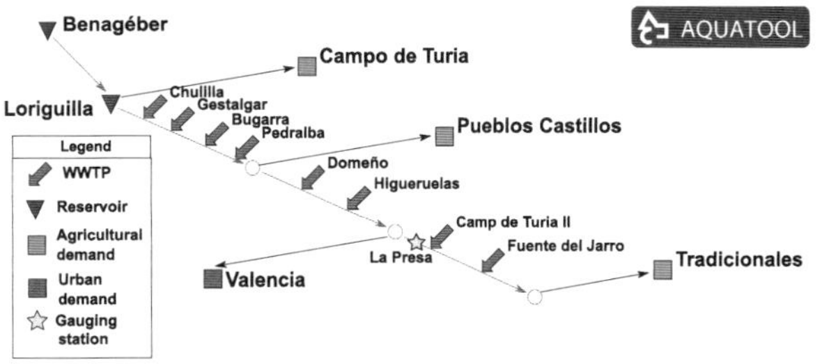


**Figure 2.** Scheme of the simulation model. The fundamental elements characterizing the Jucar River Basin.

The main water demands in the basin are agrarian and urban. Along its middle and lower stretch three agrarian demands are located: the Campo del Turia, Pueblos Castillos and Tradicionales. The Campo del Turia and Pueblos Castillo irrigation demands are located in the middle part of the system and the Tradicionales in the lower basin. The main urban demand is the supply to Valencia City, which has a population of 1 million inhabitants, and its intake is located near the Tradicionales agricultural intake.

The basin has different problems related with water quantity and quality. One of the main problems is the allocation of water among agrarian demands during drought periods. Droughts are common in this basin and during these periods there is not enough water for the three above-mentioned demands. Therefore, reductions in the water supplies have to be applied. Under this consideration, flows are lower because management tries to keep water in the reservoirs and consequently an increased concentration of pollutants is measured in the river. This causes a significant problem in the lower stretch of the river due to high concentrations of nitrates and pathogens. It is not just an environmental problem; also human health is affected since the intake of water for Valencia City is located in the lower zone. The origin of nitrates pollution is the return of flows from agrarian activities in the middle river stretch. On the other hand, urban runoff and effluents from Waste Water Treatment Plants (WWTPs), also located in the middle river stretch, are responsible for the pathogen pollution. Therefore, there is an increment of pathogens and other compounds in the river, among others TC, FC, and Escherichia Coli.

## 3.2 Model building

The necessary data to develop the water resource quality model are:

- Streamflow time series measured in gauging stations and reservoirs levels. Time series of water supplies to agrarian and urban demands. A naturalization of streamflow has been done taking into account streamflow time series measured in the gauging stations, reservoirs levels and water supplies to agrarian and urban demands.
- Environmental flows have also been considered, i.e. the model includes $0.7\,m^3\,s^{-1}$ downstream the Benagéber Reservoir, $0.5\,m^3\,s^{-1}$ downstream the Loriguilla Reservoir, and $1\,m^3\,s^{-1}$ upstream the Repartiment Weir.
- Turbined flows in hydropower stations are also taken into account.
- Water volume and pollutant concentrations treated in WWTPs, which discharge in the river, time series of pollutant concentrations recorded in the quality stations located along the river and water temperature in different locations in the river.

Figure 2 shows the scheme of the SIMGES and GESCAL models used to estimate the water quality in which the different elements of the system are identified.

## 3.3 Calibration of the model

The water allocation model has been calibrated and validated. An easy way to do this is by comparing the real volumes measured in the main gauging stations of the system with the simulated volumes in the same points of the model. This was done for the period October 2004 to September 2014. The observed and simulated series were very similar, and thus the water allocation model was considered valid to conduct the proposed drought impact and management analysis (Fig. 3).

The water quality model was developed in order to simulate the concentrations of nitrates, TC and FC along the river. The calibration of the water quality model consisted of adjusting the parameters, which control the physical, chemical and biological processes causing variations in the concentrations of the considered compounds. Notice that whenever the measurements were done, they were considered as representative values of a whole month because GESCAL works on monthly stationary basis. Thus, the calibration focused on the similarity of observed and simulated tendencies instead of looking for a perfect match. The parts of the river, which required a bigger calibration effort were the lower sections due to the high concentration of urban discharges.

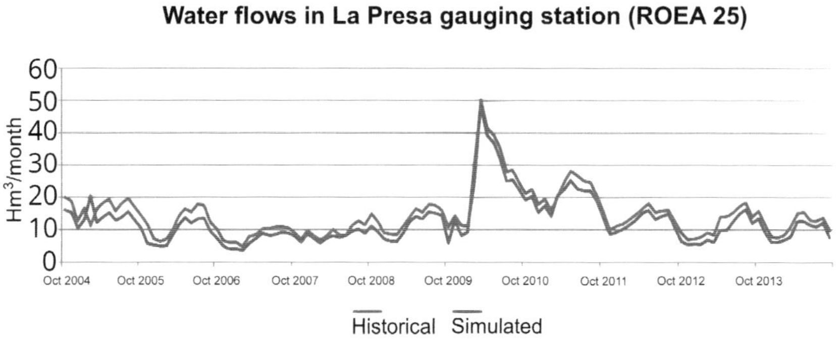


**Figure 3.** Comparison of water flows observed in La Presa and simulated by the model.

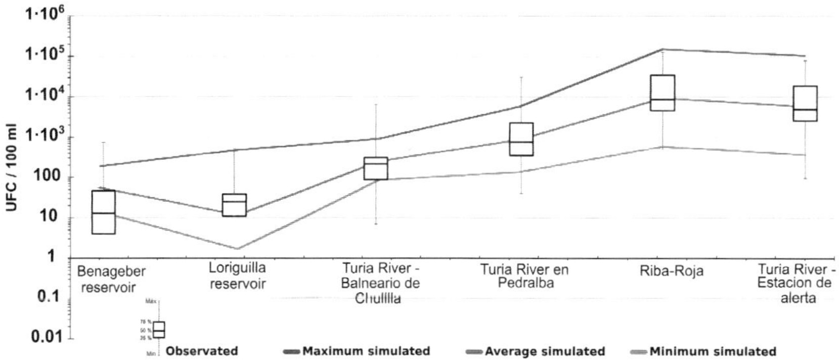


**Figure 4.** Adjustment between maximum, mean and minimum concentrations of TC observed and simulated along the Turia River.

Maximum, average and minimum concentrations observed and simulated along the Turia River are presented in Fig. 4. The outcome from the calibration shows a good agreement between the observed and simulated values. So, finally, the water quality model was also ready to perform the simulations required for the scenarios, i.e. exploration of measures.

### 3.4 Simulation of scenarios

Model outcome was used to explore improvement of the water quality of the river in the vicinity of the Valencia City intake, while maintaining reliability of agrarian water demands. In order to achieve this goal, several scenarios have been considered that impose different minimum flows just upstream of the Valencia City intake. The imposed minimum flows range from $8\,m^3\,s^{-1}$ (current historical minimum flow) to $20\,m^3\,s^{-1}$.

A traditional measure of reliability for agrarian water demand is the maximum supply deficit of the demand in one year and the maximum deficit in two consecutive years. On the other hand, resiliency is measured as the maximum number of consecutive months with deficit.

As presented in Table 1, imposed minimum flows up to $12\,m^3\,s^{-1}$ can be adopted without significant increments of agrarian deficits (compare MINFLOW12 with current situation). Flows of $14\,m^3\,s^{-1}$ or more imply a substantial increase of deficits, which would mean important economic consequences for farmers.

It is important to notice that imposing high river flows near the drinking water intake (16, 18 or $20\,m^3\,s^{-1}$) implies that in some drought periods there is not enough water to maintain such minimum flows, since it involves releasing too much water in periods when it is not necessary. This is reflected in Table 1 by the difference between "Imposed" and "Model" figures for each scenario. The "Imposed" minimum flow is the boundary condition predefined in the model, while the "Model" minimum flow is the flow at the analysed river stretch resulting from the simulation.

Regarding water quality aspects, Fig. 5 shows that the increase of river flow decreases concentrations of TC and FC overall in the critical periods. According to Spanish legislation, drinking water treatment plants, which apply intense physical and chemical treatment, refining, and disinfection (A3 type), should not use raw water with TC concentration above 50 000 UFC/100 mL. This is the case of the plant supplying Valencia City. Results show that the increase of minimum flows reduces TC concentrations to or below the legal threshold all along the simulated period, except for October 2007. Even imposing an envi-

**Table 1.** Summarised results for the reliability (Max Def 1 year (%): maximum annual water supply deficit as % of the annual water demand; Max Def 2 consecutive years (%): maximum accumulated water supply deficit in two consecutive years as % of the annual water demand) and resiliency (maximum number of consecutive months with deficit) of the three agricultural water demands for the current situation and for different imposed minimum flows at river stretch where the Valencia City water intake (VWI) is located (MINFLOW: 12, 14, 16, 18 and 20 $m^3 s^{-1}$).

| | Current SITUATION | | | MIN FLOW 12 | | | MIN FLOW 14 | | |
|---|---|---|---|---|---|---|---|---|---|
| | Minimum Flow in VMI | | | Minimum Flow in VMI | | | Minimum Flow in VMI | | |
| | Imposed | Model 8 | | Imposed 12 | Model 12 | | Imposed 14 | Model 14 | |
| | Max Def 1 year (%) | Max Def 2 consecutive years (%) | Resiliency | Max Def 1 year (%) | Max Def 2 consecutive years (%) | Resiliency | Max Def 1 year (%) | Max Def 2 consecutive years (%) | Resiliency |
| Campo del Turia | 25.5 | 50.2 | 24.0 | 25.5 | 50.2 | 24.0 | 25.5 | 50.2 | 24.0 |
| Pueblos Castillos | 29.5 | 48.3 | 24.0 | 30.0 | 48.7 | 24.0 | 35.5 | 54.3 | 24.0 |
| Tradicionales | 16.1 | 30.6 | 24.0 | 16.2 | 30.7 | 24.0 | 17.3 | 31.8 | 24.0 |
| | MIN FLOW 16 | | | MIN FLOW 18 | | | MIN FLOW 20 | | |
| | Minimum Flow in VMI | | | Minimum Flow in VMI | | | Minimum Flow in VMI | | |
| | Imposed 16 | Model 8.2 | | Imposed 18 | Model 8.2 | | Imposed 20 | Model 8.2 | |
| | Max Def 1 year (%) | Max Def 2 consecutive years (%) | Resiliency | Max Def 1 year (%) | Max Def 2 consecutive years (%) | Resiliency | Max Def 1 year (%) | Max Def 2 consecutive years (%) | Resiliency |
| Campo del Turia | 25.8 | 51.4 | 28.0 | 25.8 | 51.4 | 28.0 | 29.0 | 56.1 | 34.0 |
| Pueblos Castillos | 56.5 | 87.0 | 28.0 | 74.3 | 125.6 | 28.0 | 78.0 | 148.1 | 34.0 |
| Tradicionales | 17.8 | 33.0 | 28.0 | 16.4 | 31.6 | 28.0 | 19.0 | 36.0 | 34.0 |

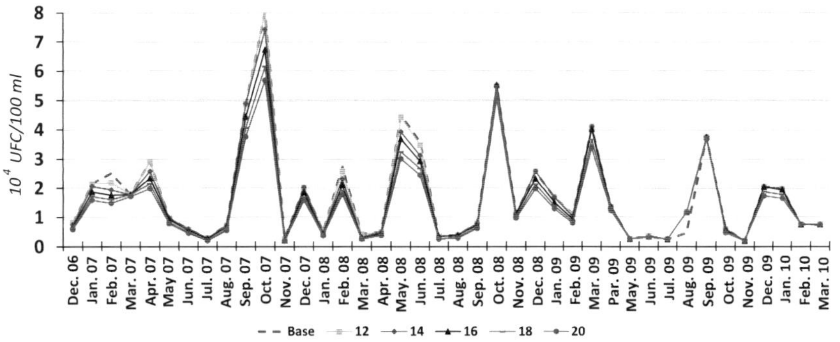


**Figure 5.** Concentrations of simulated Total Coliforms (TC) near the intake of Valencia City water for different environmental flow (imposed minimum flow) conditions.

ronmental flow of 20 $hm^3 month^{-1}$, concentrations remain above 50 000 UFC/100 mL. However, the global situation improves. Moreover, according to the results, it is not necessary to maintain the minimum flow all the time since most of the simulated period TC concentrations stay between 20 000 and 30 000 UFC/100 mL, which are more than acceptable concentrations. Hence it would be better to define a variable minimum flow regime, which allows keeping water in the reservoirs when it is not necessary for water quality issues, and which ensures increased river flows when pollution peaks occur.

Other alternatives have also been taken into consideration, such as disinfection treatments in different WWTPs, but even in the optimal scenarios, an increase of water releases from the reservoirs is necessary to reduce pathogens in the Valencia City water intake.

## 4 Conclusions

In this study a water quantity-quality model for the Turia River has been developed. This model allowed us to analyse the influence of total (TC) and faecal coliforms (FC) in the raw water arriving to the Valencia City water supply intake. The integral model allows exploring integral solutions taking into account water quantity and quality aspects. The obtained results have demonstrated the efficacy of applying environmental flows as a measure to reduce concentration of water pollutants during drought conditions. According to the results, it is not necessary to maintain a minimum flow continuously, but during drought conditions, an increase in river flow may help to improve the water quality.

## Data availability

All raw data is available on the Jucar River Basin Webpage (http://www.chj.es/). More specific information can be found in appendix 6 of river basin management plan 2015–2021 (http://www.chj.es/Descargas/ProyectosOPH/Consultapublica/PHC-2015-2021/PHJ1521_Anejo06_SE-Balances_151126.pdf).

**Acknowledgements.** The authors wish to thank the European Community's Seventh Framework Program in financing the project ENHANCE (FP7-ENV-2012, 308438), as well as the support of the "Pioneers into Practice Programme" of Climate-KIC for creating opportunities, and the framework for innovators to address climate change and shape the world's next economy. The authors acknowledge the support of the Polytechnic University of Valencia (grant agreement number 4234) through its Support Programme for Research.

## References

Andreu, J., Capilla, J., and Sanchís, E.: AQUATOOL, a generalized decision-support system for water-resources planning and operational management, J. Hydrol., 177, 269–291, doi:10.1016/0022-1694(95)02963-X, 1996.

Barros, M. C., Mendo, M. J. M., Negrão, F. C. R.: Surface water quality in Portugal during a drought period, Sci. Total Environ., 171, 69–76, doi:10.1016/0048-9697(95)04671-7, 1995.

Caruso, B. S.: Temporal and spatial patterns of extreme low flows and effects on stream ecosystems in Otago, New Zealand, J. Hydrol., 257, 115–133, doi:10.1016/S0022-1694(01)00546-7, 2002.

Chang, F.-J. and Wang, K.-W.: A systematical water allocation scheme for drought mitigation, J. Hydrol., 507, 124–133, doi:10.1016/j.jhydrol.2013.10.027, 2013.

Confederación Hidrogràfica del Júcar: River basin management plan 2015–2021, available at: http://www.chj.es/Descargas/ProyectosOPH/Consultapublica/PHC-2015-2021/PHJ1521_Anejo06_SE-Balances_151126.pdf, 2015 (in Spanish).

Mishra, A. K. and Singh, V. P.: Drought modeling – A review, J. Hydrol., 403, 157–175, doi:10.1016/j.jhydrol.2011.03.049, 2011.

Momblanch, A., Paredes-Arquiola, J., Munné, A., Manzano, A., Arnau, J., and Andreu, J.: Managing water quality under drought conditions in the Llobregat River Basin, Sci. Total Environ., 503–504, 300–318, doi:10.1016/j.scitotenv.2014.06.069, 2015.

Paredes, J.: Integración de la modelación de la calidad del agua en un sistema de ayuda a la decisión para la gestión de recursos hídricos, Universitat Politècnica de València, Valencia (Spain), doi:10.4995/Thesis/10251/1895, 2004.

Paredes, J., Andreu, J., and Solera, A.: A decision support system for water quality issues in the Manzanares River (Madrid, Spain), Sci. Total Environ., 408, 2576–2589, doi:10.1016/j.scitotenv.2010.02.037, 2010.

Paredes-Arquiola, J., Andreu-Álvarez, J., Martín-Monerris, M., and Solera, A.: Water quantity and quality models applied to the Jucar River Basin, Spain, Water Resour. Manag., 24, 2759–2779, 2010.

Sušnik, J., Vamvakeridou-Lyroudia, L. S., Savić, D. A., and Kapelan, Z.: Integrated System Dynamics Modelling for water scarcity assessment: Case study of the Kairouan region, Sci. Total Environ., 440, 290–306, doi:10.1016/j.scitotenv.2012.05.085, 2012.

Van Vliet, M. T. H. and Zwolsman, J. J. G.: Impact of summer droughts on the water quality of the Meuse River, J. Hydrol., 353, 1–17, doi:10.1016/j.jhydrol.2008.01.001, 2008.

Whitworth, K. L., Baldwin, D. S., and Kerr, J. L.: Drought, floods and water quality: Drivers of a severe hypoxic blackwater event in a major river system (the southern Murray-Darling Basin, Australia), J. Hydrol., 450–451, 190–198, doi:10.1016/j.jhydrol.2012.04.057, 2012.

Proc. IAHS, 374, 193–199, 2016
proc-iahs.net/374/193/2016/
doi:10.5194/piahs-374-193-2016


Open Access
PIAHS

# Decadal oscillations of the aquatic chemistry of river waters in Latvia

**Dmitry Porshnov and Maris Klavins**

Faculty of Geography and Earth Sciences, University of Latvia, Riga, Latvia

*Correspondence to:* Dmitry Porshnov (dmitrijs.porsnovs@lu.lv)

Published: 17 October 2016

**Abstract.** Water quality changes of surface waters can be used to assess human impact intensity, but of importance is to consider also impacts of climate change/variability and naturally occurring changes of environmental quality. In Latvia, during the recent decades a major reduction of anthropogenic pressure has happened due to restructuring of economy and industrial production, resulting in major decrease of loading of many groups of pollutants. However, trends and driving factors for other groups of substances have not been much studied. Long term (1980–2012) results of hydrochemical monitoring, performed in rivers of Latvia, are analysed during this study in connection with long-term sets of hydrological and heliophysical data, using standard statistical approaches. Our results indicate that variation of some hydrochemical values, for example COD and total Fe, show clearly visible decadal oscillated character, while variation of some other values, for example phosphate P and total P, show some individual characteristics of decadal oscillations. These results indicate the presence of a large scale, geochemical and geophysical significant process: multiannual pulse of catchment, driven by variation of solar irradiance through complex interactions between global atmospheric circulation, groundwater and surface waterbodies. The process described in our study is significant from a geochemical point of view and must be taken into account in prediction of water quality and quantity. Impacts of natural processes should be considered in the planning of environmental policy.

## 1 Introduction

The runoff of dissolved substances by riverine waters is an important factor in evaluation of not only water quality, but more important, of the character of the processes in waterbody basins (Henriksen et al., 1998). Rivers act as conveyers transporting dissolved substances from continents to seas. Thus, the integral dimensions of adverse processes in water body basins and the increased geochemical mobility of water ingredients can be identified by studying surface-water-driven geochemical flows (Stumm and Morgan, 1996). Increased flows of dissolved substances can affect the living environment of aquatic biota, especially in the receiving water bodies, such as the Gulf of Rīga and the Baltic Sea, in the case of Latvia.

Chemical composition of surface waters is determined by numerous factors and it depends, at first, on natural conditions (physico-geographical character of the region, climate, hydrology, hydrogeology etc.), but also on the level and extent of human impact. Considering many different sources of water ingredients, the study of their contribution to the aquatic chemistry is at utmost importance in evaluation of water quality (Hounslow, 1995). Studies of aquatic chemistry have also been used to reveal the dimension of the human activity-related processes in water body basins which have changed water composition (both at global, regional and local level), to mention industrial production (wastewater and solid waste dumping) and agricultural activities (non-point pollution sources), as well as impacts of hydro-technical constructions and changes of land-use patterns. Also, impacts of climate change of water composition and flows of substances might be considered as major actuality (Whitehead et al., 2009).

Commonly, temporal changes of geochemical flows of dissolved substances, have been done in terms of linear trend analysis, even though the character of variance of these flows can be much more complex (Libiseller and Grimvall,

Published by Copernicus Publications on behalf of the International Association of Hydrological Sciences.

2002). At the same time river discharge regimes, surface water composition and flows of dissolved substances have not only linear trend of changes, but well expressed periodicity (Pekarova et al., 2003). Oscillations of the discharge intensity, with main frequency of about 20 and 13 years in rivers of Latvia, were reported by Klavins et al. (2002a). Periodic oscillations of groundwater levels and levels of deeper underground aquifers in Latvia, with a main frequency of about 11 years, were reported by Levina et al. (1995). So, it is very credible, that similar oscillational patterns are characteristic also for fluxes of dissolved substances with riverine waters.

The results for the 35-year-period of systematic observations of the concentrations of analysed substances allow estimating more or less credible values of flows of these substances in river waters. The aim of this study is to investigate the oscillation patterns in the temporal variation of flows of dissolved substances in surface waters in Latvia.

## 2 Physio-geographical characteristics of the study area

Latvia is a country situated on the eastern coast of the Baltic Sea in North Eastern Europe. Chemical composition of surface waters of Latvia is influenced by geographical, geological, and geochemical processes. At first, possible impacts on geographical factors should be mentioned: the surface is characterised by flat surface topography (57 % of Latvia's territory is located below 100 m above the sea level, and only 2.5 % of the area reaches 200 m). The surficial geology has been formed by Quaternary glacial and ancient sea sediments and by recent exogenous processes. The most widely distributed parent soil materials are moraine loam and sands which are resistant to weathering processes. More than one half of Latvian soils are humic podsols with comparatively high content of organic matter (Klavins et al., 2002b). Latvia has a comparatively dense network of rivers. The climatic conditions of Latvia are dominated by transport of cyclonic air masses from the Atlantic Ocean, leading to comparatively high humidity, uneven distribution of atmospheric precipitation through the year, mild winters and moist summers (Klavins et al., 2003). The study area covered the major river basins of Latvia (Lielupe and Daugava) which supply the dominant part of the water runoff (24 $km^3$ or 69 % of the total runoff).

Daugava, also known as Western Dvina is the largest river of Latvia and also one of the biggest rivers in the Baltic region. It is 1005 km long and its basin size is 87 900 $km^2$, including territories of Latvia, Belorussia and Russia. Mean water runoff in Daugava constitutes 20,4 $km^3$. The mean precipitation ranges from 600 to 900 mm per year (Statistical Yearbook of Latvia, 1997). More than half of the river discharge takes place during spring floods, and the water discharge pattern is characterised by steep fluctuations of water discharge intensity. Characteristic soil bedrocks in Daugava basin are quaternary sand, loam and sandy loam of glacial origin. Podzolic soils, stagnosols and peat soils are widely distributed. The area covered by forests (mostly boreal) is 37 800 $km^2$ (43 %), and bog area about 4400 $km^2$ (5 %). Depth of groundwater active exchange zone is 200–350 m, aeration zone, mostly consisted of sand and is 1–15 m deep. In the previous studies, a 26-year periodicity of the flow of the Daugava River was considered as the main period, which includes smaller cycles of 2, 6 and 13 years (Glazacheva, 1988).

Lielupe is the second biggest river of Latvia, with strongly different catchment characteristics. Length of Lielupe is 119 km, basin area 17 600 $km^2$, mean annual runoff 3.6 $km^3$. Annual precipitation amounts in Lielupe basin are lower than average in Latvia (500–700 mm), due to dominantly downward direction of air mases in this region (Anonymous, 2009). Lielupe receives the major part of its discharge from direct surface run-off, spring floods dominate, and the role of permanent water discharge during the year is comparatively low ($\sim 40$ %) (Klavins et al., 2003). The geological structure of the catchment is characterised by several meters of thin layer of quaternary deposits (mainly limnoglacial clay). Underlying Devonian deposits, mainly dolomite, limestone, sandstone and clay embed very close to surface and often expose in river valleys. Sod alkaline soils and brown soils on clay and sand bedrock are characteristic of this area, stagnasols and podzols also are prevalent (Anonymous, 2009). Bog areas (3 %) are present in Lielupe catchment, but they are situated downstream from monitoring point used for this study, therefore impacts of wetlands on geochemical fluxes from Lielupe catchment are not taken into account. Soils of Lielupe catchment are fertile and valuable for agriculture, so agricultural areas are widely distributed here and take 71 % of total catchment area, forest areas (mostly nemoral) are limited and cover only 22 % of area (Anonymous, 2009).

## 3 Materials and methods

Hydrochemical and hydrological data used in this study were collected by the Centre for Geology, Hydrometeorology and Environment within the National monitoring program. Monthly records of river discharge from gauging stations on Daugava (Jēkabpils) and Lielupe (Kalnciems) were used. Sampling and analytical methods followed the standard methods during all the study period (Tsirkunov et al., 1992). From 1980 till 1993, water ingredients were analysed according to standard methods used in USSR, as summarised in Standard Methods (1973). After 1993 standard methods described in APHA (1992) were used. During 1992–1994, inter-calibration programmes, both at national and international levels, have demonstrated general acceptability of the used methods (Stålnacke, 1996). Average concentrations of chemical substances (Fetot, TOC, COD, P-$PO_4$, N-$NH_3$ and N-$NO_3$) in rivers were calculated using the data

**Figure 1.** Monitoring points under analysis.

obtained at the following sampling stations: Piedruja, Daugavpils, Jekabpils, Lipshi and Ogre on the Daugava River and Bauska, Jelgava 1, Jelgava 2 and Kalnciems on the Lielupe River (Fig. 1). The used Total Solar Irradiance (TSI) data (2015) was obtained by the VIRGO project as described by Frohlich (2006). All the data under analysis was recalculated, using 24 month moving average function. Microsoft Excel was employed for data management and the construction of curves. NIWA Time Trends software was taken for linear trend analysis by Mann Kendall test, as described by Hirsh and Slack (1984), as well as for linear regression analysis.

Flux of total iron and nutrients were estimated, with the help of available raw monitoring data, while flux of organic matter, taking into account limits of data availability, was evaluated using COD values for the period 1980–2003 and TOC values for the period 2003–2012. TOC values for the first 23 years of study were calculated from COD concentrations using formula proposed by Klavins et al. (2012).

$$\mathrm{TOC} = (0.2928 \times \mathrm{COD}) + 7.9503 \quad (1)$$

The export of substances was determined as:

$$R = Wc \quad (2)$$

where $W$ – volume of the water flow per day in a specific period m$^3$, (km$^3$); $c$ – average concentration of a substance in this period in river waters, mg L$^{-1}$ (g m$^{-3}$). The result values were recalculated as tons per day (24 h). Descriptive characteristics of the obtained data sets are described in Table 1.

**Table 1.** Descriptive characteristics of data sets under analysis.

| Daugava | TOC | Fe Tot | N-$NO_3$ | N-$NH_4$ | P-$PO_4$ | Color |
|---|---|---|---|---|---|---|
| Sample size | 311 | 250 | 369 | 368 | 370 | 199 |
| Min. | 103.7 | 0.16 | 0.06 | 0.13 | 0.1 | 16 |
| Max. | 5190.3 | 71 | 336 | 56.4 | 12 | 300 |
| Mean | 937.1 | 12.7 | 41.6 | 5.3 | 1.9 | 85.6 |
| Median | 650.6 | 7.6 | 26.8 | 3.9 | 1.5 | 83 |
| Stand. dev. | 826.6 | 13.7 | 45.1 | 5.7 | 1.65 | 38 |
| **Lielupe** | **TOC** | **Fe Tot** | **N-$NO_3$** | **N-$NH_4$** | **P-$PO_4$** | **Color** |
| Sample size | 342 | 274 | 353 | 354 | 362 | 313 |
| Min. | 12.9 | 0.1 | 0.02 | 0.03 | 0.01 | 14 |
| Max. | 1091.2 | 2.0 | 307 | 21.2 | 9.1 | 521 |
| Mean | 161.5 | 0.7 | 29.1 | 1.6 | 0.6 | 82.8 |
| Median | 93.6 | 0.6 | 9.5 | 0.7 | 0.4 | 75 |
| Stand. dev. | 163.5 | 0.3 | 43.9 | 2.7 | 1.0 | 49 |

## 4 Results and discussion

As it can be seen in Fig. 2, the character of the variation of discharge values in the studied rivers obviously can be described as oscillated. Signs of oscillations with 20, 11, 8–9, 3–5 and 2 years long periods can be observed, that generally corresponds to the results of earlier studies (Glazacheva, 1988; Klavins et al., 2002a). Increase of water color, known also as brownification, was frequently reported in many lakes and rivers of the Northern hemisphere (Driscoll et al., 2003; Frey and Smith, 2005; Monteith et al., 2007; Erlandsson et al., 2008; Haaland et al., 2010), including Latvian rivers as well (Klavins et al., 2012). Our data show that despite low data quality and short period of observations (1980–2003), statistically significant upward trends of color values can be found in both rivers (Table 3). However, a closer look at obtained curves indicates also an oscillated character of this process in both cases (Fig. 3). The drivers behind brownification are heavily debated and have been ascribed to hydrological factors (Hongve et al., 2004; Erlandsson et al., 2008), as well as an increase in temperature, changes in land-use and reduced acid deposition (Correll et al., 2001; Freeman et al., 2001; Monteith et al., 2007).

Although it has been difficult to find out a single factor causing brownification, there is an implicit consensus that the direct factor causing this process is the increased concentrations of terrestrially derived organic matter (OM) in the water (Kritzberg and Ekström, 2012).

As we see in Fig. 4, temporal variations of TOC flow in both studied rivers are very similar. Mann-Kendall test results do not show any positive trend in fluxes of organic carbon in the rivers under analysis. While the oscillated charac-

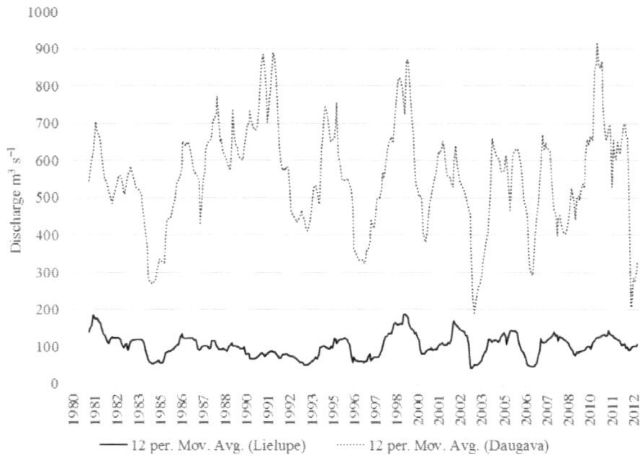


**Figure 2.** Variation of discharge in studied rivers (24 monthly moving average).

**Figure 3.** Variation of water color in studied rivers.

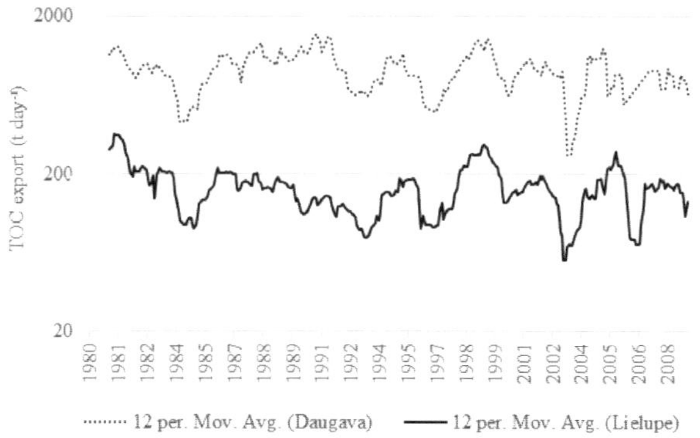


**Figure 4.** Variation of TOC export in studied rivers.

**Figure 5.** Variation of total iron export in studied rivers and total solar irradiation (TSI).

ter of data variance is evident, decadal oscillations, as well as oscillations with shorter period (3–6 years) can be observed in both rivers. The character of variation is similar to the character of variation of discharge data, while statistically significant correlation ($r = 0.18$) between TOC concentration and discharge can be observed only in the case of Daugava, but such correlation for the Lielupe river is insignificant (Table 2).

However, it is necessary to admit that the increase in water color cannot be explained only by changes in the quantity and quality of OM, iron also must be mentioned as a potential source of color (Canfield et al., 1984; Kritzberg and Ekström, 2012; Ekström, 2013). Considering low solubility of iron compounds it becomes evident that iron compounds in surface waters exist in form of complexes with OM. Dissolved OM is known to bind to the surface of ferrihydrite and stabilize against aggregation by giving the particle surface a net negative charge (Tipping, 1981), so the flux of organic matter and the flux of iron are strongly connected amongst themselves.

Variation of total iron flux in Lielupe indicates clearly visible decadal oscilations, while period of oscillations in Daugava is shorter (3–6, mainly 4 years). Observed oscillations are highly similar to the cyclic signal of changes in total solar irradiance (TSI) with a periodicity of $\sim 11$ years (Schwabe cycle) between a sunspot minimum and a sunspot maximum, measured instrumentally in the last 3 decades (Frohlich, 2006; Engels and van Geel, 2012) (Fig. 5). Pair regression analysis of total iron fluxes in both catchments in a connection with TSI data shows statistically significant correlations between these values: $R = 0.15$ in the case of Lielupe and $R = 0.16$ in the case of Daugava. Correlation amongst total iron concentrations and river discharge can be observed in Daugava. In the case of Lielupe these values are not correlative (Table 2).

Differences in both curves can be explained by the difference of iron sources in the studied areas. In the catchment of Daugava as the most significant source of iron, the leaching from soils must be mentioned, as indicated also by a very

**Table 2.** Linear correlations between hydro-chemical values and river discharge.

| | Daugava | | Lielupe | |
|---|---|---|---|---|
| | $n$ | $R^2$ | $n$ | $R^2$ |
| Color | 308 | 0.173 | 310 | 0.088 |
| $N\text{-}NH_4$ | 368 | 0.047 | 354 | 0.016 |
| $N\text{-}NO_3$ | 369 | 0.004 | 353 | 0.249 |
| $P\text{-}PO_4$ | 370 | 0.032 | 362 | 0.030 |
| $Fe_{tot}$ | 253 | 0.14 | 274 | 0.014 |
| TOC | 312 | 0.034 | 394 | 0.001 |

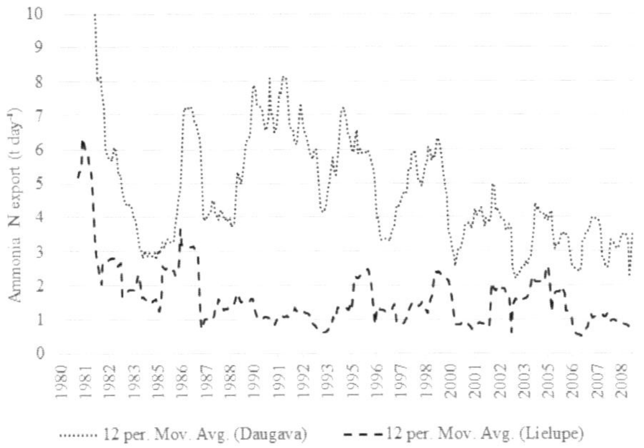


**Figure 6.** Variation of ammonia nitrogen export in studied rivers.

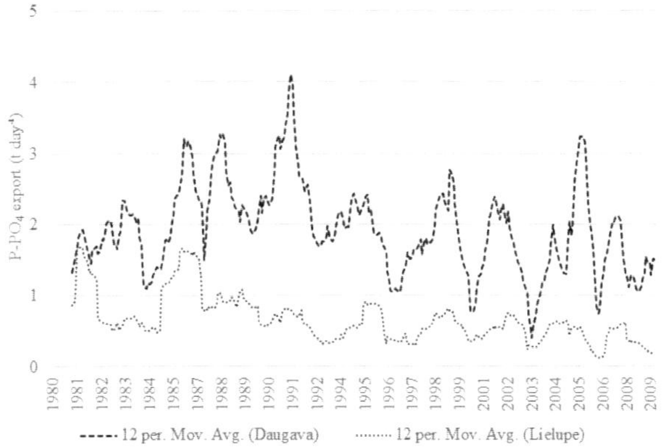


**Figure 7.** Variation of phosphate phosphorous export in studied rivers.

strong correlation with features of covariation between fluxes of TOC and Total Iron ($R = 0.91$) demonstrating similar sources of both substances (Kritzberg and Ekström, 2012). Meanwhile, in the Lielupe basin the correlation is much less pronounced ($R = 0.32$), indicating difference of sources. It is very likely that the main source of iron in the case of Lielupe is groundwater from Devonian aquifers. It is necessary to add that similar ideas were expressed also by Klavins et al. (2002b) and Delina (2007). So we can make an assumption, that short (3–6 years long) periods of oscillations represent impact of changes in surface and soil runoff, while decadal oscillations represent a hydrodynamic response of deeper underlying aquafers.

The results of Mann Kendall test (Table 3) show that the decrease of nutrient export amounts can be observed in both rivers. These tendencies can be explained by significant reduction of nutrient loads that took place in last decade of XX century, after the collapse of the Soviet economics (Stalnacke, 1996). However, taking into account the oscillational character of nutrient flows, downward trends in Daugava also can be explained with the minimum in cycle with approximately 20 years long period. In this case, upward trends in nutrient export values can be predicted for Daugava catchment, irrespectively to the reduction of loads, in the next few years. The reduction of phosphorous and ammonia nitrogen exports in Lielupe appears to be more reliable, representing self-purification process of over fertilised agricultural area after sufficient cuts in fertiliser loads.

## 5 Conclusions

Oscillational patterns that can be defined as multiannual pulse of catchment are obviously characteristic of export values variation of the dissolved substances from the studied catchments. The oscillational character of geochemical flows can be explained by impacts of multiple ranges of cyclic natural events driven by changes in solar irradiation, e.g. oscillation of precipitation and river discharge, cyclic oscillations of groundwater aquafers, changes in evaporation intensity. Cyclic character of geochemical fluxes must be taken into account researching geochemical processes, e.g. brownification of waters and self-purification of soils. 30 years long sets of geochemical data are too short for a fully credible linear trend analysis.

We have failed to explain upward trends of color in Daugava and Lielupe rivers by increase of export of OM. Our results show that iron must be mentioned as the very significant factor affecting water color in the studied rivers. The character of brownification in both rivers is oscillational and shows significant connection with solar drivers.

Successful recovery of over-fertilized agricultural area is evident in the case of the Lielupe River. Longer observation period is needed to evaluate self-purification of the Daugava River catchment area. Upward trends of nutrient export values from Daugava catchment are possible in next few years, irrespectively to reduction of loads.

## 6 Data availability

Datasets can be found in the Supplement.

**Table 3.** Results of Mann Kendall test.

| Daugava | TOC exp. | $Fe_{tot}$ exp. | N-$NH_4$ exp. | N-$NO_2$ exp. | P-PO4 exp. | Color |
|---|---|---|---|---|---|---|
| Median value | 650.6 | 7.62 | 3.85 | 26.77 | 1.46 | 83 |
| Kendall statistic | −1385 | −1795 | −10912 | −8362 | −6781 | 2985 |
| $Z$ | −0.75 | −1.36 | −4.63 | −3.53 | −2.85 | 3.18 |
| $P$ | 0.45 | 0.17 | 0 | 0 | 0.004 | 0 |
| 5 % confidence limit | −7.84 | −0.19 | −0.11 | −0.67 | −0.026 | 0.72 |
| 95 % confidence limit | 2.53 | 0.017 | −0.05 | −0.237 | −0.007 | 2.22 |
| **Lielupe** | **TOC export** | **$Fe_{tot}$ export** | **N-$NH_4$ export** | **N-$NO_2$ export** | **P-$PO_4$ export** | **Color** |
| Median value | 93.56 | 0.59 | 0.7 | 9.53 | 0.4 | 72 |
| Kendall statistic | −837 | 2013 | −7555 | −3130 | −13 535 | 9103 |
| $Z$ | −0.4 | 1.33 | −3 | −1.4 | −6 | 5.4 |
| $P$ | 0.69 | 0.18 | 0 | 0.16 | 0 | 0 |
| 5 % confidence limit | −0.97 | −0 | −0 | −0.2 | −0 | 1 |
| 95 % confidence limit | 0.59 | 0.008 | −0 | 0.01 | −0 | 1.8 |

**The Supplement related to this article is available online at doi:10.5194/piahs-93-193-2016-supplement.**

## References


Anonymous: Management plan for Lielupe basin district, Latvian Center of Environment, Hydrometeorology and Geology, 2009.

APHA: Standard Methods for the Examination of Water and Wastewater, 18th edition, American Public Health Association, Washington DC, 1992.

Canfield, D. E., Linda, S. B., and Hodgson, L. M: Relations between color and some limnological characteristics of Florida lakes, Water Res. Bull., 20, 323–329, 1984.

Correll, D. L., Jordan, T. E., and Weller, D. E.: Effects of precipitation, air temperature, and land use on organic carbon discharges from Rhode river watersheds, Water Air Soil Poll., 128, 139–159, 2001.

Delina, A.: Quaternary groundwater of Latvia: Disertationes Geologicae Universitas Latviensis Nr. 8, 2007.

Driscoll, C. T., Driscoll, K. M., Roy, K. M., and Mitchell, M. J.:Chemical response of lakes in the Adirondack Region of New York to declines in acidic deposition, Environ. Sci. Technol., 37, 2036–2042, 2003.

Ekström, S.: Brownification of freshwaters-the role of dissolved organic matter and iron, Doctoral dissertation in biology, Lund University, 2013.

Engels, S. and Van Geel, B.: The effects of changing solar activity on climate: contributions from palaeoclimatological studies, Journal of Space Weather and Space Climate, 2, A09, 1–9, 2012.

Erlandsson, M., Buffam, I., Folster, J., Laudon, H., Temnerud, J., Weyhenmeyer, G. A., and Bishop, K.: Thirty-five years of synchrony in the organic matter concentrations of Swedish rivers explained by variation in flow and sulphate, Glob. Change Biol., 14, 1191–1198, 2008.

Freeman, C., Evans, C. D., Monteith, D. T., Reynolds, B., and Fenner, N.: Export of organic carbon from peat soils, Nature, 412, 785–785, 2001.

Frey, K. E. and Smith, L. C.: Amplified carbon release from vast West Siberian peatlands by 2100, Geophys. Res. Lett., 32, 1–4, doi:10.1029/2004GL022025, 2005.

Frohlich, C.: Solar Irradiance Variability since 1978: Revision of the {PMOD} Composite during Solar Cycle 21, Space Sci. Rev., 125, 53–65, 2006.

Glazacheva, L.: Long-term trends of the river run-off, air temperature in the Baltic region and atmospheric circulation in the Euro-Atlantic sector, in: The factors of regime formation, hydrometeorological conditions and hydrochemical processes in the seas of USSR, Leningrad, Hydrometeorological Agency, 227–241, 1988 (in Russian).

Haaland, S., Hongve, D., Laudon, H., Riise, G., and Vogt, R. D.: Quantifying the Drivers of the Increasing Colored Organic Matter in Boreal Surface Waters, Environ. Sci. Technol., 44, 2975–2980, 2010.

Henriksen, A., Skjelvale, B. L., Mannio, J., Wilander, A., Harriman, R., Curtis, C., Jensen, J. P., Fjeld, E., and Moiseenko, T.: Northern European lake survey, Ambio, 27, 80–91, 1998.

Hirsch, R. M. and Slack, J. R.: A Nonparametric Trend Test for Seasonal Data with Serial Dependence, Water Resour. Res., 20, 727–732, 1984.

Hongve, D., Riise, G., and Kristiansen, J. F.: Increased colour and organic acid concentrations in Norwegian forest lakes and drinking water – a result of increased precipitation?, Aquat. Sci., 66, 231–238, 2004.

Hounslow, A. W.: Water quality data: analysis and interpretation, Lewis PublBoca Raton, 1995.

Klavins, M., Briede, A., Rodinov, V., Kokorite, I., and Frisk, T.: Long-term changes of the river runoff in Latvia, Boreal Environ. Res., 7, 447–456, 2002a.

Klavins, M., Rodinovs, V., and Kokorite, I.: Chemistry of surface waters in Latvia, Riga, LU, 2002b.

Klavins, M., Kokorite, I., Ansone, L., Eglite, L., Rodinov, V., and Springe, G.: Spectrofluorimetric study of dissolved organic matter in River Salaca (Latvia) basin waters, Knowledge and Management of Aquatic Ecosystems, 404, 1–14, 2012.

Kritzberg, E. S. and Ekström, S. M.: Increasing iron concentrations in surface waters – a factor behind brownification?, Biogeosciences, 9, 1465–1478, doi:10.5194/bg-9-1465-2012, 2012.

Levina, N., Levins, I., Prols, J., and Straume, J.: Use and monitoring of drinking groundwater resources in Latvia (in Latvian), Geological Service of Latvia, Riga, 1995.

Libiseller, C. and Grimvall, A.: Performance of partial Mann-Kendall test for trend detection in the presence of covariates, Environmetrics, 13, 71–84, 2002.

Monteith, D. T., Stoddard, J. L., Evans, C. D., de Wit, H. A., Forsius, M., Høgasen, T., Wilander, A., Skjelkvale, B. L., Jeffries, D. S., Vuorenmaa, J., Keller, B., Kopacek, J., and Vesely, J.: Dissolved organic carbon trends resulting from changes in atmospheric deposition chemistry, Nature, 450, 537–541, 2007.

Pekarova, P., Miklanek, P., and Pekar, J.: Spatial and temporal runoff oscillation analysis of the main rivers of the world during the 19th–20th centuries, J. Hydrol., 274, 62–79, 2003.

Stålnacke, P.: Nutrient loads to the Baltic Sea, PhD. Thesis, Linköping University: Linköping, 1996.

Standard Methods for Chemical Analysis of Surface Waters: Gidrometeoizdat: Leningrad, 1973 (in Russian).

Statistical yearbook of Latvia: 1996 Central Statistical Bureau of Latvia, Rīga, 1997.

Stumm, W. and Morgan, J. J.: Aquatic chemistry, J. Wiley, NY, 1996.

Tipping, E.: The absorption of aquatic humic substances by iron oxides, Geochim. Cosmochim. Acta, 45, 191–199, 1981.

Tsirkunov, V. V., Nikanorov, A. N., Laznik, M. M., and Dongwei, Z.: Analysis of long-term and seasonal river water quality changes in Latvia, Water Res., 26, 1203–1216, 1992.

Whitehead, P. G., Wilby, R. L., Battarbee, R. W., Kernan, M., and Wade, A. J.: A review of the potential impacts of climate change on surface water quality, Hydrol. Sci., 54, 101–123, 2009.

# AUTHOR INDEX